British Economy and Society

British Economy and Society
1870–1970

Documents · Descriptions · Statistics

Selected and Edited by

R. W. BREACH and R. M. HARTWELL

Principal Lecturer
King Alfred's College of
Education Winchester

Professorial Fellow
Nuffield College
Oxford

OXFORD UNIVERSITY PRESS
1972

Oxford University Press, Ely House, London W.1
GLASGOW NEW YORK TORONTO MELBOURNE WELLINGTON
CAPE TOWN IBADAN NAIROBI DAR ES SALAAM LUSAKA ADDIS ABABA
DELHI BOMBAY CALCUTTA MADRAS KARACHI LAHORE DACCA
KUALA LUMPUR SINGAPORE HONG KONG TOKYO

© Oxford University Press 1972

Printed in Great Britain by
Richard Clay (*The Chaucer Press*), Ltd.,
Bungay, Suffolk

PREFACE

This book is intended for teachers and students in the sixth form and in the first years of higher education. We have had three main aims in preparing it: to provide a selection of documents (widely chosen from official publications, novels, pamphlets, critics of society, social scientists, etc.) to illustrate the economic and social history of the century 1870 to 1970; to comment on these documents, both individually, and also in four long essays which put the documents into the context of an historical summary of the whole period; to provide a statistical appendix, with an explanation of economic terms, to enable the reader to strengthen his grasp of the period with firm facts and analysis. We have tried to meet the challenge which faces all editors of documentary collections: in a limited space to present a broad picture of the period, and yet to provide substantial material on the great themes and events. We have tried also to balance the vivid and descriptive with the sober and analytical. We have tried to include some words from some of the more important people of the period, both those who made history and those who commented on it.

We believe that an understanding of the history of modern Britain can perhaps best start with a consideration of economy and society. Unfortunately, though much has been written on the economic history of modern Britain, very little has been written on her social history. In this volume, therefore, the weighting is greater on those sections entitled 'general social conditions', 'work', and 'opinion' than on 'economy'. To an important extent, however, there is redress for those who prefer economics and statistics in the statistical appendix, and at all times the reader, in considering any topic, should proceed from the document, to the explanatory introductions, to the statistics.

R.W.B.
R.M.H.

ACKNOWLEDGEMENTS

The Authors gratefully acknowledge permission to reproduce extracts from the following copyright works.

P. W. S. Andrews and E. Brunner: *The Life of Lord Nuffield*. Reprinted by permission of Basil Blackwell.
C. R. Attlee: *As it Happened*. Reprinted by permission of William Heinemann Ltd.
Automation in Perspective. Reprinted by permission of the Controller of Her Majesty's Stationery Office.
Lady Bell: *At the Works*. Reprinted by permission of Edward Arnold (Publishers) Ltd.
Aneurin Bevan: *In Place of Fear* (William Heinemann Ltd.). Reprinted by permission of David Higham Associates, Ltd.
W. H. Beveridge: *Full Employment*. Reprinted by permission of George Allen & Unwin Ltd.
C. Booker: *The Neophiliacs*. Reprinted by permission of Collins Publishers.
Asa Briggs: *The Birth of Broadcasting*, volume one of *The History of Broadcasting in the United Kingdom*. Reprinted by permission of Oxford University Press.
The British Economy: Key Statistics 1900–1970. Times Newspapers Ltd., 1971. Reprinted by permission.
G. M. Carstairs: *This Island Now* (The Reith Lectures, 1962). Reprinted by permission of the author and the Hogarth Press Ltd.
S. J. Chapman: *The Cotton Industry and Trade*. Reprinted by permission of Methuen & Co. Ltd.
S. J. Chapman: *Work and Wages*. Reprinted by permission of Methuen & Co. Ltd.
W. S. Churchill: 'The Crux of the Whole War' from *The War Speeches* vol. 1. Reprinted by permission of Cassell & Co. Ltd. and the Canadian Publishers, McClelland and Stewart Ltd., Toronto.
J. H. Clapham: *The Woollen and Worsted Industry*. Reprinted by permission of Methuen & Co. Ltd.

H. A. Clegg and R. Adams: *The Employees' Challenge*. Reprinted by permission of Basil Blackwell.

K. Coates and R. Silburn: *Poverty: The Forgotten Englishmen*. Copyright © Ken Coates and Richard Silburn, 1970. Reprinted by permission of Penguin Books Ltd.

Commercial Policy in the Inter-War Period: International Proposals and National Policies. Reprinted by permission of the United Nations office at Geneva.

Co-operative Independent Commission Report. Reprinted by permission of the Co-operative Union Ltd.

R. H. S. Crossman: *New Fabian Essays*. Reprinted by permission of the Turnstile Press Ltd.

W. W. Daniel: *Racial Discrimination in England*. Reprinted by permission of Political and Economic Planning.

P. Deane and W. A. Cole: *British Economic Growth, 1688–1959*. Reprinted by permission of Cambridge University Press.

A. V. Dicey: *Law and Opinion in England*. Reprinted by permission of The Macmillan Company of Canada and Macmillan London and Basingstoke.

T. A. Fyfe: *Employers and Workmen under the Munitions Acts*. Reprinted by permission of William Hodge & Co. Ltd.

T. R. Fyvel: *Intellectuals Today*. Reprinted by permission of the author and Chatto & Windus Ltd.

Growth in the British Economy. Reprinted by permission of George Allen & Unwin Ltd.

W. Hannington: *Unemployed Struggles 1919–1936*. Reprinted by permission of Lawrence & Wishart Ltd.

H. D. Henderson: *The Inter-War Years and other Papers* edited by H. Clay. Reprinted by permission of The Clarendon Press.

U. K. Hicks: *British Public Finances: Their Structures and Development, 1880–1952*. Reprinted by permission of The Clarendon Press.

J. A. Hobson: *Imperialism*. Reprinted by permission of George Allen & Unwin Ltd.

R. Hoggart: *The Uses of Literacy*. Reprinted by permission of Chatto & Windus Ltd.

B. Jackson: *Working Class Community*. Reprinted by permission of Routledge & Kegan Paul Ltd.

Hilda Jennings. *Brynmawr* (Allenson & Co.). Reprinted by permission of James Clarke & Co. Ltd.

J. Jewkes: *Ordeal by Planning*. Reprinted by permission of the Macmillan Company of Canada and Macmillan London and Basingstoke.

A. C. Jones: *Working Days* edited by M. A. Pollock. Reprinted by permission of Jonathan Cape Ltd.

D. C. Jones: *The Social Survey of Merseyside*. Reprinted by permission of Liverpool University Press.

Keesing's Contemporary Archives (1960), pp. 17269–70. Reprinted by permission of Keesing's Publications Ltd.

J. M. Keynes: *The Economic Consequences of the Peace*. Reprinted by permission of the Macmillan Company of Canada and Macmillan London and Basingstoke.

J. M. Keynes: *The End of Laissez-Faire* (Hogarth Press). Reprinted by permission of the Trustees of the Keynes Estate.

George Lansbury: *My Quest for Peace* (Michael Joseph Ltd.). Reprinted by permission of the Literary Executor of the Lansbury Estate.

R. Lewis and Angus Maude: *The English Middle Class* (Phoenix House). Reprinted by permission of David Higham Associates, Ltd.

R. Lewis and Angus Maude: *Professional People* (Phoenix House). Reprinted by permission of David Higham Associates, Ltd.

G. I. H. Lloyd: *The Cutlery Trades: An Historical Essay on the Economics of Small-Scale Production*. Reprinted by permission of Longmans, Green & Co., Ltd.

H. W. Macrosty: 'The Trust Movement in Great Britain' from *British Industries* by W. J. Ashley. Reprinted by permission of Longmans, Green & Co., Ltd.

B. Mallett: *British Budgets 1887–88 to 1912–13*. Reprinted by permission of the Macmillan Company of Canada and Macmillan London and Basingstoke.

C. F. G. Masterman: *The Condition of England*. Reprinted by permission of Methuen & Co. Ltd.

Sir E. Mellanby: *Recent Advances in Medical Science*. Reprinted by permission of Cambridge University Press.

George Orwell: *The Lion and the Unicorn*. Reprinted by

permission of A. M. Heath & Co. Ltd. on behalf of Miss Sonia Brownell and Secker & Warburg.

Harold Owen: *The Staffordshire Potter* (The Richards Press). Reprinted by permission of John Baker Publishers Ltd.

Alan T. Peacock and Jack Wiseman: *The Growth of Public Expenditures in the United Kingdom*. Reprinted by permission of the National Bureau of Economic Research.

J. C. Philip: *What Science Stands For*. Reprinted by permission of George Allen & Unwin Ltd.

M. A. Pollock: *Working Days*. Reprinted by permission of Jonathan Cape Ltd.

J. B. Priestley: *Angel Pavement*. Reprinted by permission of the author and William Heinemann Ltd.

Problems of Progress in Industry No. 10—Woman, Wife and Worker. Reprinted by permission of the Controller of Her Majesty's Stationery Office.

R. Robson: *The Cotton Industry in Britain*. Reprinted by permission of Macmillan London and Basingstoke.

R. Seebohm Rowntree: *Poverty: A Study of Town Life*, *Poverty and Progress* and *Poverty and the Welfare State* (Longmans, Green & Co., Ltd.). Reprinted by permission of The Joseph Rowntree Charitable Trust.

J. Simon: *Three Speeches on the General Strike*. Reprinted by permission of the Macmillan Company of Canada and Macmillan London and Basingstoke.

Osbert Sitwell: *Laughter in the Next Room* (Macmillan). Reprinted by permission of David Higham Associates, Ltd.

John Strachey: *The Menace of Fascism* (Victor Gollancz Ltd.). Reprinted by permission of Mrs. Celia Strachey.

Flora Thompson: *Lark Rise to Candleford*. Reprinted by permission of Oxford University Press.

R. M. Titmuss: *Essays on the Welfare State*. Reprinted by permission of George Allen & Unwin Ltd.

H. Tout: *The Standard of Living in Bristol* (J. W. Arrowsmith Ltd.). Reprinted by permission of the University of Bristol.

Wages, Profits and Prices: A Report. Reprinted by permission of the Labour Research Department.

Beatrice Webb: *My Apprenticeship* (Longmans, Green & Co.,

Ltd.). © The London School of Economics and Political Science, and reprinted with their permission.
H. G. Wells: *Experiment in Autobiography*. Reprinted by permission of A. P. Watt & Son on behalf of the Estate of H. G. Wells and Victor Gollancz Ltd.
E. W. Williams: *The Case for Protection* (G. Williams). Reprinted by permission of the Hutchinson Publishing Group Ltd.
H. T. Williams: *Principles for British Agricultural Policy*. Reprinted by permission of the Clarendon Press.
Leonard Woolf: *The Journey not the Arrival Matters*. Reprinted by permission of the Author's Literary Estate and the Hogarth Press.
F. Zweig: *The Worker in an Affluent Society*. Reprinted by permission of Heinemann Educational Books Ltd.

While every effort has been made to secure the permission of all copyright holders, we were unsuccessful in a few cases, and we apologize for our apparent negligence.

CONTENTS

	page
List of Documents	xv
General Introduction	1
Part I: 1870–1918	21
Introduction	23
Documents	39
Part II: 1918–1945	131
Introduction	133
Documents	146
Part III: 1945–1970	229
Introduction	231
Documents	243
Part IV: Statistical Appendix	365
List of Tables and Charts	369

LIST OF DOCUMENTS

PART I, 1870–1918

GENERAL SOCIAL CONDITIONS
page
1. Social classes in the countryside — 39
2. Society: (a) the London season, (b) changes in manners — 42
3. The new suburbans — 44
4. The Potteries — 47
5. The East End of London — 50
6. Poverty in York, 1899 — 54
7. The rise in living standards — 59
8. The Poor Law — 62
9. The framework of town life, 1889 — 67
10. Schools: (a) the village school, (b) the public schools — 69

WORK
11. Farm workers: (a) Hiring (1870), (b) Conditions (1900) — 73
12. Domestic servants — 75
13. The Gantrymen, Middlesbrough — 77
14. The workshop of Joseph Brown, Birmingham — 80
15. Basic conditions of factory work, 1901 — 82
16. The Professions: the law — 84
17. War work — 86

OPINION
18. The monarchy — 89
19. Religion: (a) the sceptical intellectual, (b) the Salvation Army — 90
20. State education — 94
21. Ideas of empire — 95

		page
22.	The origins of socialism	100
23.	Women's rights	102
24.	The Golden Age: three points of view	104

ECONOMY

25.	The survival of small-scale industry after 1900: (a) general, (b) Sheffield	108
26.	The organization of the staple industries: (a) cottons, (b) woollens	112
27.	The combination movement	116
28.	International competition: (a) iron and steel, (b) cotton textiles	119
29.	Growth of welfare expenditure	124
30.	Trade Unions: (a) transport strike (1907), (b) the Russian Revolution (1917)	127

PART II, 1918–1945

GENERAL SOCIAL CONDITIONS

31.	The end of the war, 1918	146
32.	Broadcasting, programmes, and audiences	148
33.	The cinema	151
34.	The lower-paid worker in the twenties	154
35.	The General Strike, 1926	158
36.	The urban environment, Merseyside 1934	163
37.	Unemployment in South Wales, 1934	166
38.	Violence in Belfast	168
39.	Working-class standards of living in Bristol, 1938	171
40.	The changing social structure	176
41.	The effect of the social services	179

WORK

42.	Office work in the twenties: the cashier and his children	183
43.	Workers in the ports, 1926	186

LIST OF DOCUMENTS xvii

                                                                page
44. Other workers: (a) the miner, (b) the shirt-maker
    machinist   189
45. New industries: cars and electricity   192

OPINION

46. The end of *laissez-faire*   195
47. The British Fascists   198
48. The abdication of Edward VIII   200
49. Pacifism   201
50. Science and Society   203
51. The War, 1939–1945   208

ECONOMY

52. An American view of the British economy in 1927   212
53. The new industrial revolution   214
54. The decline of overseas trade: cottons and coals   218
55. The difficulties of the gold standard   221
56. The financial crisis, 1931   222
57. International influences on commercial policy   225

PART III, 1945–1970

GENERAL SOCIAL CONDITIONS

58. The Crisis of 1947   243
59. The National Health Service   246
60. Working-class standards of living, 1950   249
61. The quality of working-class life   251
62. Women after the war   254
63. Influences on family size   256
64. Retail shopping   259
65. Poverty in the sixties   261
66. Educational needs   266
67. Racial discrimination   271

		page
68.	The urban environment	277
69.	The Aberfan disaster	280

WORK

70.	The manufacture of biscuits	285
71.	The Anglican clergyman	288
72.	The cotton industry	290
73.	The factory worker	292
74.	The effects of automation on labour	295
75.	The Civil Service	299
76.	Dockers	303

OPINION

77.	A Social revolution?	309
78.	The end of Imperialism	312
79.	The role of the middle classes after the war	315
80.	A changing morality	317
81.	The cult of youth	319
82.	A new kind of society?	322

ECONOMY

83.	The post-war balance of payments problem (1950)	325
84.	The problem of inflation	327
85.	A National plan for economic growth	329
86.	The issue of planning	333
87.	The development of nuclear power	335
88.	The special case of agriculture	337
89.	Monopoly and competition	340
90.	Wages and earnings in shipbuilding and engineering	343
91.	Industrial relations at Fords	345
92.	Collective bargaining	348
93.	The low-paid worker	353
94.	Overseas aid	356
95.	Britain and Europe	361

GENERAL INTRODUCTION

I

This book is concerned with a hundred years of the social and economic history of the United Kingdom, and is divided into three chronological periods, 1870–1918, 1918–45, 1945–70. The starting date, 1870, was in many ways a turning-point in British history. The franchise had been extended in 1867, and this marked the beginning of mass democracy and the emergence of a political labour party. From 1873 to 1896 there was a long period of falling prices, known as 'the great depression', and of a slower rate of growth of the economy, known as 'the climacteric', which were the first set-backs the economy had suffered since the beginning of the industrial revolution in the eighteenth century. The 1870s also saw a resurgence of empire-building, a remarkable expansion which added massively to Britain's empire in many parts of the world. The policy of *laissez-faire*, which had been virtually unopposed for a long period, came under attack, and growing doubts about the effectiveness of the market economy and especially about free trade, led to increasing intervention by the government in economic matters, and ultimately, after 1900, to the laying of the foundations of the modern welfare state. Abroad, also, 1870 marked a significant turning-point: the face of Europe changed with the triumph of Bismarck's Germany over France; the American Civil War had recently preserved the political unity of the U.S.A.; Japan had just achieved a political revolution of momentous future importance; these three countries—Germany, U.S.A., and Japan—were to industrialize rapidly before the end of the century, surpassing Britain in some fields and destroying once and for all Britain's claims to be 'the workshop of the world'; the great burst of European expansion known as 'the new imperialism' was about to take place, during the course of which most of the non-European world was to come under the control or influence of Europe, through conquest, trade, investment, and emigration. This extraordinary expansion of European society, in which Britain played a leading role, created a situation by 1900 which was the starting-point of so many developments of the twentieth century.

But if 1870 began an era, 1918 certainly ended one. The old-world economy and society suffered shocks between 1914 and 1918 from which it never recovered. In Britain, the well-ordered society of the middle and upper classes, whose status and income seemed so secure in 1913, was changed permanently, so too was the pre-war economy whose great staple industries, on whose prosperity Britain had so long depended, were either permanently or temporarily harmed by the war. There is little wonder, then, that Keynes, with many others, recalled the Edwardian age with such nostalgia and affection (see p. 104). In one respect only, Britain seemed to benefit from the war: in the increase in empire which resulted from the additions of ex-enemy colonies as mandates under the League of Nations.

The period 1918–45 had an obvious unity, enclosed as it was by two world wars of unparalleled dimensions. The inter-war years have been seen, retrospectively, as years of failure, when Britain was unable to cope with serious economic and political problems, and when the world drifted, hopelessly and inevitably, into conflict. The great problems of Britain were to restructure an economy, to solve a massive unemployment problem (at no time between 1921 and 1939 were less than 10 per cent of British workers unemployed), and to begin to carve away some of the social and economic privileges which characterized British society. As the British people, and their governments, grappled unsuccessfully with these problems, there was an air of tension and disruption, and a lack of confidence, which contrasted markedly with feelings before 1914. Attempts to return to 'normalcy'—i.e. to the pre-1914 conditions—were bound to fail, and too much of thinking and policy in these years was backward- rather than forward-looking. Nevertheless, in spite of declining industries and unemployment, this was a period of technological advance and structural change, and, for those in employment, a period of rising living standards.

The Second World War for Britain produced 'the moment of truth'. There were still those in 1939 who, encouraged perhaps by the survival of a vast empire, thought of Britain in nineteenth-century terms. Nobody after 1945 could realistically include Britain among the super-powers, and the emergence of the U.S.A. and Russia as the great powers of the world marked a new stage in Britain's position in world politics and economy. Moreover the economic strength of Britain, so long dependent on trade, services, and the returns from foreign investment, was sadly diminished: there was greater competition; the returns from services and foreign investment were relatively less important. Britain after 1945 had to rehabilitate the economy and to depend increasingly on her trade to balance her commitments with the rest of the world. Britain was still a dependent economy, particularly dependent on food and raw material imports, and trade was never more important for survival. But the lessons of the inter-war years and war years were not forgotten, and Britain was to benefit from the revival of world trade after 1945, and especially after 1950. The war had increased the productive capacity of the economy and had nurtured desirable economic and social change. The prosperity which finally ensued was not without its problems, and, indeed that prosperity seemed always to poise on a knife-edge, with continual balance of payments difficulties and 'stop–go' economic policies. The advance of technology after 1945 was extraordinarily rapid; and much of the discussion of the period was on how to make the most of the new mastery of production, and to a lesser extent on how it should be distributed. But there was also social advance, especially in education, and in the emergence of Britain's young as an obvious and articulate new force in society, not a class force, but an age and income group of considerable power and influence.

The periods we have chosen are convenient rather than sacrosanct. Different criteria will produce different dates. 1900 was a turning-point in the history of social attitudes as well as in the history of real wages. 1926 was a turning-point in the history of industrial relations. 1951, or even 1955, was the beginning of the period of mass high consumption rather than 1945. Again the whole period could be divided by wars—the Boer War, the two world wars—or that remarkable economic set-piece, the Great Depression of 1929–34. However, remembering that history is a continuum, our turning-points do mark significant discontinuities beyond which the history of Britain was significantly different from what it had been before.

II

Even with discontinuities, there are certain social and economic developments which have persisted over the hundred years since 1870. The most striking phenomenon has been the continuous rise in the standard of living of the mass of the British population, along with a continuous increase in total numbers. Poverty and unemployment—the specific social evils of our first and second periods—have, to a large measure, been overcome since 1945, and Britain has now reached what can be reasonably called an age of high mass consumption. That has been accomplished in the face of a rising population makes it all the more remarkable. The second most striking phenomenon of the last hundred years has been the expansion of the role of government in social and economic life, and, in particular, the growth of the welfare state whereby the government has assumed the responsibility of making life tolerable for the masses. The working classes of Britain have increased their real wages, and they have received increasing welfare payments from the State, especially in the forms of health, education, and social security. The achievements of the State, socially, have two dimensions: on the one hand, to provide greater equality of opportunity, so that everybody has the chance to make the most of his talents, both for himself and for the benefit of society; on the other hand, to help the individual (and his dependents) over the humps of life, the misfortunes caused by bad health, unemployment, accidents, death, etc. Both dimensions of State activity have been motivated by theories about, and desires for, egalitarianism, theories, and desires which have become increasingly explicit and conscious. The Victorians accepted men as equal in the sight of God as democrats do in the presence of the ballot box; the modern Elizabethans (if we may so describe ourselves) take this feeling of equality into society at large. The uncertainties and inequalities of life, whch were accepted stoically as inevitable in 1870, have been significantly modified, and, moreover, are now seen generally as social evils to be remedied further rather than as unchangeable facts of life for the mass of people.

Just as there has been a narrowing of the gap of economic privilege so has there been a move towards cultural unity. A national society has been created out of many regional societies, and out of many social classes. Sometimes this process has been called the *embourgeoisement* of the working classes; it could easily be called the proletarianization of the middle and upper classes (which followed on such social change as the drastic decline in servants). In any case, a broadly unified culture—more than merely literate and urbanized—has been consolidated from an earlier industrial–urban society which was already showing something of these characteristics. Of course regional and class differences still exist: regional accents persist in spite of education and mass media; class interests are reflected in entertainment and sport; in Scotland and Wales nationalism still flourishes. But such differences have been softened, and often obliterated, for example, as national rather than local newspapers increasingly have dominated popular reading, and as national television has dominated viewing and has consolidated folk-heroes like the Beatles. Other aspects of this more unified society can be seen in the scientific and technological dominance of modes of thought, the general decline of religion and the growth of secularism, the increasing proportion of working-class children in institutions of higher learning. If we could put ourselves back to 1870 the most striking contrasts (physical environment apart) between now and then would probably appear to be in social attitudes, which were marked in the earlier period by deference and servility, a lack of understanding, class hostility, and prejudice. Yet we would also find much in common to remind us of historical continuity in our attitudes and habits. For example attitudes towards the Irish and the Jews then were very similar to the prevalent attitudes to coloured immigrants today. To find a world we have lost completely it is necessary to go back beyond the industrial revolution of the eighteenth century.

What have been the continuities and changes in the economy since 1870? We have already emphasized rising living standards and the growth of government. In 1870 the government controlled less than 10 per cent of the economy, today more than 50 per cent. In the hundred years since 1870 Britain passed from a predominantly *laissez-faire*, free enterprise, market economy to a mixed government-private, largely controlled economy. The economic achievements (and problems) of 1870 were the result of a host of individual decisions operating through the market; the economic achievements (and problems) of today are the result as much of government decisions, often inspired by political and social objectives, as of individual choice. This change in the importance of government in the economy came gradually, not by revolutionary jumps, although both world wars lifted government expenditure on to new plateaux, raising both expectations and what was considered to be tolerable level of taxation. At first government action was mainly for social welfare—to solve the problems of inequality, insecurity, and poverty—but since the 1930s government has been increasingly concerned with controlling and directing the economy. The two basic beliefs which motivated government action—or rather, lack of action—before 1914 were allegiance to free trade and to the gold standard. These, it is important to remember, involved fundamental adherence to a self-regulating system; free trade posited that, in the long run, the economy should adjust itself to international specialization and division of labour; the gold standard implied that, in the short run, the economy adjusted itself, by deflation or inflation, to balance of payments disequilibria. Belief in free trade allowed the remarkable decline of wheat growing after 1880; belief in the gold standard facilitated the establishment of London as the financial capital of the world before 1900. But all this ended with the First World War, and finally, in 1931 and 1932, Britain abandoned both gold and free trade. The momentous implication for government was that henceforth it had to have both a monetary policy and a trade policy, whereas in the past governments had acted as though money and trade looked after themselves.

A survey of the economy since 1870 reveals two distinct periods, with a turning-point also about 1930. After the great growth of the industrial revolution, that rate of growth after 1870 was less fast, as also was the rate of structural change. The declining rate of growth has been called 'the climacteric' and its causes have been widely sought: the disadvantages of an early start; dependence on a narrow range of ageing staple industries coupled with neglect of important new industries like electricity and chemicals; a social and educational system (particularly the lack of technical education) which discouraged talent from entering industry and encouraged amateurism and rule of thumb methods in industry; increasing international competition and tariffs tending towards a slowing-down in the rate of growth of international trade; too much foreign investment at the cost of home investment. The great structural change which from the beginnings of the industrial revolution had transferred resources from agriculture to industry and services was completed by 1900; thereafter the proportion of national resources devoted to agriculture was so low that it could no longer afford further transfers; and thereafter the service sector would grow at the expense of industry. Even so, Britain in 1913 seemed rich and confident, whereas in 1926 the country seemed much poorer and very unconfident. Between the wars Britain had a regular 10 per cent unemployment problem as the staple industries spent twenty years in the doldrums. Only after 1930 were there the beginnings of a great new change in the British economy: the advent of a new technology of revolutionary characteristics; the growth of new industries and the decline of old; a change in educational facilities which greatly expanded the general base of education and strategically increased scientific and technological education; the vigorous growth of export industries with a new resurgence of trade. By the mid-1950s Britain had transformed the depressed and inappropriate economy of the 1920s into one that could compete successfully in an increasingly competitive world trade.

III

These social and economic changes took place within a political framework itself changing and interacting with economy and society. Fundamental, though historians tend to take it for granted, has been the survival of the democratic state with its central and local institutions generally distinguished by a high level of honesty and efficiency. The long and successful history of democracy in Britain, the continuing though weakened strength of local government, a largely bi-party political system respresenting broadly the class interests of society, parliamentary and cabinet government (combining legislative and executive functions) which allows governments once elected to govern (without, for example, the frustrations of an American president), the strength of common law, have all ensured the survival and continuing development of British political institutions. This mature political organism survived two world wars and great social and economic changes without a civil war (such as occurred in the U.S.A., at the beginning of the period), without alternation from Republic to Empire and back again via a revolutionary Commune (as in France), without resort to autocracy (as in Germany and Russia), or ineffectiveness (as in Italy). An Empire was won and lost; an important international role was maintained, against reality, and then gracefully surrendered. Traditional institutions were modified to accommodate change and did so, not simply to avoid revolution, but also positively to put into effect new ideals. Men of conscience and moral stature led the way to increasing democracy and social justice. Of course there is another way of putting it: the cynical argue that the ruling classes yielded enough, just quickly enough, to give an illusion of democracy while hanging on effectively to power and privilege. But whatever the motives the political machinery did effectively cope with change, and even in a book primarily devoted to social and economic history it is worth while giving some consideration to how it was done.

Great political changes had occurred in the century before 1870, so that by 1870 the new middle classes, created by the industrial revolution, had already achieved the conditions for the peaceful taking over of political power from the aristocracy and the old landed gentry. Such acts as the Great Reform Bill of 1832 and the Municipal Corporations Act of 1835 had made this possible. This had had the effect of producing some Tory radicalism, and with it support for controls on industrialism and urbanization, and even for the extension of democracy; for example, it was the aristocratic Shaftesbury who pioneered factory legislation, and Disraeli who put through the second Reform Bill. At the same time some of the middle classes had begun to envisage and work for a sharing of new authority with the working classes, under the powerful stimulus of middle-class conscience with its Christian roots, disturbed also by the revelations of poverty and abuse of power which were revealed by the great reports of special committees, royal commissions (and later, social surveys), and, after 1870, by a weakening faith in *laissez-faire*. The second Reform Bill was not just the product of a party manoeuvre; by the time of its passing both political parties had come to think that proletarian enfranchisement in some form was important. However, just as it took the middle classes a long time to replace their predecessors as the actual wielders of power after it became politically possible, so the working classes allowed power to elude them long after the Reform Bill of 1884 had brought it within their grasp. The Labour Party had its origins in the 1880s, was formed in 1906, adopted socialist objectives only in 1918, formed two inadequate [minority] governments in 1924 and 1929, shared in the government of 1939 to 1945, but only came into real power in 1945.

The working classes were not only slow to take power with their own party, they often resisted good things done for them by Tories and Liberals. There is a history of working-class opposition to compulsory elementary education, health, and sickness insurance, and better housing, because all were seen to be, to some extent, at the cost of, rather than as additions to, wages. Indeed the working classes, and their leaders, placed more trust in industrial than in political action, and so much of their effort went into trades unions, which, except for a brief Syndicalist period from 1909 to 1913, were more concerned with wages and conditions of work than with political power. Nevertheless as the Labour members increased in the House of Commons, political power did come, and in the 1920s Labour took over from the Liberals as the great second party. After the great Liberal landslide of 1924, Labour became the party of alternative government, and this is a turning-point of the utmost significance in the history of Britain. And before this the Labour Party, by adopting specific socialist objectives in its 1918 platform, had officially endorsed the acceptance of the role of the State in changing even the economic framework of society to assist the workers. The eclipse of the Liberals cannot be attributed to short-term causes, for example, to the quarrel between Lloyd George and Asquith. Rather it was the result of long-term social and economic changes which had united Britain geographically, at the same time as dividing the population in terms of economic or class interests. As social and status differentials became less important, and as the franchise extended, then it became natural for the voter to try and match his vote with his economic interest. The structure of politics became more democratic, but at a time when the decline of the staple industries condemned so much of Britain's working population to idleness, and, hence, to political discontent. Economic ills between the wars were seen increasingly by the working classes as remedial political ills, which were not remedied because of the wickedness or stupidity of governments. Nevertheless the State did increasingly respond to the problems of the inter-war years, and the thirties was a decade of increasing though ineffective State intervention in economic life.

The war after 1939 was fought under a national government, and with a tightly controlled economy, and there were many who thought that the post-war problems should also have been faced by a national government. But Labour was swept into power in 1945 by an electorate who vividly remembered the Conservative failure to provide employment in the 1930's. The State was seen clearly as reflecting social and economic needs, and the political parties as supporters of particular class interests; for the first time in British history most of the working classes identified their interests with those of a Labour Party and voted for it. The 1945 election was won under the leadership of Clement Attlee, the middle-class son of a well-to-do miller, educated at Haileybury College (a school long devoted to educating Imperial district officers) and Oxford. As leader of post-war Labour Attlee was succeeded, first by Hugh Gaitskell, who in background and education resembled Attlee, and second by Harold Wilson, more proletarian in origin (though not working class) and more technocratic in style (moved as much by doctrines of efficiency as of equality). These three men have led a party which has derived its strength from the working-class vote, but the fact that that party has not been continuously in power means still that many working-class voters vote Conservative or Liberal, or do not vote at all. Indeed, as increasing affluence extends down the income scale, it may be that more of the working classes will cease to identify themselves automatically with the Labour Party. It may be that this source of identification was a temporary incident in British political life, the direct result of a period of high unemployment and social discontent. It is, even now, inappropriate to make a simple alignment of social, economic, and political forces; the affluent State may make such an alignment even more inappropriate. And when the policies and actions of post-war Conservative governments are considered it becomes obvious that the Tories have not stood for exclusive class interests and have not been prepared to die in any last ditches for old privileges. As politics were adapted to changing conditions, so were institutions, although the general framework of British institutions remained unaltered; just as there was no revolution in politics, so there was none in institutions. However, a very long list of peaceful and gradual changes can be drawn up. The Church of England surrendered wealth and establishment in Ireland, and later in Wales. In England the Church of England remained the established Church in a society in which allegiance to any religion or creed was no barrier to social, political, or economic advancement. The monarchy increasingly gave up any pretence of an independent political role. The Army gave up the selling of commissions; the Civil Service opened its ranks to successful examinees; the Commons accepted the widening of the franchise, first to all adult males and then to all adult females; the Lords blustered but accepted the Commons' right to finance without interference; the wealthy accepted death duties, and the middle class progressive income tax; local government increased its activities in a wave of 'municipal socialism'. The political foundations of these various moves had been laid before 1870, beginning particularly with the first great reform of the franchise in 1832. But whence came the impetus to reform? An important part of it came from persons who were in public office, as administrators or politicians: members of parliament, civil servants, members of local government, members of commissions of inquiry. Many like Lloyd George were waging a private 'war against poverty'. The independence of civil servants from ministerial control, and the independence of ministers from parliamentary control, allowed considerable independent initiative. But whatever the impetus to change, the nineteenth-century process of remedy followed a pattern: *identification* of a social, economic, or political problem; *propaganda* for an inquiry; *inquiry* in the form of special committee or commission; parliamentary debate and *legislation*; *administrative machinery* for enforcement. And once an administration had been established for a particular problem, its inquiries and tasks inevitably expanded. The inspectors' reports on factories, under the 1833 Factory Act, made clear the need for further reforms. Investigation of the cholera epidemics of 1848 and 1854 led to a series of reports by Dr. Simon; a preliminary Act of 1866 established the principle of compulsory health control over local authorities; the Public Health Act of 1875 established a sanitary code which lasted until 1936. By 1870 Britain already had a large Civil Service with multifarious activities and duties, making it possible to historians to refer to 'the Victorian origins of the welfare state'. When A. V. Dicey published *Law and Opinion in England* (its substance had been given in lectures in 1898) he considered that a period of Collectivism (see p. 94) had already replaced the period of Benthamism or Individualism as the main current of public opinion about 1865. This seems an exaggeration, but it does point to the danger of thinking that our generation is the first to think of itself as living in an age of collectivism. Certainly the gulf between Gladstone's political philosophy and Attlee's is immense, but even Gladstone accepted a positive role for the State. If civil servants and politicians gave impetus to reform, so also did independent critics of society, for example, the Fabians. It is important to note, however, that British social criticism owed little to Marx, and that there has never been a strong Marxist following in this country. Perhaps a more important agent in growing collectivism was the social pressure on an increasingly large and complex industrial society, which was already half-urban by 1851. And this pressure was felt as much locally as centrally; hence the municipalities promoted 'gas and water socialism', and some School Boards provided extra education which amounted to secondary education. In two world wars the State increased its powers over the economy and the individual, amounting in the second war to total direction and control, and more completely and successfully than the German enemy. It is for this reason that some economists and historians have attributed much of the growth of the State to the demands of total war, which led to total government. After the wars, it is argued, the powers of the State were permanently increased. The Great Depression after 1929 was another potent force for State control, creating problems so vast that it seemed obvious that more than individual initiative was necessary to solve them.

When Dicey wrote of public opinion he reckoned it to be an important social force producing social and political change.

But public opinion can be manipulated; newspapers, for example, both form and are formed by public opinion. In a society with many national newspapers, and a great variety of magazines, most views, however perverse, have had a chance to be heard. But with a wealth of communications, actions can follow from a particular interest group persuading society to look on its needs favourably. Of course, even with such good communications, public opinion and public approval can be misjudged and misinterpreted. A classic case of political misjudgement can be traced in the events leading up to the General Strike, and the unions' defeat, in 1926. In British democracy, however, in which there has been public and constant advocacy of a very wide variety of interests, the tyranny of any one interest has been made very difficult to achieve. The countervailing forces against tyranny are many and, combined, are very powerful indeed.

There is one other political theme of persistent importance in our period: imperialism. J. A. Hobson's thesis on *Imperialism* (1902), which was taken up by Lenin, was that the European imperialist powers exploited the rest of the world as a result of the inevitable contradictions in their own economies. A declining rate of profit at home and increasing competition abroad resulted in increasing imperialist tensions and in war. This theory has been largely discredited in its simple economic analysis, and explanations of imperialism more in socio-political terms are more convincing. A main motive in Britain's imperialism of the 1880s, nevertheless, was a fear of European economic rivals; Seeley, Dilke, Shadwell, and Chamberlain were, between them, afraid of the Germans, the Americans, and even the Chinese. Actually Britain's industrial and commercial supremacy depended more on Europe and America than on the Empire. The Empire was important, but its economic importance can be easily exaggerated. Emigration was perhaps a real safety valve for Britain, rather than the export of capital. Certainly some frustration and discontent was siphoned off by emigration, but the largest stream went, not to the Empire, but to the United States of America. British public opinion was certainly influenced by imperialism: the public cheered imperial contingents at royal jubilees; popular music-hall ditties were often imperialist in sentiment; public schools and universities trained administrators for overseas territories; schoolboys rejoiced to see, in the exaggerated contours of Mercator's projection, that so much of world maps was coloured the red of 'the Empire on which the Sun never sets'. But did interest in imperial affairs reach the masses? Imperialism, like church-going, was for the Victorians and Edwardians generally a middle-class rather than a working-class interest. Perhaps the most surprising feature of British imperial history was the disbanding of Empire after 1945. In 1939 the British Empire, in its full extent and apparent magnificence, was intact; by 1960 it had practically disappeared. This disbanding, though not casual, was done without great soul-searching or publicity. In 1947, it was a fuel crisis and not Indian independence which preoccupied the public. As J. K. Galbraith has written, the British, as usual, had shed their inconvenient luggage just at the right time. An incautious generalizer might indeed conclude that Britain prospered as she shed her imperial burden. But this would be incorrect: over the brief period of Britain's great empire, from 1870 to 1950, the Empire was less significant than the rest of the world for Britain's economic fortunes, and, in any case Britain's trading and investment in the areas of Empire had commenced before they became part of the Empire and have continued after they have become independent. The Empire, to a very great extent, has been an irrelevancy economically, although very important politically.

IV

Can anything useful be said about 'the quality of life' over the hundred years? What did the various sections of the community expect from life, and how did expectations match with achievements? Were life-choices enriched and extended in this period? The quality of life, unlike the standard of living, cannot be measured, which is what the social and economic historian would like to do. In evaluating the quality of life, we are concerned, often, with goods and ills not normally measured in the market-place. What we are looking for is the sort of evaluation that G. M. Young's *Portrait of an Age* gives of the quality of life before 1914. History is too often concerned with politics and diplomacy, with industries and depressions, and too little with literature, architecture, painting, and even science. The confidence of the Edwardians can be seen, perhaps most explicitly, in their opulent and decorative buildings in which, presumedly, they delighted. The quality of the First World War was most vividly delineated by the poetry and prose of its victims, or in the grim landscapes of its artists. For the whole of our period, photography has immortalized evidence in striking detail, whether it be London slums or aristocratic households. There is much social history in the *Illustrated London News*, on the one hand, and, on the other, in *Punch*, which tells us so much of changing fashions in humour, as well as of social snobberies and contemporary preoccupations. The newspapers, which expanded after 1890 to meet the demands of a literate populace, provide excellent source material for a study of tastes as well as of politics. And there are, in addition to these less formal sources, important social surveys, made with specific ends in view, and packed with relevant details about how the poor lived. For most of the period we are dependent for information about the poor on the very small minority of articulate poor, and on commentators mainly from other, superior income groups. We know much more about the articulate than the inarticulate and about the literate than the illiterate. Much of our information about the submerged tenth, the lowest 10 per cent of the working class, comes from special inquiries or surveys, initiated by private or public enterprise, in the last twenty years of the century.

Of the Victorians after 1870 it can be said that they worked harder and longer than we do (for we have measures of this), lived their lives by stricter ideals than ours (this can be derived from their writings), and inhabited a social environment much harsher than ours (this also being partly measurable by such social indices as mortality rates). Theirs was a rational and materialist society (rational in its adherence to science, materialist in its unending quest for material success), yet one in which the irrational survived (for example, the opposition to Darwin's theory of evolution) and in which adherence to Christianity was loudly and almost universally proclaimed (except among the working classes). Their British world was continually expanded by conquests and settlements which were adornments rather than burdens. The Industrial Revolution, which had brought Britain to the foremost economic place in the world, had not yet worked out its rich vein of reward and, although other economies threatened, Britain retained a remarkable world trading and financial hegemony right through to 1914. There had been no major war since 1815, and the Victorians looked forward to further and strengthened peace as international trade entangled the world in mutual dependence. Poverty still existed, but its presence was known and analysed, and there was hope that it could be eventually reduced to humane proportions. Virtue favoured the hard-working, and in a highly competitive world success went, not only to those who inherited wealth, but to the hard-working, imbued with ideas of self-help and thrift.

C. F. G. Masterman plotted the alleged decline of these virtues in his *Condition of England*, written in 1908. He noted private opulence and public squalor; a monarchy more interested in pleasure than duty; the rich more interested in consolidating their wealth as *rentiers* than in increasing it as entrepreneurs. Certainly there were great trials ahead, precipitated in part by the war. But there was no inevitability in Britain's decline. It is fanciful to see Britain as a sort of modern Rome, her strength eaten away by internal weaknesses. There were weaknesses, undoubtedly, but those, as we have already shown, would be remedied or partly remedied. Similarly it is difficult to accept the thesis that the great blow to European civilization that occurred as a result of the First World War was due to a minor political affray in Serbia, or simply to the greed and folly of the Austrian foreign minister. Great events have large and complex causes, and 'the decline of the West' in more than relative terms is to be found in the complex of economic and political forces that constitute the history of Europe in the fifty years before 1914. No one can look at Britain before 1914 without seeing considerable virtues (a concern for the poor, the aged, and the illiterate, for example) as well as defects (insensitivity about the rights of women, for example).

It is important to think of Britain as carried into the First World War by the nature of international politics, for which, certainly, she must share some responsibility; but no large nor even significant blame attaches to Britain as a maker of the conditions which made the war possible. The war was a great tragedy for the people of Britain; it killed part of one generation, as monuments to the dead in every town and village testify, and left those who survived to face a world which was certainly not the world they fought to save. The war destroyed the conviction that reason, hard work, and good will on the one hand, and increasing international economic contacts on the other, could solve the problems of international politics. And the fate of the internationalization of political institutions, in the form of the League of Nations, proved that the elusive aim of world peace would continue to be frustrated by national ambition. Physically a good deal of pre-war life survived the holocaust. For the mass of the people the standard of living was rising between 1918 and 1939, but the nagging problem of unemployment persisted as the specific social evil of the period. No matter what changes in the economy, it took the Second World War finally to absorb the unemployed. In such circumstances, with democracy obviously failing to provide a satisfactory solution to social and economic problems, many of the intellectuals of the thirties became blindly enamoured of Soviet Communism, to a degree that even the Stalinist purges of the late thirties failed to unhinge their convictions. Even that great rationalist and scorner of follies, George Bernard Shaw, is to be found in the thirties among the admirers of Soviet Russia. Some even looked towards the Continental solutions provided by the Fascism of Mussolini and Hitler, who solved the problem of unemployment by road-building and soldiering. But British governments, left and right, and the National Government of the thirties, failed to solve the problem of the workless, or to greatly modify the problems of wealth and privilege; they continued to soften poverty and misfortune, but there was no radical reform, of either economy or society.

The Second World War produced a genuine reformation in British society. Even more than in the First World War, it was fought with a sense of common purpose, much inspired by the imaginative leadership of Winston Churchill. It was important that the first durable consumer good of the new industries to reach the mass of the people in Britain was the radio, so that Churchill and the other war leaders had instant and effective access to almost all English homes. Common hardships produced more sympathy and understanding between social groups than had ever occurred before in British history. There was more determination during the war to make post-war Britain a place fit for heroes than had been the case during the First World War. Indeed this determination ensured the election of a Labour government in 1945. The electorate, with memories of the Tory inter-war failures, returned a government pledged to reform. This government succeeded in preserving in the post-war world the feeling of struggle and survival which was indeed almost necessary in a Britain facing a peace-time situation almost as grim as that of the war years. As equality of sacrifice was accepted during the war, so equality of opportunity was expected after it. It was not until the 1950s, however, that post-war austerity gave way to affluence. The sort of society which affluence produced is still being investigated and judged. The real quality eludes us, in spite of sociologists' surveys of how the British live and behave, in spite of consumer research and public opinion polls. The slogans of the day, however, do reveal something, especially in the phrase 'the permissive society' (the Pill and greater sexual freedom; less discipline for the young; easier divorce; etc.). But no aspect of modern British society has been more striking than the willingness and ability of the young to express themselves. Just as the workers of the nineteenth century were able, for the first time in history, to organize themselves for the furtherance of their own ends and influence society and the economy, so the young of all social classes in the 1950s and 1960s have been able to influence society in many ways, and often beneficially; they have not only brought different fashions, in clothes and music, they have also brought a sharper moral judgement to bear on the elders, with disturbing results for those who have always assumed their rights without always examining the moral bases of those rights.

V

What of today? What judgements should we make of Britain after the century of change and adjustment we have been describing? Can we agree with the popular commentator who wrote in 1967: 'England is immeasurably the best place in the world in which to live. Its people are more tolerant and more talented than even they believe. It has a uniquely important part in the modern world.' Or are we to believe those who see decline in all aspects of British life? Crime and violence, immorality, drug-taking, social prejudice—are these the social irrationalities of modern Britain? Balance of payments difficulties, troubled industrial relations, slowly rising productivity—are these the economic rewards of affluence? Opposition to the Common Market, partisan politics with doctrinaire theories and prejudices, a creaking parliamentary procedure—are these the fruits of political advance? Certainly failures and unsolved problems should be opposed to achievements and problems solved or being solved. But as we pointed out in this Introduction, and as the texts will verify, the Britain of the 1970s is in almost all ways superior to the Britain of the 1870s. In material terms the population is healthier, richer, and better educated; in social terms, it is less servile and degraded; in political terms, it is more democratic. To criticize comprehensive schools and the National Health Service, for example, is not to deny their considerable merits, and their advance on what went before. If housing is still inadequate, it was even more inadequate a century ago. It is fashionable among a particular class of middle-class commentators, especially on television, to express fastidious disgust for the present. Their memories are short and their reading of history is inadequate. Britain certainly has never had it so good. And even now what is perhaps the greatest single problem of modern industrial society—pollution—is being carefully analysed with a view to control and remedy. This is not a complacent society; it has great capacity for self-criticism and great ability to face and solve problems. It may not be quite the best country in the world in which to live but it is certainly one of the best. In a world of continuing conflict it is gratifying to dwell in a country where the police are not armed and where the great issues of public policy are still minutely and intelligently debated, not only in Parliament, but in the newspapers, in the weeklies, and on radio and television.

PART I, 1870–1918

PART ONE: 1870–1918*

In 1870 British society could be described as hierarchical, religious, and masculine: it was very unequal in its distribution of wealth and firmly stratified socially; it was ostensibly very religious, with some of the great public controversies of the day centring in religion; it was masculine, with women in an inferior social and economic position, married women, for example, not being able to own property in their own right until 1882. By 1918 society was less hierarchical, less religious, and less masculine. The franchise, the growth of a political labour party, and the expansion of trade unions to the unskilled, had extended democracy, economically and politically; increasing taxation and social welfare had reduced income inequality; educational opportunities had widened, although secondary education was still denied to all but a small proportion of the working classes; the old social order of the countryside,[1] that order which had long dominated the British way of life, was breaking up, and in the towns the working classes were taking a firmer grip on their own lives, and on the ordering of both their living and their working conditions. Society generally had become more secular. Perhaps most important, women had gone far towards formal equality politically, though economic equality had not been achieved even fifty years later. The State had been forced to play a more active role in social and economic life,[2] before 1914 to provide a modicum of social security and welfare for the very poor, and during the war to mobilize the economy's resources more efficiently. The State's role in the economy would never again fall to the insignificant levels of the nineteenth century. The greatest, saddest, and most influential experience of the period was undoubtedly the First World War itself,[3] which affected the

---

*The footnotes in the introductions refer to documents which illustrate points made in the text.

[1] 1.   [2] 29.   [3] 31.

whole nation, and which changed British society and economy permanently. It was not just the slaughter of a generation of young men; the war, an engineer's war in many ways, hastened technical change and also stimulated social and economic change (for example, votes for women). Less obvious at the time, but perhaps equally important, was Britain's changed economic position in the world after the war. The growth of other industrial powers,[4] coupled with a slowing-down in the rate of growth of the British economy, heralded the end of Britain's world economic hegemony, an end hastened by the war. Britain's century was over; the twentieth century would see the rise of other great European powers, and the emergence into the world political power of the non-European peoples.

Who were the agents and what were the forces which produced these changes? At the political–administrative level the State (through Parliament and the growing Civil Service) and the municipalities[5] were powerful instruments of change. At the economic level, the growing power of trade unions on the one hand, and the pervasive power of industry on the other— industry which was increasing in unit size,[6] and becoming more and more concentrated and more and more under pressure from foreign competition—were also instruments of change. At the level of ideas there were complementary forces at work, the inspection effect of social surveys and investigations (like those of Booth and Beveridge), and the impact on fundamental beliefs of changing concepts in economics, sociology, and philosophy (such as the philosophies of Green and Sidgwick, the sociology of Spencer, the economics of Marshall) which stressed the need for ideals of social responsibility beyond those of individual economic gain, and which questioned the social efficiency of the market economy. At the social level, the suffragette movement was important in focusing attention on a particular type of inequality, and thus in influencing social change. Other people, like the Fabians, formed themselves into interest and pressure groups, as the women had done, and talked, wrote, agitated, argued, and generally had their influence on decision-making at the political level. As a sort of overriding condition, influencing all decisions and change, were Britain's external commitments, not only the vast Empire but also the worldwide network of trading and

[4] 28.   [5] 9.   [6] 27.

financial connections which could, and did, change autonomously with important consequences for Britain. Much of what Britain could do was externally determined, because Britain was in 1914 the most dependent economy in the world, importing about 50 per cent of her food needs and even more of her industrial raw material needs (excluding coal); thus changes in the rest of the world inevitably meant changes in Britain. For example, British agriculture was transformed between 1880 and 1914 because of American ability to produce cheap wheat; and similarly the great Lancashire cotton industry, for so long the backbone of Britain's export strength, was never to recover from the war which prompted already growing cotton industries elsewhere in the world into greater production and more effective competition. But, however subtle and complex the mechanism of change, two factors by themselves were undoubtedly powerful engines of reform: universal manhood suffrage and rising living standards. And with all the forces making for change there was no revolution; the British (with some notable exceptions) were for peaceful change, not for violence and discontinuity.

The stages and periods of change cannot be so clearly defined as the agents. There are several important turning-points, but even these can be misleading: there was a fundamental Education Act in 1870,[7] but already there was an 80 per cent literacy rate in England and Wales; the battle for 'the dockers' tanner' in 1889 gave the unions a great victory, but the strength of the trade union movement had been building up gradually since 1850; the Public Health Act of 1875 had been prefaced by a half century of municipal concern for health. But quite new ground was broken with the Old Age Pension Act of 1908 and the National Insurance Act of 1911 (against sickness and unemployment). Some historians divide the period 1870 to 1918 at 1901–2, which marked the death of Queen Victoria (22 January 1901) and the signing of the Peace of Vereeniging (May 1902) which concluded the Boer War, the war described by Smuts as 'the beginning of the end of Imperialism'. Certainly the new monarch considerably relaxed the formalities of the court; Victoria was the last British monarch to have exerted considerable political influence, and Edward was the first almost completely constitutional monarch. Again

[7] 20.

the Edwardian period has a character of its own, partly determined by the economic trends of the period 1900 to 1914 (a slowing-down in the rate of growth of the economy, a slowing-down in the rate of increase of real wages with consequent industrial trouble, increasing international competition), and partly by its political problems (a near civil war in Ireland, the suffragettes' nagging campaign, a constitutional conflict between Commons and Lords). The writers of this Edwardian period have more in common with us than do the Victorians; it is the modernity of Shaw which attracts us still today, while it is the historical character of Dickens, a social mirror of his times, which holds us to him. Shaw saw the significance of technical change quite clearly; Straker, the chauffeur in *Man and Superman*, relies on his technical expertise and has little respect for his social superiors. Marconi transmitted signals across the Atlantic in 1901; the Wright brothers flew a heavier-than-air machine in 1903; the discovery of the mosquito as the transmitter of malaria was made in 1899; these were the achievements of the scientist rather than of the engineer, and the twentieth century was to be the century of incredibly rapid scientific discovery and its application to industry and warfare.

According to the decennial census of 1871 there were 27,431,000 people in the United Kingdom, of whom well over half lived in urban areas, as they had done since 1851. Between town and country there was still manifest antagonism. In Charles Kingsley's *Alton Locke*, of 1850, country men regarded townsmen as sly and corrupt, while townsmen in turn saw country dwellers as oafish yokels. This attitude was still prevalent in 1870, as was the strength of local loyalties and interests. But, of course, as town dwellers increasingly dominated in numbers, so did their attitudes and manners. In the countryside, it is true, there were still great landowners who ruled what Escott has called 'their principalities' like feudal lords, and who continued to do so until 1914. Of the towns, the great ones, like Birmingham, Manchester, and Glasgow, had a distinctive provincial quality and a distinctive urban society, while the small ones remained market towns which reflected the farming communities and landed society around them. Local accents, customs, folk-lore, songs, and other cultural characteristics, and also local industries, formed a social and economic framework which was distinctive and in which family connections were strongly maintained. In the working-class areas of the towns an intricate web of conventions, relationships, and status controlled the family's social existence. Son followed father in occupation, and social and occupational mobility was still limited. The migration of families from the countryside to the town or city in search of betterment, a migration which continued, though decreasingly, up to the war, did not destroy the extended family system, and urban groupings tended to be more stable than today. The move from country to town—one of the great social facts of British life since the eighteenth century—meant a constant flow of country habits and values into the new industrial society. But that great migration was practically complete by 1900; thereafter urban society developed more or less autonomously.

In the countryside the hierarchy of British society could still be seen clearly, and Joseph Arch recalled in his autobiography how as a child he had watched the labourers in their smocks wait till last before taking their sacrament in the parish church, and how the parson's wife exercised a social despotism in his Warwickshire village. The 'good' parson married, baptized, visited, and buried, sat on school boards, charities, and the Board of Guardians of the Poor Law; 'bad' parsons squabbled with Dissenters, talked down to their parishioners, quarrelled with the farmers, and were often absent from their duties. Even Escott recognized that it was possible to take a Marxist view of the parson (not that he used that term) as 'a class burden' on the village community, along with the squire. The standard of living of the village labourer depended much on the district: the declining cereal regions of the South and East paid poorer wages (barely adequate for physical efficiency) than the mixed farming areas of the North and the West; and wherever industry was competing for labour, agricultural wages were higher. The more unpleasant work practices of agriculture were gradually being stifled by legislation; for example, the gang system whereby an entrepreneur took a gang of labourers across the country working on contract. Nevertheless the traditional agricultural society—stable and hierarchical—was in 1870 on the eve of its profoundest upset since the agricultural revolution. The massive influx, after 1880, of cheap imported grain led to the decline of a whole sector of British agriculture, and to the ruin of many farmers.

The effect was largely regional—on the South and East—but the social consequence generally was a large decline in the agricultural labour force. The effect also was a changed structure of agriculture, with greater concentration on fruit and vegetable growing, on meat and milk production, and, generally, a swing away from cereals. In the village, the parson's income and status both suffered, and the traditional confidence in land as a social trust was inexorably weakened as rents fell and landed incomes were diminished. Agricultural labourers, in the long period of falling prices which continued almost to the end of the century, did relatively better than the other landed classes,[8] so that the Royal Commission on Agriculture of 1893 noted a marked improvement in their position over that of 1867. In the 1890s, also, agricultural trade unions revived on a more solid basis, so that working conditions and wages were both improved. But the backward-looking attempts to create a country of smallholders were a failure, partly because labourers did not want to become peasants, and partly also because county authorities failed to put into operation such legislation as the Smallholdings Act of 1892. It was the war which once again boosted agriculture generally, with the demand for home-produced food at an unprecedented level. Cereal production, for example, which had been declining during the thirty years before, increased almost a third over its pre-war level. Both labourer and tenant profited from the inflation of food prices, and conditions of work so improved that the agricultural labourer, like the town worker, often finished his week's work at 4 p.m., or even at noon, on Saturday. In the resultant prosperity, landowners perhaps profited less than tenants who were often able to purchase farms and set up as independent producers.

For an earlier period of British history, at the beginning of the eighteenth century, it is possible to write about 'the British' confident that the majority of the inhabitants of the islands shared enough of the same agricultural way of life to make the generalization valid; this is again true of the mid-twentieth century and beyond, when they shared an urban way of life. In the 1870's, however, there was still enough of the old agricultural way of life surviving to make it prudent, when writing about 'the British', to write separately about town and

[8] 11.

country. And, also, there were other groups with a separate and distinctive way of life; for example the life of the colliers, in the mining villages of the north or on the straggling hillsides of South Wales, was quite unique. By 1870 the first raw phase of English town-living was over. Grand public buildings, libraries and art galleries, schools, and even university colleges had already been well established. Thus even in a comparatively new industrial town like Middlesbrough there was a town hall, a free library, the various offices of the Corporation, churches, schools, a fine park, and a large public square. Readers of Charlotte Brontë's *The Professor* will recall that the rich and ruthless capitalist Edward each morning drove in his gig from his country house to his office, and after five, to his club. Escott, writing in 1879, noted the grim exteriors of the 'towns of business', but also that in them there was 'no lack of humanising influences at work'; and he contrasted their stable and law-abiding communities with the lawlessness and wild conditions of a hundred years before. He saw that industrialization had its goods as well as its discomforts; in all, it was a boon for the working classes. Evidence of 'the middle-class conscience', of humanitarianism, of Christianity, of prudence, of justice—or whatever of the many motives which prompted the wealthy to help the poor— is plentiful both before and after 1870. In Birmingham Joseph Chamberlain was Mayor from 1873 to 1876 and his efforts for the city were more like an attempt to match work already done in cities like Glasgow than to set the pace for a new generation of improving Corporations. Many studies of Victorian cities have made us familiar with the vigorous provincial culture of what could properly be described as 'provincial capitals'. In some this culture was associated also with political and economic causes, like that of free or fair trade. Not that reform was always successful. Many of the houses that replaced the slums cleared under the powers of the Artisans Dwelling Act of 1875 were nearly as bad as those they replaced.

Was there a town type? What reality was there in the descriptions of the diminutive mill-hands of the textile towns of Lancashire and Yorkshire, the big and tough foundry hands of Middlesbrough,[9] the small operatives of the workshops of Birmingham? Physical differences undoubtedly existed, as they

[9] 13.

do today, between those whose occupations demanded markedly different physical strength, but the noted differences were rather those of occupation, ethos, speech, social habits, and pay; in other words, differences in standard-of-living and way-of-life. Cockneys talked and behaved differently from the Yorkshire mill-hands. Escott wrote of Northern bluntness, lack of servility, pride in calling and in work skills; and there is much testimony of the characteristics of the Londoner. Uniformity in a general way of life had not yet submerged regional differences. Indeed, it must be remembered that in 1870 there was little to soften the sharpness of regional differences: no national popular newspapers, little mobility, and little education. Even London did not assume that peculiar national eminence which belonged to her as the seat of government, of British finance, and of the new imperialism, until the 1890s. Certainly the towns of England in 1870 represented a new kind of civilization, and one in which the problems of urban living were increasingly recognized and partly solved. Given the predominant *laissez-faire* attitudes of the nineteenth century, given the inexperience of dealing with the problems of cities, given the technical ignorance (about medicine and engineering, for example), it is reasonable to argue that the city and town authorities had done a tolerable job, even by 1870. Much more was to follow. The later Victorians were eminently practical, and municipalized services at the same time as they talked about freedom of enterprise; they helped the destitute while proclaiming the truth of Social Darwinism. Affluence came to many of the working classes,[10] even though class distinctions remained relatively rigid. Time has even made the homes they lived in acceptable, and a modern poet has hymned the plum brick villa and its idea of Gothic.

Nevertheless, whatever the achievements, the problem of Victorian poverty remained as the most profound and insoluble of problems. We know a great deal about this poverty, because there was, especially in the eighties, a quickening interest in the poor and in the problems of poverty. We have to go back to the forties—for example, to the great reports of Edwin Chadwick and the passionate pleading of Friedrich Engels— to find a crop of social writing which compares with Charles Booth's remarkable investigation into the life and

[10] 7.

labour of London, and Seebohm Rowntree's similar study of York.[11] If Booth and Rowntree were impressive and convincing in the sobriety of their evidence, more strident and emotional work, like the Rev. Andrew Mearn's *Bitter Cry of Outcast London* (1883), Jack London's *On the Edge of the Abyss* (1901), and William Booth's *In Darkest England and the Way Out* (1890),[12] had as much impact on public opinion, and deeply influenced the sensitive and responsible middle classes. And, it must be remembered, desperate poverty was not confined to the city. There was in the countryside dire poverty which only rose-covered cottage sentimentalists could excuse, and which was well documented in Parliamentary Reports (for example, those of 1867 and 1893) and in the writings of James Caird, Rider Haggard, and Joseph Arch. Such writings, reflecting a continued interest in poverty and the problems of society, continued in the Edwardian period in the work of William Beveridge and Charles Masterman.[13] Indeed there was at all times between 1870 and 1914 influential writers who challenged the conventional wisdom of the time, that wisdom which had a vision of the world as an unfolding drama of Whiggish progress—ensured by constitutional government and freedom of enterprise. Statistics and reporting show that about one-third of the population of the towns, and perhaps more in the countryside, lived at subsistence level, and that 10 per cent at all times were well below a standard of life which would have allowed the barest physical well-being. Even above the poverty line few working men enjoyed any security in times of sickness or unemployment, and the facilities of Trades Unions and Friendly Societies offered little more than temporary protection. Against sickness, industrial accident, drink, laziness, old age, and all the other hazards of life, the only shields were one's family and, ultimately, the workhouse. Under these conditions it is little wonder that much was made of the character of the extended family and its responsibilities among the working classes. Between a respectable, happy, and comfortable life and cruel poverty, deprivation, and even disgrace was a thin line, and the simplest of events could push a family across it. Their food, shelter, and entertainment, at best, were simple and unvaried; their education was minimal; their hopes of enrichment small. Even the skilled artisan, with

[11] 6.   [12] 19.   [13] 24.

a good standard of living, was quite sharply cut off from the society of his 'betters', and his children were segregated in schools which gave them little opportunity of social improvement. In the century of industrialization, an industrial working class had been formed from an agricultural and small-town labouring populace. But whereas in the early stages of the industrial revolution many artisans had become wealthy and even famous, there was by 1870 less social mobility. The industrial revolution industries were now well-established, and industrial dynasties dominated them rather than new entries; sons of existing industrialists rather than aspiring artisans were the entrepreneurs of the 1870s. There were, of course, many exceptions, and especially in new industries and services, but there is no doubt that social mobility was less by the last quarter of the nineteenth century than it had been in the first. On the other hand there was, between 1870 and 1918, a gradual improvement in the lot of the working classes, both in real wages, and in the provision of some security for old age, insurance for some poorly paid workers, an educational ladder (though it was climbed mainly by the middle classes) legislation ensuring safety standards in factories and workshops,[14] more effective trade union organization, and the beginnings of a Labour party. This resulted in some redistribution of income, although the working classes continued to pay more in indirect taxation, in aggregate, than they received in social benefits. The distribution of wealth remained, as it always had been, most unequal.

To what extent was the First World War a great discontinuity in British history? It has been suggested that the war left few permanent marks on British life; and it has been argued, on the contrary, that the war promoted vast economic and social change from which Britain never recovered. Certainly the State had exercised an economic control during the war that simply could not be forgotten. Soldiers had learnt lessons, similarly, which could not be forgotten, including the birth-control techniques which contributed to the declining inter-war birth rate. 750,000 men had been killed, and one and a half millions left with war-time injuries, a killing and wounding that cut across classes but, proportionately, affected the rich and educated more than the poor and working classes.

[14] 15.

Women would never again be asked to accept the inferior political status of the pre-1914 period; too many of them had been accepted as men's equals during the war to go back to the simple pre-war male dominance of politics. The rich had contributed heavily to the war—up to 30 per cent of their incomes—and taxation would never again drop to the pre-1914 low. Inequalities remained, but more money went into the public purse, and more into social services.

The intellectual changes of the period are as profound as the political and economic. The Anglican version of Protestant Christianity was embodied in an established church under the nominal control of the State, and wielding enormous intellectual and social influence. In 1870 the Church of England was rich, assured, and powerful. In the village the parson claimed equality with the squire, often controlled what education there was, and seemed little concerned with Dissenting or Roman Catholic rivals which twenty years before had seemed to threaten the Anglican hegemony. Nevertheless this apparent strength was already evaporating, not so much by the growth of other religions, as by the growth of secularism. In particular the urban working classes had never been as Christian as the agricultural working classes; and, where their religion was strong, as in the Northern towns or in the Welsh villages, it was the Chapel rather than the Church which dominated. But by the end of the century a larger proportion of the urban workers and their families attended church not at all, or only to participate in vestigial ceremonies like baptism, marriage, and burial. If religion had once given some emotional, social, and cultural outlet to the working classes, some interest outside their drab lives of toil, there were now more exciting and satisfying activities; the working classes were on the point of assuming political and economic influence more commensurate with their numbers, their increasing literacy enabled them to read an increasingly popular press, their slowly increasing affluence enlarged their interests. In these circumstances religion was on the decline, and after 1918 in rapid decline. Other factors also contributed. The Churches lost the intellectual battle with the scientists which followed the publication of *The Origin of Species* (1859), the book that reinforced an attack on the literal interpretation of the scriptures already made by the geologists. The State began to share, and to take over, the provision of education, and there was, once again, a stirring up of the embers of the sectarian dispute between those who believed that education was essentially religious, and those who believed equally passionately that education should be entirely secular. The literary intellectuals like Matthew Arnold had already deserted the Church, and even Gladstone had betrayed it from within by agreeing to the disestablishment of the Irish Anglican Church in 1869. But the most convincing evidence is in figures: Lady Bell reckoned in 1907 that 70,000 of Middlesbrough's total population of 97,000 went to no place of worship.

The changes in formal education were among the most spectacular of social changes in this period. The Education Act of 1870 allowed for the building and running of elementary schools by ratepayers' school boards where no adequate denominational school was available. The provision of public schooling had been bedevilled by inter-religious squabbling, and although Britain in 1870 was, at least for males, a largely literate country, much of existing schooling was being provided in religious schools, and the provision of public education had been delayed because of the unwillingness of the churches to surrender their educational role and privileges. After 1870, however, and especially after attendance became compulsory in 1880, and free in 1891, it meant that all children were likely to acquire a modicum of education in part or all of eight years of required school attendance in State or State-supported schools. W. E. Foster, in introducing his Bill in the Commons on 17 February 1870, pointed out that only two-fifths of working-class children between the ages of six and ten and only one-third of the ten to twelve-year-old age group were even nominally at schools aided by government grants. The situation changed rapidly after 1891, and the board schools, once firmly established, began to run classes beyond the statutory leaving age. At the same time the county councils began to set up technical schools, and some of the old-established Grammar Schools (when they were not taken over by the middle classes and made into fee-paying Public Schools) revived to help develop a rather chaotic system of secondary education. University education also expanded, with more colleges in London, women's colleges being established in Oxford and Cambridge, and expansion also in the north where colleges in Manchester, Leeds, and Liverpool combined in 1884 to award their own degrees, and separated again in 1903-4 to form autonomous universities. Other 'redbrick' universities were also founded around this time, for example the University of Birmingham in 1900. More significant were the attempts by universities to provide education for the working classes: Ruskin Hall was founded in Oxford in 1899, and there was a great blossoming of the University Extension Movement and the Workers' Educational Association. This first great experiment in adult education attracted men of the calibre of G. D. H. Cole and R. H. Tawney.

With the firm establishment of compulsory elementary education, the next obvious advance was in secondary education which was provided by the Act of 1902. By the Act the County Councils took over the primary schools in their areas, and were empowered to provide secondary schools. Such schools, however, were not a continuation of existing primary schools, but parallel institutions, covering the ages seven to fifteen, and charging fees. In 1907 the government laid down that 25 per cent of places in these secondary schools should go to the primary schools, candidates being chosen by a competitive examination, 'the Scholarship', an earlier version of the eleven-plus exam. But most working-class children received their schooling in the same school, even after twelve, on a part-time basis. Genuinely comprehensive schooling at the secondary level was not provided for the working classes until after 1945. The effects of these educational opportunities can be traced in various ways. In 1901 in Great Britain 5,843,000 children of the age-group five to fourteen were at school—but only 52,000 for higher ages were at school. In 1913 88,000 children were in secondary schools and over 28,000 at university. There was by 1914 an educational ladder which could be climbed from any class, but the difficulties for the talented working-class child were still very great indeed. For the well-to-do a regular system of privately owned preparatory schools provided for boys up to thirteen, and the Public Schools, under boards of governors, provided for secondary education. The Public Schools were influential social institutions, still imbued with the ethical ideas of Arnold of Rugby, but with broadened curricula which included science (practically none of which had been taught when the Clarendon

Commission reported on them in 1864),[15] art, and music. And there were a few co-educational schools, like Bedales (founded by J. H. Badley in 1893) and progressive schools which sought both to mitigate the rigours of the traditional boarding schools and also to be experimental with teaching and curricula. Education, however, was not as yet a spearhead of social change. Educational facilities tended to reflect social classes, and although they did provide mobility for the most talented, generally speaking secondary and tertiary education were for the already privileged. Elementary education, which all working-class children had, did not fit its recipients for middle-class occupations, and, indeed, as late as the debate on Fisher's Education Act of 1918 there were its opponents who argued that the working classes were happier when ignorant. Nevertheless the breach had been made, and the trickle of working-class children into higher education would thereafter continuously increase.

The great public debates of the period centred on two main themes: Britain's position in word politics and economy; and the state of the people. The first of them embraced such questions as: Had free trade outlived its usefulness? Should the country take on further burdens of Empire?[16] How could Britain combat the increasing trade rivalry of Germany and the United States? The second theme continued 'the condition of England' question. Had capitalist society succeeded, with progress, or should it take into account Fabian, syndicalist, and other critics? What was a reasonable living wage? How much alcoholic liquor should be consumed, and how was its distribution to be controlled? These and other questions—for example, the ever-present problem of Ireland—were worrying, but there was no feeling of impending crisis, and certainly no expectation of the catastrophe which the First World War proved to be. Indeed there was behind all the debate a generally held assumption that progress was inevitable, and almost continuous, and that while the end was obvious and desirable, there would be argument only about means. Even with the damage done to high Victorian society and ideals—for example, by the decline of religion—there is no picture before 1914 of an unsure society. The nostalgia of the well-to-do of the 1920s for the pre-war period was not just sentimentality

[15] 10.   [16] 21.

for a golden age that never existed.[17] It was nostalgia for an age of greater certainties, when taxes were lower, when the working classes knew their place, when women could not interfere in politics and male superiority was assured, when Britannia ruled the waves, and when the idea of Empire conjured up feelings of greatness and adventure.

The British economy, also, presented a generally flourishing appearance in the forty years before the First World War: rising productivity, huge overseas investment, increasing exports and imports, a comfortable balance of payments. Indeed Britain held a dominant place in the world economy, not only in trade but in finance. The City of London was the financial centre of the world. Britain accounted for a third of the trade in world primary products, as well as 30 per cent of world manufactures in 1913. Living standards had risen throughout the nineteenth century. Nevertheless there were a number of underlying weaknesses in the economy, of which variants are still recognizable in the 1970s. The rate of growth of the economy slowed down at some time between 1870 and 1900 and was certainly slower than the growth rates of U.S.A., Germany, and Japan. Real wages ceased to rise in the decade before 1914. In some industries, like coal, productivity was either stagnant or even falling. Why was this? Argument still rages around these facts, which are indisputable. Was Britain's domestic investment too low because too much capital went abroad? Were some industries, like coal mining, suffering from technical obsolescence? Were managers inefficient? Were the unions traditionalist and obstructive? Were entrepreneurs unenterprising? All these have been used to explain Britain's slower performance. While the causes of the declining rate of growth are elusive, the period undoubtedly marks a turning-point in British history, when Britain lost the economic pre-eminence which the industrial revolution had given her.

The period 1873 to 1896 is known as 'the great depression'. Its main characteristic is not declining production or trade, but the long price-fall of up to 40 per cent which almost certainly reduced profits and undoubtedly raised real wages. The squeeze on profits produced much protest, the beginnings of a protectionist lobby, the questioning of the economy's com-

[17] 24.

petitiveness, and more unemployment. One part of the economy was really depressed, agriculture, or rather that part of it which grew wheat. This was the result mainly of cheap American imports, which, at the same time as embarrassing the wheat farmer, resulted in cheaper bread for the worker. The period 1896 to 1914 produced an upturn in prices and stationary wages. At the same time there was something of an industrial boom, with an expansion in exports of the staple industries. This was accompanied by considerable emigration and a large export of capital. Because wages were not rising, and because trade unions and political labour were becoming better organized, this was a period also of considerable social unrest and some notable strikes. Finally there was the period of the war, dominated by increasing war production and the killing of so many young men. It took some time to break away from the idea that a large-scale war could be fought on the slogan 'Business as usual', but gradually the government took over the economy in the interest of winning the war, so that by 1918 Britain had an almost fully-controlled economy. The lesson was not lost when Britain was faced again with a large war in 1939.

Social history and economic history do not move in this period in that close harmony which the historians prefer. The rise of a Socialist party owed more to middle-class consciences than to changes in real wages. The reforms of the great Liberal ministry of 1906 to 1914 cannot be simply explained by the demands of those who benefited from them. Nor can the First World War be explained any longer as the result of economic imperialism.

GENERAL SOCIAL CONDITIONS

1.   Social Classes in the Countryside
James Caird, *The Landed Interest and the Supply of Food* (London, Cassell, 1880), pp. 56–64.

Sir James Caird, F.R.S. (1816–92) is one of the main sources on Victorian farming from his *English Agriculture in 1850–51* to his evidence to the *Royal Commission on Agricultural Interests* of 1881–2. His insistence helped bring about the compulsory collection of agricultural statistics, the annual *Agricultural Returns*, from 1866.

The landowners are the capitalists to whom the land belongs. . . . He takes a lead in the business of his parish, and from his class the magistrates who administer the criminal affairs of the county, and superintend its roads, its public buildings, and charitable institutions, are selected. . . .

This class in the United Kingdom comprises a body of about 180,000, who possess among them the whole of the agricultural land from 10 acres upwards. The owners of less than 10 acres each hold not more than one-hundredth part of the land, and in regard to farming may be regarded as householders only. The property of the landowners, independent of minerals, yields an annual rent of sixty-seven millions sterling, and is worth a capital value of two thousand millions. There is no other body of men in the country who administer so large a capital on their own account, or whose influence is so widely extended and universally present. From them the learned professions, the church, the army, and the public services are largely recruited.

The tenant-farmers are the second class, and a much more numerous one. Their business is the cultivation of the land, with a capital quite independent of that of the landowner. They occupy farms of very various extent, 70 per cent of them under 50 acres each, 12 per cent between 50 and 100 acres, and 18 per cent farms of more than 100 acres each. 5,000 occupy farms of between 500 and 1,000 acres, and 600 occupy farms exceeding 1,000 acres. His intimate knowledge of the condition of the labourer, and constant residence in the parish, fit him best for the duty of Overseer of the Poor, member of the Board of

Guardians, Churchwarden, and Surveyor of the Roads. He is frank and hospitable to strangers, as a rule; in favour of the established political institutions of the country; loyal as a subject; often available in case of need as a mounted yeoman; and constantly in requisition as a juryman in the Courts of Law.

The farmers are six times as numerous as the landowners, there being 560,000 in Great Britain, and 600,000 in Ireland, the holdings there being on a smaller scale. They employ a vast capital in the aggregate, upwards of four hundred millions sterling, and, unlike that of the landowners, much of it is in daily use, circulating among tradesmen and labourers.

Between the landlords and farmers there is an intermediate class, the land-agents, to whom on most large estates the details of transacting business with the farmers, and looking after the cultivation and buildings and general condition of the property, are committed. . . .

The Labourers

The third class comprises the agricultural labourers, who are necessarily much more numerous than both landowners and tenants. They cannot be said to have any other capital than the furniture of their dwellings, their well-acquired experience in all the details of husbandry, and the bodily strength to use it. The English labourer, of the southern counties especially, has hitherto had but little education, except in his business.

The state of the agricultural labourer of the southern counties has long been the subject of reproach, and, till a recent period, not without good reason. In many parishes the average rate of wages was below the means of maintaining a man's bodily strength adequate to good work, and the result was that two men at low wages was kept to do the work of one well-paid labourer. The employer was a loser by this; and though he might be aware of it, he could not help it, for there was a redundancy of labour seeking employment, and which had to be maintained either by wages or poor-rates. The labourer himself was uneducated, having little knowledge of any district outside his own parish, no means of moving beyond it, while he risked the loss of his legal right to the parish relief in illness or old age if he left it. In such circumstances it was hardly possible for the agricultural labourer to attain any degree of independence. There was no margin for saving, no surplus out of which an enterprising man could make the venture of moving his labour to places in which it would command a better return. And during the long period that this continued, his condition was low, and still shows itself in his small stature and slow gait. From the pressure of this system he was at last emancipated by the extension of his legal right of relief from the parish to the Union, a district much more extensive, and by the simultaneous increase in the demand for labour arising from the rapid development of the other industrial resources of the country. The great extension of steam communication with America, and the encouragement thereby afforded to emigration, drew off rapidly the surplus agricultural population of Scotland and Ireland; wages in both countries quickly increased, and this soon extended its influence southwards. Agricultural labourers' unions were formed in the depressed districts just when this wholesome feeling was spreading throughout the country, and to their efforts much of the natural effect of other causes in producing a rise of wages has been ascribed. This increase of wages was attended by a most useful result, for it forced upon farmers the more extensive use of machinery, and, in the end, brought about a higher scale of wages to the labourer, while the additional cost to the farmer is met to some extent by superior skill, and greater economy in the application of labour. It is worthy of note that the increase of agricultural wages has been greatest in Scotland, where labourer's unions have not taken root, but where the parish school is an old Institution.

The general condition of the agricultural labourer was probably never better than it is at present.

2. Society: (a) the London Season, (b) Changes in Manners

(a) Beatrice Webb, *My Apprenticeship* (London, Longmans, 1924), pp. 47, 50.
(b) Algernon West, *The Nineteenth Century* (April 1897)

Beatrice (1858–1943) and her husband Sidney Webb (1854–1947) were involved in many of the characteristic movements of two of our periods, and some of their work, like Beatrice's share in the *Minority Report* of the *Poor Law Commission* (1909), produced successful results only in the third. Beatrice's father, Richard Potter, was a railway magnate, one of the new class of entrepreneurs who came, as a business aristocracy, to social and political influence.

(a)

Now the first and foremost characteristic of the London season and country-house life, a characteristic which distinguished it from the recreation and social intercourse of the rest of the community, was the fact that some of the men and practically all the women made the pursuit of pleasure their main occupation in life. I say advisedly *some* of the men, because the proportion of functionless males, I mean in the economic sense, varied according to whether the particular social circle frequented was dominated by the Cabinet and ex-Cabinet or by the racing and sporting set. Among my own acquaintances (I except mere partners at London and country-houses dances, for dancing men in my time were mostly fools) there were very few men who were not active brain-workers in politics, administration, law, science or literature. In the racing set, which I knew only by repute, I gathered that the professional brain-workers, whether speculators or artists, bookmakers, trainers or jockeys and the like, rarely belonged to 'society'; in their social and economic subordination these professional workers of the world of sport did not differ materially from other providers of entertainment—gamekeepers, gardeners, cooks and tradesmen. But about the women there was no such distinction. . . . Riding, dancing, flirting and dressing up—in short, entertaining and being entertained—all occupations which imply the consumption and not the production of commodities and services, were the very substance of her life before marriage and a large and important part of it after marriage. And my own experience as an unmarried woman was similar. How well I recollect those first days of my early London seasons: the pleasurable but somewhat feverish anticipation of endless distraction, a dissipation of mental and physical energy which filled up all the hours of the day and lasted far into the night; the ritual to be observed; the presentation at Court, the riding in the Row, the calls, the lunches and dinners, the dances and crushes, Hurlingham and Ascot, not to mention amateur theatricals and other sham philanthropic excrescences. There was of course a purpose in all this apparently futile activity, the business of getting married; a business carried on by parents and other promoters, sometimes with genteel surreptitiousness, sometimes with cynical effrontery.

(b)
One of the most remarkable changes of manners has been that familiarities have taken the place of formalities. In my early days few elderly ladies addressed their husbands by their Christian names in public. I never heard my mother call my father by his Christian name. I recollect that Lady ——'s fame was imperilled because, after some great man's death, a letter from her to him was discovered beginning with his Christian name. I think I am right in saying that at Eton we never recognized the existence of such a thing. Even boys who 'knew each other at home' never divulged them. Letters between friends often began 'My dear Sir', and many boys in my time addressed their fathers always as 'Sir'. A friend of mine, Gerald Ponsonby, dining with Lady Jersey, heard her say that she never recollected her father, Lord Westmorland, though specially attached to his sister, Lady Lonsdale, call her anything but Lady Lonsdale; and Henry Greville, who was present at the same dinner, said he remembered his mother, Lady Charlotte, and her brother, the Duke of Portland, meeting in the morning at Welbeck and saying, 'How is your Ladyship this morning?' and her replying with all solemnity, 'I am quite well, I am obliged to your Grace.'

All shopkeepers are now 'young gentlemen' and 'young ladies'. The Duchess of Somerset, on making inquiry about something she had purchased at Swan & Edgar's, was asked if she had been served by a young gentleman with fair hair. 'No,' she said meditatively, 'I think it was by an elderly nobleman with a bald head.'

Photography was in its infancy early in the fifties, and had just begun to be common in the hideous daguerreotypes and talbotypes of that time. The witty Lady Morley used to say in reply to any complaint of the dulness of the weather, 'What can you expect when the sun is busy all day taking likenesses in Regent Street?'

3. The New Surburbans

C. F. G. Masterman, *The Condition of England* (London, Methuen, 1909), pp. 57–60, 67–8.

Masterman was a junior minister in the Liberal government brought to power by the electoral landslide of 1906. On 'the condition of England' question he wrote in the great Victorian tradition of Carlyle, Matthew Arnold, and Ruskin, arguing for reform and the extension of the power of the State. In his brilliant, impressionistic description of English society, he saw the following groups: the Conquerors, the rich ruling class; the Suburbans, the emergent middle classes; the Multitude, who made up the inarticulate 80 per cent of the nation; the Prisoners, the worst victims of the system; and the countryside which was feudal, poor, and decaying.

(a)

They are the creations not of the industrial, but of the commercial and business activities of London. They form a homogeneous civilization—detached, self-centred, unostentatious—covering the hills along the northern and southern boundaries of the city, and spreading their conquests over the quiet fields beyond. They are the peculiar product of England and America; of the nations which have pre-eminently added commerce, business, and finance to the work of manufacture and agriculture. It is a life of Security; a life of Sedentary occupation; a life of Respectability; and these three qualities give the key to its special characteristics. Its male population is engaged in all its working hours in small, crowded offices, under artificial light, doing immense sums, adding up other men's accounts, writing other men's letters. It is sucked into the City at daybreak, and scattered again as darkness falls. It finds itself towards evening in its own territory in the miles and miles of little red houses in little silent streets, in number defying imagination. Each boasts its pleasant drawing-room, its bow-window, its little front garden, its high-sounding title— 'Acacia Villa' or 'Camperdown Lodge'—attesting unconquered human aspiration. There are many interests beyond the working hours: here a greenhouse filled with chrysanthemums, there a tiny grass patch with bordering flowers; a chicken-house, a bicycle shed, a tennis lawn. The women, with their single domestic servants, now so difficult to get, and so exacting when found, find time hang rather heavy on their hands. But there are excursions to shopping centres in the West End, and pious sociabilities, and occasional theatre visits, and the interests of home. The children are jolly, well-fed, intelligent English boys and girls; full of curiosity, at least in the earlier years. Some of them have real gifts of intellect and artistic skill, receiving in the suburban secondary schools the best education which England is giving today. You may see the whole suburbs in August transported to the more genteel of the southern watering places; the father, perhaps, a little bored; the mother perplexed with the difficulty of cramped lodgings and extortionate prices. But the children are in a magic world, crowding the seashore, full of the elements of delight and happy laughter.

(b)
Yet in the crumbling and decay of English rural life, and the vanishing of that 'yeoman' class which in Scotland provides a continuous breeding ground of great men, it would seem that it is from the suburban and professional people we must more and more demand a supply of men and women of capacity and energy adequate to the work of the world. Sufficiently vulnerable to criticism as they appear today, finding no one who will be proud of them because they are not proud of themselves, they yet offer a storehouse of accumulated physical health and clean simplicities of living. Embedded in them are whole new

societies created by legislation and a national demand, whose present development is full of interest, whose future is full of promise. Here is, for example, the new type of elementary teacher—a figure practically unknown forty years ago—drawn in part from the tradesmen and the more ambitious artisan population, and now, lately, in a second generation, from its own homes. It is exhibiting a continuous rise of standard, keen ambitions, a respect for intellectual things which is often absent in the population amongst which it resides. Its members are not only doing their own work efficiently, but are everywhere taking the lead in public and *quasi*-public activities. They appear as the mainstay of the political machine in suburban districts, serving upon the municipal bodies, in work, clear-headed and efficient; the leaders in the churches and chapels, and their various social organizations. They are taking up the position in the urban districts which for many generations was occupied by the country clergy in the rural districts; providing centres with other standards than those of monetary success, and raising families who exhibit sometimes vigour of character, sometimes unusual intellectual talent. A quite remarkable proportion of the children of elementary schoolmasters is now knocking at the doors of the older Universities, clamouring for admittance, and those who effect entrance are often carrying off the highest honours. This process is only in its beginning; every year the standard improves; these 'servants of the State' have assured to them a noteworthy and honourable future. Again, there is no doubt that the conception of social service is making progress against the resistance of whatever is solid in the suburban tradition of individualism and indifference. Even the Socialist no longer turns from the Middle Classes in disgust. He is coming to regard them as the most fruitful field for his propaganda. The women— or a remnant of them—are finding outlet for suppressed energy and proffered devotion in an agitation for the vote. Sixpenny reprints of proof or disproof of religion, the world's classics in neat shilling volumes, sevenpenny novels, and a variety of printed matter are irrigating the suburbs with a fresh flood of literature. It is not impossible to conceive of a time when a Middle Class will definitely build up a standard of its own: no longer turning to a wealthy and leisured company above it for effective imitation of a life to which it is unsuited. Becoming conscious, for the first time, that it possesses elements to contribute to the stream of natural life which can be provided neither by the rich nor the poor, it may gain that collective respect and pride in itself which it has not yet achieved. Abandoning its panic fear of the industrial peoples, it may find itself treating with them as an equal, exacting terms in return for its alliance. At best it may even resist the stampedes of those who find the support of the 'Middle Classes' always easily obtainable for an agitation against the Income Tax or in favour of municipal reaction, or for any system which will 'broaden the basis of taxation' by shifting it from the shoulders of the rich to the shoulders of the poor.

4. The Potteries

Harold Owen, *The Staffordshire Potter* (London, Grant Richards, 1901), pp. 337–42.

The area of the Potteries was typical of the specialized regional character of the nineteenth-century industry. Such areas were generally unbeautiful, but housed distinctive communities with strong local feeling. The Potteries were recorded also in the novels of Arnold Bennett (1867–1931). The extract here quoted is concerned with 'where and how the workers live'.

Starting from Longton, which rejoices in a Town Council, you run into the arms of Fenton Local Board, pass thence through the domain of the Corporation of Stoke-upon-Trent, are handed on to the County Borough of Hanley, and having thus reached the zenith of municipal glory, pass with a gentle descent through the incorporate town of Burslem, to end the journey within the realm of the Tunstall Local Board.

This seven-mile main road—which has two intervals, each of half-a-mile's length, of villa houses—cuts through the centre of the towns and townships in which the pottery workers work and live. The footpaths are fairly well paved, the roads are possibly all that the principles of Macadam and the ministrations of a steam-roller can make them; but the clayey subsoil keeps the rain-water on the surface, and the mud of the roads is blackened blacker by the droppings from the jolting coal-carts which pass along the thoroughfares to the manufactories. These coal-carts are followed by poor and ragged little boys and girls, sent out by their mothers—and often accompanied by them—to pick up in baskets, buckets, or folded dress-fronts, the small coal which falls from the carts. Occasionally, a tolerably large lump is furtively loosened, and when the next jolt in its charity sets it free, there is a scramble for the prize.

The Potteries is in its interests and definite characteristics just as much one town as Manchester, Liverpool, or Birmingham. Its powers and action, however, are split up between six separate governing bodies, and the opportunity of natural amalgamation which was offered by the formation of County Councils under the Local Government Act of 1888 was allowed to pass. Hanley did, indeed, take unto itself the dignity of a County Borough, by virtue of its superior population, but for any good that this act did to the rest of The Potteries, Hanley might just as well have set up to be a Republican State. But Hanley is in every respect, save population, acreage, and rateable value, precisely like Burslem, as Tunstall is like Longton, and as all are like each other.

The effect of thus having an industry (whose products are undoubtedly beautiful, but the production of which involves conditions that are undoubtedly not) straggling through the whole length of the district, is that the whole partakes of the character of its worst part, and there are no oases of a purely residential character. Birmingham has its Edgbaston, but The Potteries is all Soho. The residential quarters lie outside it—Wolstanton and Newcastle barely escape The Potteries (and all its 'works') by the providence of nature in setting them on a hill—and the favoured spots for those who are not compelled to live where they work lie farther out, to Trentham, Stone, and Sandon to the south, and northwards even unto Alsager and Congleton in Cheshire. Possibly the greatest advantage incidental to living in The Potteries is the hope of being able to make enough money to live out of it.

The topography of The Potteries is very varied. You may pass down a street to find that its exit is barred by a mound of ashes, higher than the houses, gathered from furnace fires long since extinct, or of shale, raised from a coal-pit. The coal-pit is probably disused, and a dome of bricks covers the shaft on a neighbouring mound. You may pass up a road, and dis-

GENERAL SOCIAL CONDITIONS   49

cover through an interval in the rows of houses that the playground of the children at the 'backs' verges on a yawning marlhole, deeper many times than the houses. These pimples of pit-mounds and these pock-marks of marl-holes scar the features—otherwise none too fair—of The Potteries. Man's faith cannot move pit-mounds, so houses are built on the lower slopes, and the monarch of the range stands an eternal sentinel at the back doors. But the deep places of the earth may be made level, and the marl-hole which yields no more marl is filled up, in the course of time, by cart-loads of broken pots and shards, with an occasional variation of cinders, from the manufactories. Dirty women and dirty little children may be seen sifting the contents of these carts before the dust of their falling has cleared away. They wait there all day for the carts, but what they hope to find is a mystery, as manufacturers sometimes pay for the privilege of thus getting rid of the rubbish of their works, and at any rate give it away. What they have taken from the earth they thus give back to it. When a marlhole is nearly filled up, it swims within the ken of the speculative builder, and a row of cottage houses soon rests upon its foundations of fifty yards of broken pottery. One wonders whether Macaulay's New Zealander will stay long on the ruins of London Bridge, to the neglect of the archæological research which excavations in The Potteries might satisfy.

Each town has thus its ups and downs of building levels. The exigency of the presence of a pit-mound or of a 'shord-ruck' in the centre of a plot of ground may demand that the space left unoccupied shall be filled up with more broken pots, until that particular piece of ground has reached an independent and exclusive level of its own, without reference to its neighbour. Its neighbour may be a plot of virgin ground, hitherto undefiled, yet on a much lower level. Or, to vary the metaphor with the matter, a few cottages may rest on mother earth, yet no longer enjoy nature's equality, for around and above them stretch acres of broken shards, whose advancing tide must one day engulf them, so that a street may pass over their chimneypots.

The Potteries is thus, and inevitably, mainly a place of muddy, squalid streets, insignificant public buildings, smoky atmosphere, pot-works, and higgledy-piggledy rows of small houses. Nine-tenths of its population are of the working-class, and its well-to-do live mostly in their little colonies outside. The character of a neighbourhood must inevitably have some effect on the dwellers within it, but, apart from a certain habit of thought which may be styled 'independence'—the result of the self-contained character of the district and its trade—the inhabitants of The Potteries bear up well against the aesthetic disadvantages of their surroundings. Each town has its free library; schools of art flourish amongst them, and remind one of the innately beautiful side of the trade in which they are engaged; musical societies, which have successfully competed with some of the best choral combinations in the country, have many adherents and supporters; and Hanley holds one of the best of the provincial musical festivals.

5. **The East End of London**

Paul de Rousiers, *The Labour Question in Britain* (translated by F. L. D. Herbertson, London, Macmillan, 1896), pp. 100–2.

Factory Acts had curbed the worst industrial excesses of the nineteenth century, but sweated workers in domestic industries, especially in London, were not protected until the late nineteenth and early twentieth century.

... For one poverty-stricken woman who refuses to work herself blind by stitching for hours to earn a few pence, there are two others still more abjectly poor who will accept the terms. In the furniture trade the man who has made a chest or a cupboard at home, and who has procured the wood and varnish on credit, and who is also hungry and in debt for his rent, is obliged to put the furniture on a truck and take it to a big emporium in Curtain Road, where it is bought and paid for in ready money at a low rate—and this too is sweating. Another victim is the small shoemaker with no custom, who, when the leather-seller refuses further credit, or an apprentice asks his trifling wage, is obliged to take half a dozen pairs of boots to the wholesale shop to satisfy his creditors.

It would be easy to multiply examples, but these are enough to show that sweating is not confined to a single trade. They also point to the second characteristic I alluded to, the chronic nature of the system. The man without a fair start, who is driven by necessity to carry his work to the wholesale dealer as fast as it is finished, is at the mercy of every vicissitude, and will often find his way back to the wholesale shop, where he pawns rather than sells his goods. The needlewoman who is perpetually competing with all the other needy women of a large town will never get a remunerative price for her work. Further, the sweating system is organised and in uninterrupted swing; it is provided with organs, it has its sale-rooms and its recognised middlemen. Curtain Road, in London, is lined with general shops for the sale of furniture and boots and shoes, while in Paris the huge *maisons de nouveautés*, and the weekly Saturday fair of the Avenue Ledru-Rollin, play the same part. Sweating is as much a chronic condition as improvidence and poverty in those grades where improvidence and poverty are the general rule.

God knows that this is the case in the East End. There is no need to push an inquiry very far to be assured on this point; it is quite enough to walk at broad noonday in Whitechapel or about Tower Hill. A population in abject poverty, squalid, wan, and forbidding in appearance, seethes in the narrow, dirty, muddy, reeking streets. I was at an open-air market in a lane off Commercial Street. Every sort of merchandise, especially food, was exposed for sale under temporary awnings, and was purchased by the women of the neighbourhood. It was a painful sight to see the repulsive appearance of these women, with their bloated, disfigured faces, often diseased or withered, and almost all bearing the evident brand of vice and degradation. The impression was confirmed by their dress. They wore ragged garments, third or fourth hand, which had doubtless been worn by some woman of the middle class till they were shabby, and then handed on to the servants, descending finally, when they were worn out, to their present wearers. Every article in the costume had had a different origin, and strange medleys were produced, thanks to the English love of brilliant colours. One toilette which I noticed consisted of a violet skirt and a bodice of brown plush, over which was thrown a crochet shawl of bright red, which went sufficiently badly with the rest. Notwithstanding the stains and the countless holes and the frayed edges, the violet silk skirt still shot brilliant flashes here and there. Add to the picture unkempt hair, an absence or insufficiency of under-linen which betrayed itself where a button was missing or through tell-tale rents, filthy hands and face, boots down at the heel, everything proclaiming aloud the most abject poverty and the total absence of care and self-respect. To add to the horror of the general effect, many women wore frowsy wigs, generally put on awry! I was informed that this ornament is peculiar to Polish Jewesses, whose rites oblige them to shave their heads when they marry. Jews are very numerous in the East End, and it is chiefly among them that the sweating system finds hands, more especially in the clothing trade.

About the doors of public-houses, especially on Saturdays, may be seen crowds of persons of whom a considerable number are women. Drink is a ceaseless cause of physical and moral degradation, poverty, and irremediable ruin. It is only necessary to enter any one of the lanes which branch off from the main streets to convince oneself that vice is in its element. On the doorsteps of the little, mean, narrow houses where families are crowded together higgledy-piggledy, many a wretched creature may be seen like those just described. The majority of them have no regular occupation. At a pinch they will go into a factory, but they cannot be got to stay. The least degraded go to swell the number of the sweated.

Even this is not the most pitiful sight of the East End, for it is still more heartrending to see children growing up in such surroundings. Unhappy little creatures, in rags and tatters, running barefoot on the muddy or frozen stones! In the damp climate of London this material destitution assumes a character of its own....

It is the depressing surroundings of East London that the sweating system has taken root most firmly and flourishes best. The name was first applied to the Jewish tailors of the quarter who gave out work to be done at home, but it has been extended to cover a number of other cases, for, as we have seen, a variety of different industries have become its prey.

It should be borne in mind, however, that none of the industries which suffer are universally affected. There are plenty of tailors, shoemakers, and cabinetmakers in London who escape. In the first place, there is the fashionable tailor, who obtains a high price for his cut, and who executes orders in his own workroom under his direct supervision. In the same category are included the elegant West End shoemaker and the first-class cabinetmaker. All these are producing articles of luxury. Then we have the large manufactories in the boot and furniture trades, engaged in the manufacture either of complete boots or furniture, or else of some particular parts. Here, too, sweating is unknown. The men are collected in large groups under the control of a rich employer, are assisted by machinery, and come under the type of the factory hand, whom we shall study in the third part of this work. If they think they have any grievance, they can strike suddenly and in a body, and compel attention to their demands. Further, the factory is subject to severe police regulations as to sanitary condition, and consequently they are certain of working under hygienic conditions. Legislation also protects women and children employed in factories against excessive hours of labour. In short, although no law limits the hours of work for adult males, yet, as a matter of fact, no large employer of labour keeps his hands at work more than ten or twelve hours a day, while in small tailoring establishments a day of fourteen or fifteen hours is common when there is any pressure, and the poor needlewomen, shoemakers, or cabinetmakers, who do their work at home, often work all night to finish a piece of work and buy the much-needed bread.

Thus, in the trades affected by sweating, neither the establishments producing a costly article nor the large factory are touched. Where it is felt is in the manufacture of low-class goods made in the small workshop or at home.

6. Poverty in York, 1899

B. Seebohm Rowntree, *Poverty: a Study of Town Life* (London, Longmans, 1941 edition), pp. 28–31, 132–7.

The first great social study of an English city was *Life and Labour of the People of London* (1889–97) by Charles Booth, whose methods were followed by Rowntree. Their conclusions were similar: that a third of the population lived in poverty, or below subsistence. Two of Rowntree's basic concepts were: 'primary poverty', 'families whose total earnings were insufficient to obtain the minimum necessaries for the maintenance of merely physical efficiency'; 'secondary poverty', applying to families some of whose earnings were diverted from physical necessities and who were, in consequence, pulled down to poverty.

CLASSIFICATION OF THE POPULATION OF YORK

Class	Family Income (for moderate Family)	Number of Persons in each Class	Percentage in each class calculated upon total wage earners in York (excluding domestic servants and persons in public institutions	Percentage of whole Population
A	Under 18s. per week	1,957	4·2	2·6
B	18s. and under 21s.	4,492	9·6	5·9
C	21s. and under 30s.	15,710	33·6	20·7
D	Over 30s.	24,595	52·6	32·4
E	Female domestic servants	4,296	...	5·7
F	Servant-keeping class	21,830	...	28·8
G	In public institutions	2,932	...	3·9
		75,812	100·0	100·0

If we now add up the various items of necessary expenditure under the three heads, Food, House Rent (and Rates), and Household Sundries (including clothes), we obtain the following figures:

TABLE SHOWING THE MINIMUM NECESSARY EXPENDITURE PER WEEK FOR FAMILES OF VARIOUS SIZES

Family	Food	Rent, say—	Household Sundries	Total
1 man	3s.	1s. 6d.	2s. 6d.	7s.
1 woman	3s.		2s. 6d.	7s.
1 man and 1 woman	6s.	2s. 6d.	3s. 6d.	11s. 8d.
1 man, 1 woman, 1 child	8s. 3d.		3s. 9d.	14s. 6d.
,, ,, 2 children	10s. 6d.	4s.	4s. 4d.	18s. 10d.
,, ,, 3 ,,	12s. 9d.		4s. 11d.	21s. 8d.
,, ,, 4 ,,	15s.		5s. 6d.	26s.
,, ,, 5 ,,	17s. 3d.		6s. 1d.	28s. 10d.
,, ,, 6 ,,	19s. 6d.	5s. 6d.	6s. 8d.	31s. 8d.
,, ,, 7 ,,	21s. 9d.		7s. 3d.	34s. 6d.
,, ,, 8 ,,	24s.		7s. 10d.	37s. 4d.

Having established a minimum standard of necessary expenditure, we are now in a position to ascertain what proportion of the population of York are living in 'primary' poverty.

As stated in [Chapter II] an estimate was made of the earnings of every working-class family in York. In order to ascertain how many of these families were living in a state of 'primary' poverty, the income of each was compared with the foregoing standard, due allowance being made in every case for size of family and rent paid.

Let us now see what was the result of this examination. No less than 1465 families, comprising 7230 persons, were living in 'primary' poverty. *This is equal to 15·46 per cent of the wage-earning class in York, and to 9·91 per cent of the whole population of the city.*

The above estimate, it should be particularly noted, is based upon the assumption that *every penny earned by every member of the family* went into the family purse, and was judiciously expended upon necessaries.

The statement on page 56 shows the income and estimated necessary expenditure of Section 6.

It will have been noticed that Section 6 consists of unskilled workers of various grades, 73 per cent being general labourers; whilst the others holding the lower posts in their respective occupations are employed upon work which is scarcely more difficult or responsible than that of the general labourer, and whose wages are consequently only slightly, if at all, in excess of those paid to the latter. That so many wage-earners should be in a state of primary poverty will not be surprising to those who have read the preceding pages. Allowing for broken time, the average wage for a labourer in York is from 18s. to 21s.; whereas, according to the figures given earlier in this chapter, the minimum expenditure necessary to maintain in a state of physical efficiency a family of two adults and three children is 21s. 8d., or, of there are four children, the sum required would be 26s.

It is thus seen that *the wages paid for unskilled labour in York are insufficient to provide food, shelter, and clothing adequate to maintain a family of moderate size in a state of*

Income				Expenditure			
	£	s.	d.		£	s.	d.
Weekly income of 640 families . . .	600	0	0	Weekly rent . .	112	15	5
Weekly balance—				Weekly minimum cost of food and sundries necessary to maintain 1376 adults and 2880 children in a state of physical efficiency . . .			
Deficiency . . (= 5s. 0¼d. per family)	160	17	5		648	2	0
	£760	17	5		£760	17	5

bare physical efficiency. It will be remembered that the above estimates of necessary minimum expenditure are based upon the assumption that the diet is even less generous than that allowed to able-bodied paupers in the York Workhouse, and that *no allowance is made for any expenditure other than that absolutely required for the maintenance of merely physical efficiency.*

And let us clearly understand what 'merely physical efficiency' means. A family living upon the scale allowed for in this estimate must never spend a penny on railway fare or omnibus. They must never go into the country unless they walk. They must never purchase a halfpenny newspaper or spend a penny to buy a ticket for a popular concert. They must write no letters to absent children, for they cannot afford to pay the postage. They must never contribute anything to their church

		s.	d.
Food	.	12	9
Rent—say	.	4	0
Clothes—two adults at 6*d*.	.	1	0
three children at 5*d*.	.	1	3
Fuel	.	1	10
All else—five persons at 2*d*.	.	0	10
	Total	21	8

or chapel, or give any help to a neighbour which costs them money. They cannot save, nor can they join sick club or Trade Union, because they cannot pay the necessary subscriptions. The children must have no pocket money for dolls, marbles, or sweets. The father must smoke no tobacco, and must drink no beer. The mother must never buy any pretty clothes for herself or for her children, the character of the family wardrobe as for the family diet being governed by the regulation, 'Nothing must be bought but that which is absolutely necessary for the maintenance of physical health, and what is bought must be of the plainest and most economical description.' Should a child fall ill, it must be attended by the parish doctor; should it die, it must be buried by the parish. Finally, the wage-earner must never be absent from his work for a single day.

If any of these conditions are broken, the extra expenditure involved is met, *and can only be met*, by limiting the diet; or, in other words, by sacrificing physical efficiency.

That few York labourers receiving 20*s*. or 21*s*. per week submit to these iron conditions in order to maintain physical efficiency is obvious. And even were they to submit, physical efficiency would be unattainable for those who had three or more children dependent upon them. It cannot therefore be too clearly understood, nor too emphatically repeated, *that whenever a worker having three children dependent on him, and receiving not more than* 21*s*. 8*d*. *per week, indulges in any expenditure beyond that required for the barest physical needs, he can do so only at the cost of his own physical efficiency, or of that of some members of his family.*

If a labourer has but two children, these conditions will be better to the extent of 2*s*. 10*d*.; and if he has but one, they will be better to the extent of 5*s*. 8*d*. And, again, as soon as his children begin to work, their earnings will raise the family above the poverty line. But the fact remains that every labourer who has as many as three children must pass through a time, probably lasting for about ten years, when he will be in a state of 'primary' poverty; in other words, when he and his family will be *underfed*.

The life of a labourer is marked by five alternating periods of want and comparative plenty. During early childhood, unless his father is a skilled worker, he probably will be in poverty; this will last until he, or some of his brothers or sisters, begin to earn money and thus augment their father's wage sufficiently to raise the family above the poverty line. Then follows the period during which he is earning money and living under his parents' roof; for some portion of this period he will be earning more money than is required for lodging, food, and clothes. This is his chance to save money. If he has saved enough to pay for furnishing a cottage, this period of comparative prosperity may continue after marriage until he has two or three children, when poverty will again overtake him. This period of poverty will last perhaps for ten years, i.e. until the first child is fourteen years old and begins to earn wages; but if there are more than three children it may last longer. While the children are earning, and before they leave the home to marry, the man enjoys another period of prosperity—possibly, however, only to sink back again into poverty when his children have married and left him, and he himself is too old to work, for his income has never permitted his saving enough for him and his wife to live upon for more than a very short time.

7. The Rise of Living Standards

D. A. Wells, *Recent Economic Change*
(New York, Longmans, 1898), pp. 354–8.

Industrialization not only changed the way of life of the workers, for the majority it raised their standards of living. At the same time as social investigators like Booth and Rowntree were discovering widespread poverty in London, others were demonstrating just how much national wealth had increased in the second half of the nineteenth century. The paradox of poverty along with increasing wealth continued into the twentieth century.

During the last twenty-five or thirty years, the aggregate wealth of Great Britain, as also that of the United States and France, has increased in an extraordinary degree. In Great Britain the increase from 1843 to 1885 in the amount of property assessable to the income-tax is believed to have been one hundred and forty per cent, and from 1855 to 1885 about one hundred per cent. The capital subject to death duties (legacy and succession taxes) which was £41,000,000 in 1835, was £183,930,000 in 1885; an increase in fifty years of nearly three hundred and fifty per cent. An estimate of the total income of the country for 1886 was £1,270,000,000; and of its aggregate wealth, about £9,000,000,000, or $45,000,000,000. Have now the working-classes of Great Britain gained in proportion with others in this enormous development of material wealth? Thanks to the labours of such men as the late Dudley Baxter? and Leoñe Levi, David Chadwick, and Robert Giffen, this question can be answered (comparatively speaking for the first time) with undoubted accuracy.

Fifty years ago, one third of the working masses of the United Kingdom were agricultural labourers; at present less than one eighth of the whole number are so employed. Fifty years ago the artisans represented about one third of the whole population; today they represent three fourths. This change in the composition of the masses of itself implies improvement, even if there had been no increase in the wages of the different classes. But, during this same period, the 'money' wages of all classes of labour in Great Britain have advanced about one hundred per cent, while the purchasing power of the wages in respect to most commodities, especially in recent years, has been also very great. Among the few things that have not declined, house-rent is the most notable, a fact noticed equally in Great Britain and France, although in both countries the increase in the number of inhabited houses is very large; the increase in the item of houses in the income-tax assessments of the United Kingdom between 1875 and 1885 having been about thirty-six per cent. But high rents, in the face of considerable building, are in themselves proof that other things are cheap, and that the competition for comfortable dwellings is great.

The Government of Great Britain keeps and publishes an annual record of the quantities of the principal articles imported, or subject to an excise (internal revenue) tax, which are retained for home consumption per head, by the total population of the kingdom. For these records the following table has been compiled. From a humanitarian point of view, it is one of the most wonderful things in the history of the latter half of the nineteenth century:

Per capita CONSUMPTION OF DIFFERENT COMMODITIES (IMPORTED OR SUBJECT TO EXCISE TAXES) BY THE POPULATION OF GREAT BRITAIN

Articles		1840	1886	1887
Bacon and hams	lbs.	0·01	11·95	11·29
Butter	,,	1·05	7·17	8·14
Cheese	,,	0·92	5·14	5·39
Currants and raisins	,,	1·45	4·02	4·34
Eggs	No.	3·63	28·12	29·37
Rice	lbs.	0·90	10·75	7·69
Cocoa	,,	0·08	0·41	0·43
Coffee	,,	1·08	0·86	0·79
Wheat and wheat-flour	,,	42·67	185·76	220·75
Raw sugar	,,	15·20	47·21	52·95
Refined sugar	,,	None.	18·75	20·25
Tea	,,	1·22	4·87	4·95
Tobacco	,,	0·86	1·42	1·44
Wine	gals.	0·25	0·36	0·37
Spirits (foreign)	,,	0·14	0·24	0·23
British (British)	,,	0·83	0·73	0·72
Malt	bush.	1·59	1·64	...
Beer (1881)	gals.	27·78	26·61	26·90

During all the period of years covered by the statistics of this table, the purchasing power of the British people in respect to the necessities and luxuries of life has therefore been progressively increasing, and has been especially rapid since 1873–'76. Converting this increase in the purchasing power of wages into terms of money, the British workman can now purchase an amount of the necessaries of life for 28s. 5d., which in 1839 would have cost him 34s. 0½d. But this statement falls very far short of the advantages that have accrued to him, for wages in Great Britain, as before stated, are fully one hundred per cent higher at the present time than they were in 1839.

The impression probably prevails very generally in all countries that the capitalist classes are continually getting richer and richer, while the masses remain poor, or become poorer. But in Great Britain, where alone of all countries the material (i.e., through long-continued and systematized returns of incomes and estates [probate] for taxation) exists for scientific inquiry, the results of investigations demonstrate that this is not the case.

In the case of estates, the number subjected to legacy and succession duties within the last fifty years has increased in a ratio double that of the population, but the average amount of property per estate has not sensibly augmented. If, therefore, wealth among the capitalist classes has greatly increased, as it has, there are more owners of it than ever before; or, in other words, wealth, to a certain extent, is more diffused than it was. Of the whole number of estates that were assessed for probate duty in Great Britain in 1886, 77.5 per cent were for estates representing property under £1,000 ($5,000).

In the matter of national income, a study of its increase and apportionment among the different classes in Great Britain has led to the following conclusions: Since 1843, when the income-tax figures begin, the increase in national income is believed to have been £755,000,000. Of this amount, the *income* from the capitalist classes increased about one hundred per cent, or from £190,000,000 to £400,000,000. But, at the same time, the number of the capitalist classes increased so largely that the average amount of capital possessed among them per head increased only fifteen per cent, although the increase in capital itself was in excess of one hundred and fifty per cent. In the case of the 'upper' and 'middle' classes, the income from their 'working' increased from £154,000,000 to £320,000,000, or about one hundred per cent; while, in the case of the masses (i.e., the manual labor classes), which have increased in population only thirty per cent since 1843, the increase of their incomes has gone up from £171,000,000 to £550,000,000, or over two hundred per cent. Between 1877 and 1886 the number of assessments in Great Britain for incomes between £150 ($750) and £1,000 ($5,000) increased 19.26 per cent, while the number of assessments for incomes of £1,000 and upward decreased 2.4 per cent. What has happened to all that large class whose annual income does not reach the taxable limit (£150) is sufficiently indicated by the fact that while population increases pauperism diminishes.

Thus, in the United Kingdom, during the last fifty years, the general result of all industrial and societary movement, according to Mr. Giffen, has been, that 'the rich have become more numerous, but not richer individually; the "poor" are, to some smaller extent, fewer; and those who remain "poor" are, individually, twice as well off on the average as they were fifty years ago. The poor have thus had almost all the benefit of the great material advance of the last fifty years.'

8. The Poor Law

Report of the Royal Commission on the Poor Laws (Cmd. 4499, H.M.S.O., 1909), *The Minority Report*, pp. 9–12, 705.

The Poor Law of 1834, still in operation at the beginning of the twentieth century, forbade 'outdoor relief' to the 'able-bodied poor', requiring such people to enter the workhouse and to work for wages less than those paid to the lowest-paid employed workers. In this way the workhouse was supposed to act as a deterrent to the lazy and irresponsible. Such a system was inoperable in times of trade depression and it was attacked as inhumane and impractical. The Royal Commission of 1909 was divided in its recommendations, and *The Minority Report*, by the Radicals led by Beatrice Webb, received wide publicity. *The Majority Report* envisaged insurance against unemployment, and both reports wanted to use the new local government machinery created by the County Councils Act (1888) to administer poor relief.

(a) THE PROMISCUITY OF THE GENERAL MIXED WORKHOUSE

We see no reason to differ from our predecessors, the Royal

Commissioners of 1834, in their decisive condemnation of the General Mixed Workhouse. We do not wish to suggest or imply that the workhouses of to-day are places of cruelty; or that their 250,000 inmates are subjected to any deliberate illtreatment. These institutions are, in nearly all cases, clean and sanitary; and the food, clothing and warmth are sufficient— sometimes more than sufficient—to maintain the inmates in physiological health. In some cases, indeed, the buildings recently erected in the Metropolis and elsewhere have been not incorrectly described, alike for the elaborateness of the architecture and the sumptuousness of the internal fittings, as 'palaces' for paupers. In many other places, on the other hand, the old and straggling premises still in use, even in some of the largest Unions, are hideous in their bareness and squalor. But whether new or old, urban or rural, large or small, sumptuous or squalid, these establishments exhibit the same inherent defects. We do not ignore the zeal and devotion by means of which an exceptionally good Master and Matron, under an exceptionally enlightened committee, here and there, for a brief period, succeed in mitigating, or even in counteracting, the evil tendencies of a general mixed institution. But these evil tendencies, exactly as they were noted by the Commissioners of 1834, are always at work; and sooner or later they have prevailed in every Union of which we have investigated the history. After visiting personally Workhouses of all types, new and old, large and small, in town and country, in England and Wales, in Scotland and Ireland, we find that the descriptions of the Workhouses of 1834, so far as we have quoted them above, might be applied, word for word, to many of the Workhouses, of to-day. The dominant note of the institutions of today, as it was of those of 1834, is their promiscuity. We have ourselves seen, in the larger Workhouses, the male and female inmates, not only habitually dining in the same room in each other's presence, but even working individually, men and women together, in laundries and kitchens; and enjoying in the open yards and long corridors innumerable opportunities to make each other's acquaintance. It is, we find, in these large establishments, a common occurrence for assignations to be made by the inmates of different sexes, as to spending together the 'day out', or as to simultaneously taking their temporary discharge as 'Ins and Outs'. It has not surprised us to be informed that female inmates of these great establishments have been known to bear offspring to male inmates and thus increase the burden on the Poor Rate. No less distressing has it been to discover a continuous intercourse, which we think must be injurious, between young and old, innocent and hardened. In the female dormitories and day-rooms women of all ages, and of the most varied characters and conditions, necessarily associate together, without any kind of constraint on their mutual intercourse. There are no separate bedrooms; there are not even separate cubicles. The young servant out of place, the prostitute recovering from disease, the feeble-minded woman of any age, the girl with her first baby, the unmarried mother coming in to be confined of her third or fourth bastard, the senile, the paralytic, the epileptic, the respectable deserted wife, the widow to whom Outdoor Relief has been refused, are all herded indiscriminately together. We have found respectable old women annoyed by day and by night by the presence of noisy and dirty imbeciles. 'Idiots who are physically offensive or mischievous, or so noisy as to create a disturbance by day and by night with their howls, are often found in Workhouses mixing with others both in the sick wards and in the body of the house.' We have ourselves seen, in one large Workhouse, pregnant women who have come in to be confined, compelled to associate day and night, and to work side by side, with half-witted imbeciles and women so physically deformed as to be positively repulsive to look upon. In the smaller country Workhouses, though the promiscuity is numerically less extensive, and, in some respects, of less repulsive character, the very smallness of the numbers makes any segregation of classes even more impracticable than in the larger establishments. A large proportion of these Workhouses have, for instance, no separate sick ward for children, and, in spite of the ravages of measles, etc., not even a quarantine ward for the constant stream of newcomers. Accordingly, in the sick wards of the smaller Workhouses, with no constraint on mutual intercourse, we have more than once seen young children in bed with minor ailments, next to women of bad character under treatment for contagious disease, whilst other women, in the same ward, were in advanced stages of cancer and senile decay. Our Children's Investigator reports, after visiting many Workhouses in town and country, 'that children when detained in the Workhouse always come into contact with the ordinary inmates. Certainly, in a country Workhouse this seems impossible to avoid. Paupers are always employed to help with the rough scrubbing and cleaning, and though Matrons invariably try to send the more respectable women into the children's quarters, often the only women available are the mothers of illegitimate babies'. In many Workhouses we have ourselves found the children having their meals in the same room and at the same times as the adult inmates of both sexes, of all ages, and of the most different conditions and characters. Even the imbeciles and the feeble-minded are to be found in the same dining-halls as the children. In some Workhouses, at any rate, the boys over eight years of age have actually to spend the long hours of the night in the same dormitories as the adult men. In all the small Workhouses and in many of the larger ones, the infants are wholly attended to by, and are actually in charge of, aged, and often mentally defective, paupers; the able-bodied mothers having, during the first year, daily access to their own babies for nursing, and, subsequently, such opportunities for visiting the common nursery as the Master may decide. In the better managed, and in the largest establishments the nursery is, it is true, in charge of a salaried nurse, but even here the handling of the babies is mostly left to pauper inmates. However desirable may be the intercourse between an infant and its own degraded mother, it is not the advantage of the scores of infants in the nursery to be perpetually in close companionship for the first three or four years of their lives, with a stream of mothers of various types that we have mentioned. Such a nursery embedded in the midst of an institution containing, not merely hundreds, but thousands of paupers of the most diverse classes is impregnated through and through with the atmosphere of pauperism.

(*b*) SCHEME OF REFORM

91. That the Boards of Guardians in England, Wales and Ireland, and (at any rate as far as Poor Law functions are concerned) the Parish Councils in Scotland, together with all combinations of these bodies, should be abolished.

92. That the property and liabilities, powers and duties of these Destitution Authorities should be transferred (subject to

the necessary adjustments) to the County and County Borough Councils, strengthened in numbers as may be deemed necessary for their enlarged duties; with suitable modifications to provide for the special circumstances of Scotland and Ireland, and for the cases of the Metropolitan Boroughs, the Non-County Boroughs over 10,000 in population, and the Urban Districts over 20,000 in population, on the plan that we have sketched out.

93. That the provision for the various classes of the non-able-bodied should be wholly separated from that to be made for the Able-bodied whether these be Unemployed workmen, vagrants or able-bodied persons now in receipt of Poor Relief.

94. That the services at present administered by the Destitution Authorities (other than those connected with vagrants or the able-bodied)—that is to say, the provision for:

(i.) Children of school age;
(ii.) The sick and the permanently incapacitated, the infants under school age, and the aged needing institutional care;
(iii.) The mentally defective of all grades and all ages; and
(iv.) The aged to whom pensions are awarded—should be assumed, under the directions of the County and County Borough Councils, by:

(i.) The Education Committee;
(ii.) The Health Committee;
(iii.) The Asylums Committee; and
(iv.) The Pension Committee respectively.

9. The Framework of Town Life, 1889

Report of the Metropolitan Board of Works, 1888 Parliamentary Papers, vol. LXVI (London, Eyre and Spottiswoode, 1889), p. 10.

Towns were in general governed by elected Mayor and Aldermen under the terms of the Municipal Reform Act of 1835. London grew hugely, from three millions in 1860 to four and a half in 1901, and took on a grand imperial appearance. Its government was chaotic and extravagant, divided between the Westminster authorities, the City Corporation, the 39 local vestries, and the Metropolitan Board of Works, each defending its own interests. The County Council Act of 1888 set up the London County Council (L.C.C.) in place of the last, but not for another eleven years did the often reactionary local vestries disappear in favour of 28 Metropolitan Boroughs, each complete with Mayor and Aldermen.

(ii) POWERS AND FUNCTIONS OF BOARD
[METROPOLITAN BOARD OF WORKS]

... and the functions which at the present date devolve upon the Board, and which under the recent Act are about to pass, with others, to the new County Council of London, may be shortly stated as follows:

(1.) The maintenance of the main sewers, the interception of sewage from the Thames, its conveyance to a distance from London, and its purification before being discharged into the river.

(2.) The prevention of floods from the Thames.

(3.) The formation of new main thoroughfares through crowded districts, and the carrying out of other great improvements.

(4.) The control over the formation of new streets, and over the erection of buildings, and the construction of local sewers. The naming of streets and numbering of houses.

(5.) The construction and maintenance of highways across the Thames, whether in the form of bridges, tunnels, or ferries.

(6.) The formation and maintenance of parks and gardens, and the preservation of commons and open spaces.

(7.) The demolition of houses in areas condemned as unhealthy under the Artizans' and Labourers' Dwellings Improvement Acts, and the sale or letting of the land for the erection of improved habitations.

(8.) The maintenance of a fire brigade for the extinction of fires and the saving of life and property in case of fire.

(9.) The supervision of the structural arrangements of theatres and music-halls, with special reference to the safety of persons frequenting them.

(10.) The sanctioning of tramways.

(11.) The control over the construction of railway bridges.

(12.) The supervision to a limited extent of the gas and water supply.

(13.) The control over the sale and storage of explosives, petroleum, and other inflammable substances.

(14.) The supervision of slaughter-houses and various offensive businesses; also of cowsheds, dairies, and milk-shops; and the prevention of the spread of contagious diseases among cattle, horses, and dogs.

(15.) The supervision and control of what is known as baby-farming; and other matters of detail incidental to municipal government.

10. Schools: (a) The Village School, (b) The Public Schools

(a) F. Thompson, *Lark Rise to Candleford* (Oxford University Press, 1945 ed.), pp. 170–5.
(b) *The Clarendon Report on the Public Schools* (Parliamentary Papers, 1864, XX), p. 246.

State-provided elementary education for all came relatively late in Britain, though a network of religious and private schools had ensured a high degree of literacy among the working classes, even in the eighteenth century. While National Schools by 1900 provided a good elementary education, the public schools (i.e. the private schools) provided an excellent literary education for the middle and upper classes.

(a)

Fordlow National School was a small grey one-storied building, standing at the cross-roads at the entrance to the village. The one large classroom which served all purposes was well lighted with several windows, including the large one which filled the end of the building which faced the road. Beside, and joined on to the school, was a tiny two-roomed cottage for the schoolmistress, and beyond that a playground with birch trees and turf, bald in places, the whole being enclosed within pointed, white-painted palings.

The only other building in sight was a row of model cottages occupied by the shepherd, the blacksmith, and other superior farm-workers. The school had probably been built at the same time as the houses and by the same model landlord; for, though it would seem a hovel compared to a modern council school, it must at that time have been fairly up-to-date. It had a lobby with pegs for clothes, boys' and girls' earth-closets, and a backyard with fixed wash-basins, although there was no water laid on. The water supply was contained in a small bucket, filled every morning by the old woman who cleaned the schoolroom, and every morning she grumbled because the children had been so extravagant that she had to 'fill 'un again'.

The average attendance was about forty-five. Ten or twelve of the children lived near the school, a few others came from cottages in the fields, and the rest were the Lark Rise children. Even then, to an outsider, it would have appeared a quaint, old-fashioned little gathering; the girls in their ankle-length frocks and long, straight pinafores, with their hair strained back from their brows and secured on their crowns by a ribbon or black tape or a bootlace; the bigger boys in corduroys and hobnailed boots, and the smaller ones in homemade sailor suits or, until they were six or seven, in petticoats.

Reading, writing, and arithmetic were the principal subjects, with a Scripture lesson every morning, and needlework every afternoon for the girls. There was no assistant mistress; Governess taught all the classes simultaneously, assisted only by two monitors—ex-scholars, aged about twelve, who were paid a shilling a week each for their services.

Every morning at ten o'clock the Rector arrived to take the older children for Scripture. He was a parson of the old school; a commanding figure, tall and stout, with white hair, ruddy cheeks and an aristocratically beaked nose, and he was as far as possible removed by birth, education, and worldly circumstances from the lambs of his flock. He spoke to them from a great height, physical, mental, and spiritual. 'To order myself lowly and reverently before my betters' was the clause he underlined in the Church Catechism, for had he not been divinely appointed pastor and master to those little rustics and was it not one of his chief duties to teach them to realize this? As a man, he was kindly disposed—a giver of blankets and coals at Christmas, and of soap and milk puddings to the sick.

His lesson consisted of Bible reading, turn and turn about round the class, of reciting from memory the names of the kings of Israel and repeating the Church Catechism. After that, he would deliver a little lecture on morals and behaviour. The children must not lie or steal or be discontented or envious. God had placed them just where they were in the social order and given them their own especial work to do; to envy others or to try to change their own lot in life was a sin of which he hoped they would never be guilty. From his lips the children heard nothing of that God who is Truth and Beauty and Love; but they learned for him and repeated to him long passages from the Authorized Version, thus laying up treasure for themselves; so the lessons, in spite of much aridity, were valuable.

Scripture over and the Rector bowed and curtsied out of the door, ordinary lessons began. Arithmetic was considered the most important of the subjects taught, and those who were good at figures ranked high in their classes. It was very simple arithmetic, extending only to the first four rules, with the money sums, known as 'bills of parcels', for the most advanced pupils.

The writing lesson consisted of the copying of copper-plate maxims: 'A fool and his money are soon parted'; 'Waste not, want not'; 'Count ten before you speak', and so on. Once a week composition would be set, usually in the form of writing a letter describing some recent event. This was regarded chiefly as a spelling test.

History was not taught formally; but history readers were in use containing such picturesque stories as those of King Alfred and the cakes, King Canute commanding the waves, the loss of the White Ship, and Raleigh spreading his cloak for Queen Elizabeth.

There were no geography readers, and, excepting what could be gleaned from the descriptions of different parts of the world in the ordinary readers, no geography was taught. But, for some reason or other, on the walls of the schoolroom were hung splendid maps: The World, Europe, North America, South America, England, Ireland, and Scotland. During long waits in class for her turn to read, or to have her copy or sewing examined, Laura would gaze on these maps until the shapes of the countries with their islands and inlets became photographed on her brain. Baffin Bay and the land around the poles were especially fascinating to her.

(b)
On the general results of public school education as an instrument for the training of character, we can speak with much confidence. Like most English institutions—for it deserves to rank among English institutions—it is not framed upon a preconceived plan, but has grown up gradually. It is by degrees that bodies of several hundred boys have come to be congregated together in a small space, constantly associated with one another in work and in play; and it is by degrees that methods of discipline and internal government have been worked out by their Masters and by themselves, and that channels of

influence have been discovered and turned to account. The organization of monitors or prefects, the system of boarding-houses, and the relation of tutor and pupil have arisen and been developed by degrees. The magnitude and the freedom of these schools make each of them, for a boy of from 12 to 18, a little world, calculated to give his character an education of the same kind as it is destined afterwards to undergo in the great world of business and society. Eton, Harrow, and Rugby are the *proscholia* in this respect of Oxford and Cambridge, as Oxford and Cambridge, with their larger, but still limited freedom, are for the training of adult life. The liberty, however, which is suited for a boy is a liberty regulated by definite restraints; and his world, the chief temptations of which arise from thoughtlessness, must be a world pervaded by powerful disciplinary influences, and in which rewards as well as punishments are both prompt and certain. The principle of governing boys mainly through their own sense of what is right and honourable is undoubtedly the only true principle; but it requires much watchfulness, and a firm, temperate, and judicious administration, to keep up the tone and standard of opinion, which are very liable to fluctuate, and the decline of which speedily turns a good school into a bad one. The system, we may add, is one which is adapted for boys, and not for children, and which should not be entered upon, as a general rule, till the age of childhood is past; neither, perhaps, is it universally wholesome for boys of every temperament and character, though we believe that the cases to which it is unsuited are not very numerous. But we are satisfied, on the whole, both that it has been eminently successful, and that it has been greatly improved during the last 30 or 40 years, partly by causes of a general kind, partly by the personal influence and exertions of Dr. Arnold and other great schoolmasters. The changes which it has undergone for the better are, we believe, visible in the young men whom it has formed during that period. The great schools—which, it must be observed, train for the most part the Masters who are placed at the head of the smaller schools, and thus exercise not only a direct but a wide indirect influence over education—may certainly claim, as Mr. Hedley says, a large share of the credit due for the improved moral tone of the Universities, as to which we have strong concurrent testimony.

WORK

11.   Farm workers: (*a*) Hiring (1870), (*b*) Conditions (1900)

(*a*) T. E. Kebbel, *The Agricultural Labourer* (London, Chapman and Hall, 1870), pp. 118–20.

(*b*) Sir Henry Rider Haggard, *Rural England* (London, Longmans, 1902), pp. 544–5.

Farm labourers are to be distinguished from farm servants; the the latter lived in, were hired for perhaps a year at a time and not by the week, and had better pay. Shepherds, nearly all men, made another smaller group. Of nearly 800,000 farm labourers in 1871, 33,500 were women. Until 1873 children from an early age were employed on farms without restriction, their wages often being necessary for family subsistence. In 1876 no child under the age of ten could be set to farm work, and thereafter a child had first to reach standard four at school. By 1900 the farm labourer's lot had greatly improved as Rider Haggard, novelist and Norfolk farmer, points out.

(*a*)

All along the roads in the vicinity of the market town appointed for the ceremony, the young men and women of the neighbourhood are to be seen trooping along in their best clothes, and congregating eventually in the market-place, where they stand for hire like the labourers in the parable. The candidates indicate by a badge the peculiar service which they seek. The shepherd decorates his cap with a bunch of wool; the carter with a bit of whipcord; the housemaid with a sprig of broom; and both sexes alike, when they have been hired, pin a knot of bright-coloured ribbons on the breast or shoulder, just as if they were 'agoing for soldiers'. When the business of the day is over, the evening is devoted to amusement—in other words, dancing and drinking, which produce their natural results and are to a large extent accountable for that low standard of female honour which, according to Mr. Fraser, is characteristic of the English peasantry. The servants like the system, of course, because it gives them, at all events, one good outing in the year. The farmers like it, because, as they say, 'they get a lot to pick from', and can compare the thews and sinews of a great many candidates for service before finally engaging one. We do not mean, of course, that they feel them over as they would a horse, or as their wives would thumb a couple of fowls; but they scan them critically, as the slave merchant would have scanned a negro, and naturally regard them in no other light than that of animals. It must be understood, however, that we are speaking only of one class of farmers who stick to the old road. We are aware that there are many others of a wholly distinct character, who dislike the system as much as any one can, and would willingly abolish it could they find any practicable substitute.

(b)
To come to the third class—that of the labouring men—undeniably they are more prosperous to-day than ever they have been before. Employment is plentiful; wages, by comparison, are high—in some places higher than the land can afford to pay—food and other necessaries are very cheap.

In face of these advantages, however, the rural labourer has never been more discontented than he is at present. That, in his own degree, he is doing the best of the three great classes connected with the land does not appease him in the least. The diffusion of newspapers, the system of Board school education, and the restless spirit of our age have changed him, so that now-a-days it is his main ambition to escape from the soil where he was bred and try his fortune in the cities. This is not wonderful, for there are high wages, company, and amusement, with shorter hours of work. Moreover on the land he has no prospects: a labourer he is, and in ninety-nine cases out of a hundred a labourer he must remain. Lastly, in many instances, his cottage accommodation is very bad; indeed I have found wretched and insufficient dwellings to be a great factor in the hastening of the rural exodus; and he forgets that in the town it will probably be worse. So he goes, leaving behind him half-tilled fields and shrinking hamlets.

12. Domestic Servants

The Fortnightly (March 1888), pp. 412–13.

One of the major occupations for women was domestic service, for even poor families frequently employed 'a drudge'. In 1871, 15 per cent of the total occupied population, and in 1911, 11 per cent, were domestic servants; the proportion in 1951 was less than 3 per cent. A continuous occupational trend of the twentieth century has been the decline of domestic servants, whose absolute numbers were greatest in the Edwardian period.

'Liberty at any cost,' is the watchword of the social and political reformer, and the chief ground of discontent among servants is their deprivation of liberty. The degree of labour expected from servants varies in different households, but in all the labour is exacting and protracted. Servants, as a rule, rise at seven or soon after, and go to bed at whatever time suits the convenience of their employers. They are at work generally before the master or mistress is awake, during the fifteen or sixteen hours of the day they are supposed to be at their post, ready to do any bidding, and they are the last to bed at night. If a party is given and the servants are kept at work till far into the morning, they are nevertheless compelled to be out of bed as early as though they retired at ten, in order to take the master or mistress the morning cup of tea. Whilst the employers sleep off the effects of the previous night's entertainment, their retainers have to bestir themselves to clear up. This comparison, or rather contrast, may be invidious, but is none the less instructive. 'A servant's work,' as has been well said, 'is never done potentially, if even actually.' There is no reason why this should be so. A very little reorganisation in most London homes would enable the servant to have at least four or five hours every day, or if not every day, every other day, to herself. Let a lady or gentleman try to realise how rigorous are the laws which govern the actions of servants, and it will be seen that the intractability of their retainers is not entirely unjustified. Has it ever occurred to the master or the mistress what humiliation and distress attach to the single circumstance that the footman or the housemaid, the cook or the butler, dare not stir beyond the four walls of the house without permission except once or at most twice in a fortnight? Nothing is more bitterly resented below-stairs than to have to ask leave to 'go out' if only for a few minutes. Again, when servants are allowed out, the time at their disposal is ludicrously meagre, and, as a consequence, quarrels because the man or maid is late, are frequent. But when a mistress declaims against a servant who does not come in till eleven o'clock, she never stays to think how often that servant may have sat up for her or some member of the family. Beyond doubt little good comes of girls being allowed out very late, but, if they visit friends some distance off, it is cruelly inconsiderate to compel them to be back by half-past nine or even ten. More especially is this so when 'followers' are proscribed, and a servant never sees a personal friend from week's end to week's end. If many mistresses had their way the liberty enjoyed by servants would be smaller than it is; and not very long since a mistress publicly made a proposal that servants who transgress a prescribed limit should be birched! Her remarks, published in a newspaper devoted chiefly to the working-classes, must have conveyed to the reader an idea, equally unpleasant and false, of the disposition of the upper classes towards their social inferiors. Another matter, by which masters and mistresses insist on impressing upon their servants the fact that what is right in the drawing-room is wrong in the servants' hall, is dress and general appearance. Why, if he wishes—and for all the master can tell it may be a source of comfort to the man to do so—should the coachman, the butler, or the footman not be allowed to grow his beard and moustache? Why, if she has a particular liking for it, should the housemaid be denied the privilege of a fringe? These things are small but not insignificant. They are positive causes in the alienation from domestic service of the freedom-loving sons and daughters of the democracy. Love of finery, again, is undoubtedly strong in many servants' breasts, as love of good clothes is in the ladies'. And of this love of finery comes infinite evil. But it is not fair to level the charge of plagiarism in apparel against all servants. Some may ape their masters and mistresses, but their number is small, and the constant aim of servants who respect themselves is not to follow the lead of, but to take a line diametrically opposed to, that affected by their employers.

13. The Gantrymen

Lady Bell, *At the Works* (London, Arnold, 1907), pp. 27–31 and 96–9.

Lady Bell (1876–1930), wife of a noted ironmaster and colliery owner, devoted two fascinating chapters of her book to 'wives and daughters'. Early marriage, frequent childbirth, constant work, the environment of a town given almost entirely to iron making, brought too many to premature ageing and sometimes to degradation. Middlesbrough had a population of only 154 in 1831; ten years later the coming of the railways had pushed it above 5,000 and twenty years later, iron having been found in the Cleveland Hills in 1850, the population had risen to nearly 19,000 and to just under 100,000 by 1907.

(a)

The men who work on the 'gantry'—that is, the tall platform along which the trucks arrive containing the ore, the coal, the coke, and the limestone—are called 'gantrymen'. A gantryman has from £2 to £2 12s. 6d. a week, an income equal to that of many a curate, and that of many a junior clerk in a private, or even a Government, office. For this he works either eight hours a day or six hours a day, according to the shift, for the gantryman only has two shifts in the day, and not three, either from 6 a.m. to 2 p.m., or from 4 p.m. to 10 p.m. The daily work of the gantryman, who stands on the platform at the top of the kiln, is to superintend the calcining of the ironstone—i.e. to feed with coal the kiln into which the ironstone has been 'tipped'. The coal is tipped, not immediately into the kiln, but on to broad, flat sheets of iron on either side of the opening on to which the coal has been tipped from the truck. The gantryman at intervals shovels the coal into the kiln on to the top of the ironstone, using his own judgment how much he puts in, and how often.

The shift from 6 a.m. to 2 p.m. is the hardest, as at 6 a.m. the kiln has been left since ten the night before, so that there is a great deal to do. Then by two o'clock it is so full that it can be left until four. This second shift has not so much to do as the first, as it has only two hours interval to make up. Each of the eight-hours men is supposed to have two quarters of an hour in which to have some food. This food—his 'bait'—as he would himself call it—he takes up with him. It is usually something that can be carried in a can, and kept hot in the little 'cabins', of which one is allotted to each particular gang of men working at the same job—comfortless little sheds enough, filled with dust and dirt and the smell of the gases. For a man working on the eight-hours shift, of course, it is possible—as far, that is, as time is concerned—whatever shift the man is on, for him to have a square meal at some reasonable hour. The men who go off at 2 p.m. can have their dinner at home in the afternoon; those who come at 2 p.m. can dine at home before leaving. A man who comes to his work in the evening has had his meal at home at suitable intervals during the day, if his wife is competent.

But none of the eight-hours men when at the works can leave their job for a square meal, although they may have short spells off in which to sit down either in the cabin or in the place where they are working and have a snack. They are not for one moment free from the need of that incessant vigilance required from all the workers who share the responsibility of controlling the huge forces they are daring to use.

The gantryman, not being close to the great heat of the furnace, has not to encounter in the winter the same sudden and violent changes of temperature as those which the men standing round the foot of the furnace must endure, when they pass from the almost unbearable heat of their immediate surroundings to the biting cold of a few yards off. But still, the work of the gantryman is a task absolutely shelterless, and in the sharp airs of the north-east coast the winds that blow across the high platform pierce like a dagger; added to which they bring the burning fumes which blow across from the kilns and the furnaces, the choking sulphurous fumes which are constantly blowing down the throats of the men exposed to them.

And besides the fumes and the gases, every breath of wind at the ironworks carries dust with it, whirling through the air in a wind, dropping through it in a calm, covering the ground, filling the cabins, settling on the clothes of those who are within reach, filling their eyes and their mouths, covering their hands and their faces. The calcined ironstone sends forth red dust, the smoke from the chimneys and furnaces is deposited in white dust, the smoke from the steel-rolling mills falls in black dust: and, most constant difficulty of all, the gases escaping from the furnaces are charged with a fine, impalpable brownish dust, which is shed everywhere, on everything, which clogs the interior of the stoves and of the flues, and whose encroachments have to be constantly fought against. One of the most repellent phenomena at the ironworks to the onlooker is the process of expelling the dust from the stoves, for which purpose the valves of the stove are closed, the stove is filled with air at high pressure, and then one of the valves is opened and the air is forcibly expelled. A great cloud of red dust rushes out with a roar, covering everything and everybody who stands within reach, with so intolerable a noise and effluvium that it makes itself felt even amidst the incessant reverberation, the constant smells, dust, deposits, that surround the stoves and the furnaces. That strange, grim street formed by the kilns, the furnaces, and the bunkers, darkened by the iron platforms overhead between the kilns and the gantry, a street in which everything is a dull red, is the very heart of the works, the very stronghold of the making of iron, a place unceasingly filled by glare, and clanging, and vapours, from morning till night and from night till morning.

(b)

Pickled cabbages are a favourite form of food. A man in E. Street said he never counted he had had a meal unless he had had pickles with it, or 'to' it, as they say in Yorkshire. The favourite kind are big red cabbages, of which four large ones pickled in a glass bottle would last the family one week, as an accompaniment to the usual fare of bread-and-butter, cheese, meat, sausages, bacon, pies, according to the meal-time.

A man in K. Street, if at home for breakfast, has bacon, eggs, sausages; if for dinner, a hot meal with potatoes, and a hot supper if going on the night shift, which begins at 10 p.m. The woman in this house was cooking fried steak for her own dinner and the children's. Their income was about 24$s.$ a week.

A man in M. Street, when on the early shift, takes a sandwich of cold bacon and bread for his breakfast, and one of bread and butter. The family have bread and butter for dinner. The man has a meat dinner (hot) at 2 p.m.; and anything left over is either heated up or given cold to the children at tea-time. When the man is on the afternoon shift they all have a meat dinner at twelve. On Fridays he takes cheese to the works and has fish at home, as they are Roman Catholics.

Whatever the variations of diet, a large amount of tea almost invariably forms a part of it. This is mostly Indian tea at 1s. to 1s. 4d. a pound, and it is bought in small quantities, usually in packets of about 1 ounce, sold for 1d.

14. The Workshop of Joseph Brown, Birmingham

Paul de Rousiers, *The Labour Question in Britain* (London, Macmillan, 1896), pp. 4–6

Not all industry was large scale by the end of the century, and in Birmingham 'the workshop' persisted into the twentieth century. Generally Birmingham was a city of prosperous and skilled workmen, like those here described.

Let us enter the workshop. The ground floor is occupied by three or four forges with their anvils, at one of which Brown as working when I arrived, for though he has become an employer, he still remains a workman, and though he is fifty years of age, he still flatters himself that no smith in Birmingham can turn out better or quicker work. Ascending to the first storey we find no forges: this is the finishing department. A man is filing a pair of strong pincers in order to polish them, another is cutting unequal teeth on a steel ring meant to enter into the composition of a tool for a saddle and harness maker. Brown explained that he worked a good deal for saddlers and bootmakers. He manufactures a great number of different patterns, and declared his readiness to execute any of which a design was supplied. Six men are at work in the upper workshop, making ten perhaps in all, but he has often employed as many as two-and-twenty. At present he has not many orders in hand.

All the men are skilled workmen, and those who make the bigger tools have to be smiths. I saw Brown and his assistant forging long and stout pincers of a particular shape. These, when closed, have a hollow groove at the end, of the exact diameter of a gas-pipe. The are for gripping pipes when fixing or removing them, and are indispensable to gasfitters. They are also used for fixing and repairing water-pipes. The largest size in use is only worth a shilling; this is the one which I saw made. A dozen pairs were lying on the ground to cool at the foot of the anvil, representing a morning's work.

However, everybody works very quickly here; the men move quickly and efficiently. There is no effort wasted and no talking. French workmen, so energetic by starts, would find it difficult to equal the rate of production. I made this remark to my friends in Birmingham. One of them, of French origin, a merchant of precious stones, told me he had often tried to introduce French workmen into one of the numerous jewel manufactories in Birmingham. The experiment had never succeeded: the masters found that the work was not turned out fast enough.

Brown's workmen are paid by the piece. Work begins at seven every day except Sundays, and leave off at seven at night except on Saturdays, when it ends at two. It is a general custom in England to work only half a day on Saturday, and on the other days work is interrupted for two hours, to give the men time for meals and for a little rest. Thus there are five days of ten hours and one day of five hours, or fifty-five hours a week.

However each workman does not make up this number of hours of work every week. One, for instance, never arrives before nine o'clock; others occasionally extend the Sunday holiday until Tuesday or Wednesday, if last week's pay is not completely exhausted. Brown complains greatly of these irregularities, but bears them in silence in order to keep his men. As I have said, they are skilled workers, and not to be replaced by the first comer, and rare, for apprenticeship no longer exists in the trade. Consequently smiths profit by the situation to act as they like, knowing that the employer cannot easily do without them.

Naturally, under these circumstances, wages are high. I was shown the pay-books of the workmen, and noticed weeks at £2 : 8s., while nearly all came to £2 for regular workmen. It is highly paid work. Brown explained to me that, yielding to the request of his men, he had for several years allowed a bonus of 5 per cent on the prices fixed for piece-work. Thus, for example, if a workman has done work to the extent of £2 a week, he is paid £2 : 2s.

15. Basic Conditions of Factory Work, 1901

Factory and Workshop Act, 1901 (1 Edw. 7. Ch. 22).

The first Factory Act was in 1802, and the first effective one in 1833. Thereafter there were many, regulating hours and conditions of work, first in factories and later in workshops and shops. The 1901 Act consolidated many previous Acts.

HEALTH AND SAFETY

(i.) HEALTH
Sanitary Condition of Factory.

1.—(1.) The following provisions shall apply to every factory as defined by this Act, except a domestic factory:

(*a.*) It must be kept in a cleanly state;

(*b.*) It must be kept from from effluvia arising from any drain, watercloset, earthcloset, privy, urinal, or other nuisance;

(*c.*) It must not be so overcrowded while work is carried on therein as to be dangerous or injurious to the health of the persons employed therein;

(*d.*) It must be ventilated in such a manner as to render harmless, so far as is practicable, all the gases, vapours, dust, or other impurities generated in the course of the manufacturing process or handicraft carried on therein, that may be injurious to health.

* * *

(3.) For the purpose of securing the observance of the requirements in this section as to cleanliness in factories, all the inside walls of the rooms of a factory, and all the ceilings or tops of those rooms (whether those walls, ceilings, or tops are plastered or not), and all the passages and staircases of a factory, if they have not been painted with oil or varnished once at least within seven years, shall (subject to any special exceptions made in pursuance of this section) be limewashed once at least within every fourteen months, to date from the time when they were last limewashed; and if they have been so painted or varnished shall be washed with hot water and soap once at least within every fourteen months, to date from the time when they were last washed.

Overcrowding of Factory or Workshop.

3.—(1.) A factory shall for the purposes of this Act, and a workshop shall for the purposes of the law relating to public health, be deemed to be so overcrowded as to be dangerous or injurious to the health of the persons employed therein, if the number of cubic feet of space in any room therein bears to the number of persons employed at one time in the room a proportion less than two hundred and fifty, or, during any period of overtime, four hundred, cubic feet of space to every person.

* * *

Power of Secretary of State to Act in Default of Local Authority.

4.—(1.) If the Secretary of State is satisfied that the provisions of this Act, or of the law relating to public health in so far as it affects factories, workshops, and workplaces, have not been carried out by any district council, he may, by order, authorise an inspector to take, during such period as may be mentioned in the order, such steps as appear necessary or proper for enforcing those provisions.

(2.) An inspector authorised in pursuance of this section shall, for the purpose of his duties thereunder, have the same powers with respect to workshops and workplaces as he has with respect to factories, and he may, for that purpose, take the like proceedings for enforcing the provisions of this Act or of the law relating to public health, or for punishing or remedying any default as might be taken by the district council; and he shall be entitled to recover from the district council all such expenses in and about any proceedings as he may incur, and as are not recovered from any other person.

16. The Professions: the Law

> C. Booth, *Life and Labour in London*, second series, vol. IV (London, Macmillan, 1903), pp. 72–4 (vol. VIII of the 1896 edition).

The census of 1841 classified 'professional persons' for the first time, and the modern idea of a profession was clarified in the 1870s. The early nineteenth-century professionals *par excellence* were the clergy, doctors, and lawyers (barristers and solicitors). New professions appeared later, and most, for example engineers, architects, and accountants, had organized themselves into professional associations by 1900; for example the Institute of Chartered Accountants, 1880.

London seems the natural home of barristers, solicitors, and law clerks, and the majority of those who live in London are London born. Their headquarters are the Inns of Court and Chancery Lane, at the junction of the City and the West End, with the Courts of Justice in their midst, and a large proportion of the legal business of the whole country, contentious and non-contentious alike, is here transacted.

The factories of the law are noiseless, and many a passer-by is tempted to turn aside from the roaring streams of traffic in Holborn and the Strand into the quiet backwaters formed by the courts and gardens of Gray's Inn, Lincoln's Inn, and the Temple, where the barristers' chambers are to be found.

Solicitors are more scattered. A number of large firms have their offices in the City, and there are a good many also in the West End proper, some of high repute and active in 'family business', and others, especially in the streets off Bond Street and Regent's Street, who do business for tradesmen and money-lenders, and choose this position in order to be near their clients.

The preliminary education for both branches of the profession is expensive, and men who seek to enter them, especially as barristers, must be in a position to live on their own means for some years, or else must pick up a living in the by-paths of literature, for it is usually long before they can pay their way out of their professional earnings. The reward comes later, or may not come at all, but, if it comes, the years of earning are usually prolonged. Lawyers are not past their work until well advanced in life.

The higher branches of clerks from which solicitors are usually recruited, start by being 'articled', i.e. apprenticed to a practising solicitor. They are of higher social standing than those who work their way up from boyhood, and the distinction between 'articled' and 'unarticled' clerks in a solicitor's office is a fairly sharp one.

Unarticled clerks receive much the same pay as those employed in commerce. Starting as boys at 7s. or 8s. a week they reach 28s. to 40s., and may eventually become managing or confidential clerks at from £150 to £400 per annum. It is open to them, if they pass three qualifying examinations, and can afford to pay the fees, to become solicitors on their own account. Not many do this. As clerks their work is hard and closely sustained. In addition to the clerks, a large office will usually have two or three cashiers, rent collectors, and shorthand writers.

Barristers' clerks have more leisure. They start when fourteen or fifteen years of age under a 'senior' clerk. Between the ages of seventeen and twenty-one the boys earn 15s. to 18s. and call themselves 'junior clerks'. After a year or two at this they begin to look for a position as 'only' clerk, and in this capacity may serve four or five young men who have been lately called to the bar, and are setting up for themselves in chambers. For the clerk as well as his master there is a fee with every brief. Young barristers have but little remunerative work, and the clerk, who as soon as he becomes a chief or only clerk is properly entitled to nothing beyond the regular fees, demands a guarantee that the sum paid him shall not be less than, it may be, £70 to £100. Clerks to barristers in fair practice earn from £200 to £400 a year. The incomes of the chief clerks to barristers in the largest practice would run up to £800 or £1000, or even more. It is roughly calculated that clerk's fees are from 5 to 8 per cent of their employers' earnings. In return for this they keep the fee book, and act as intermediary between barrister and solicitor with respect to the amount of fee that will be accepted, since etiquette forbids the barrister to do this for himself. They also arrange the times of conferences, and have many other duties of a semi-confidential nature. Amiability, tact, and honesty are the qualities in greatest demand.

Office hours are from 10 a.m. to 6 or 7 p.m. or to 4 or 5 p.m. on Saturdays, excepting in the long vacation, when the hours are shorter if, indeed, the chambers are not shut up altogether.

17. War Work

T. A. Fyfe, *Employers and Workmen under Munitions Acts* (London and Edinburgh, Hodge, 1918), pp. 185–8.

War-time labour shortages had beneficial effects for at least two categories of labour: females and the unskilled. There began a process known as 'dilution', the use of unskilled or semi-skilled in jobs previously done only by skilled. This move, which threatened the job monopolies of the craft unions, resulted in a reduction of the gap between the wages of the skilled and unskilled which persisted after the war.

DIRECTIONS RELATING TO THE EMPLOYMENT AND REMUNERATION OF SEMI-SKILLED AND UNSKILLED MEN ON MUNITIONS WORK OF A CLASS WHICH PRIOR TO THE WAR WAS CUSTOMARILY UNDERTAKEN BY SKILLED LABOUR.

Cir. L. 3, Order No. 71, 24th Jan., 1917.
Order No. 667, 26th June, 1917.

[*Note.*—These Directions are strictly confined to the war period and are subject to the observance of Schedule II. of the Munitions of War Act.]

General

1. Operations on which skilled men are at present employed, but which by reason of their character can be performed by semi-skilled or unskilled labour, may be done by such labour during the period of the war.

2. Where semi-skilled or unskilled male labour is employed on work identical with that customarily undertaken by skilled labour, the time rates and piece prices and premium bonus times shall be the same as customarily obtain for the operations

when performed by skilled labour.

3. Where skilled men are at present employed they shall not be displaced by less skilled labour unless other skilled employment is offered to them there or elsewhere.

4. Piecework prices and premium bonus time allowances, after they have been established, shall not be altered unless the means or methods of manufacture are changed.

5. Overtime, night shift, Sunday and holiday allowances shall be paid to such machinemen on the same basis as to skilled men.

* * *

NOTES ON THE DILUTION OF SKILLED LABOUR PREPARED BY THE MUNITIONS LABOUR SUPPLY COMMITTEE FOR GUIDANCE OF CONTROLLED ESTABLISHMENTS. *Cir. L.* 6.

Alterations in Working Conditions

Schedule II., paragraph 7, provides:—'Due notice shall be given to the workmen concerned wherever practicable of any changes of working conditions which it is desired to introduce as the result of the establishment becoming a controlled establishment, and opportunity for local consultation with workmen or their representatives shall be given if desired.'

Procedure:—The Minister is of opinion that the following procedure should be adopted by a controlled establishment when any change is made in working conditions.

1. The Workmen in the shop in which a change is to be made should be requested by the employer to appoint a deputation of their number together with their local Trade Union representative if they desire, to whom particulars of the proposed change could be explained.
2. At the interview the employer, after explaining the change proposed and giving the date when it is to come into operation, should give the deputation full opportunity of raising any points they desire in connection therewith, so that if possible the introduction may be made with the consent of all parties.
3. Should the deputation be unable at the interview to concur in the change, opportunity should be given for fur-

ther local consultation when representatives of the Trade Unions concerned might be present.

4. It is not intended that the introduction of the change should be delayed until concurrence of the work-people is obtained. The change should be introduced after a reasonable time, and if the work-people or their representatives to bring forward any question relating thereto they should follow the procedure laid down in Part I. of the Act.

5. It is not desirable that formal announcement of the proposed change should be put on the notice board of the shop until intimation has been given as above to the men concerned or their Trade Union representative.

While this is so the Minister is of opinion that it will be consistent with prudence that every endeavour should be made by employers to secure the co-operation of their work-people in matters of this description.

Any difficulties experienced by either employers or work-people should be at once referred to the Ministry in order that an immediate endeavour may be made to find a satisfactory solution.

October, 1915.

OPINION

18.   The Monarchy
Walter Bagehot, *The English Constitution*
(London, Collins, 1963), p. 82.
No one can appreciate the social and political flavour of this period without accepting the profound influence of the monarchy. The monarchy headed the State, Church, and Empire, as well as Society. Walter Bagehot was the influential editor of *The Economist*, and produced his appreciative study of the constitution in the same year as the publication of the first volume of Marx's *Das Kapital* (1867).

The use of the Queen, in a dignified capacity, is incalculable. Without her in England, the present English Government would fail and pass away. Most people when they read that the Queen walked on the slopes at Windsor—that the Prince of Wales went to the Derby—have imagined that too much thought and prominence were given to little things. But they have been in error; and it is nice to trace how the actions of a retired widow and an unemployed youth become of such importance.

The best reason why Monarchy is a strong government is, that it is an intelligible government. The mass of mankind understand it, and they hardly anywhere in the world understand any other. It is often said that men are ruled by their imaginations; but it would be truer to say they are governed by the weakness of their imaginations. The nature of a constitution, the action of an assembly, the play of parties, the unseen formation of a guiding opinion, are complex facts, difficult to know and easy to mistake. But the action of a single will, the fiat of a single mind, are easy ideas: anybody can make them out, and no one can ever forget them. When you put before the mass of mankind the question, 'Will you be governed by a king, or will you be governed by a constitution?' the inquiry comes out thus—'Will you be governed in a way you understand, or will you be governed in a way you do not understand?' The issue was put to the French people; they were asked, 'Will you be governed by Louis Napoleon, or will you be governed by an assembly?' The French people said, 'We will be governed by the one man we can imagine, and not by the many people we cannot imagine.'

19. Religion: (*a*) the Sceptical Intellectual (*b*) the Salvation Army

(*a*) Beatrice Webb, *My Apprenticeship* (London, Longmans, 1926), pp. 54–6.

(*b*) William Booth, *In Darkest England and the Way Out* (London, The Salvation Army, 1890), pp. 242–4.

Victorian Christianity reached an apogee of social influence in the period 1850 to 1890. All Protestant denominations, including the Church of England, stressed church-going and listening to sermons, Bible reading, right conduct and strict morals, and overt attention to church and chapel. With exceptions, however, the working classes, especially in the large urban complexes, were much less influenced by religion. William Booth shocked polite society with his revelations of the moral degradation of the poor; he wanted to attack social problems systematically, with a well-disciplined Christian Army. The end of the century saw many organizations with similar objectives: Dr. Barnardo opened his first home in 1870; Toynbee Hall, the first University Settlement in the East End, was founded in 1884. At the same time, however, there was increasing intellectual scepticism about religion, with a permanent decline in church allegiance setting in by 1900.

(*a*)

For, looking back, it now seems to me that it was exactly in those last decades of the nineteenth century that we find the watershed between the metaphysic of the Christian Church, which had hitherto dominated British civilisation, and the agnosticism, deeply coloured by scientific materialism, which was destined, during the first decades of the twentieth century, to submerge all religion based on tradition and revelation. Judging by my own experience among the organisers of big enterprise, with their 'business morality' and their international affiliations, the Christian tradition, already in the seventies and eighties, had grown thin and brittle, more easily broken than repaired. When staying in the country my parents were, it is true, regular churchgoers and communicants; and my father always enjoyed reading the lessons in the parish churches frequented by the household in Gloucestershire, Westmorland and Monmouthshire. Parenthetically I may remark that it was symptomatic of the general decline of orthodoxy that one who had been brought up as a Unitarian and had never been admitted to the Anglican Church by the rite of confirmation, should have been, not only accepted as a communicant by Anglican clergymen who knew the facts, but also habitually invited, as the wealthy layman of the congregation, to take an active part in the service. Owing to personal religion, filial respect, or the joy of walking to and fro with the beloved father, one or two of the Potter girls would find themselves in the family pew each Sabbath day. But here conformity ended. No compulsion, even no pressure, was put on us to attend religious services. During the London season my father, accompanied by a bevy of daughters, would start out on a Sunday morning to discover the most exciting speaker on religious or metaphysical issues; and we would listen with equal zest to Monsignor Capel or Canon Liddon, Spurgeon or Voysey, James Martineau or Frederic Harrison; discussing on the walk back across the London Parks the religious rhetoric or dialectical subtleties of preacher or lecturer. Except for this eclectic enjoyment of varieties of metaphysical experience, the atmosphere of the home was peculiarly free-thinking. There was no censorship whether of talk in the family, or of the stream of new books and current periodicals, or of the opinions of the crowd of heterogeneous guests. Any question which turned up in classical or modern literature, in law reports or technical journals, from the origin of species to the last diplomatic despatch, from sexual perversion to the rates of exchange, would be freely and frankly discussed within the family circle. Perhaps the only expenditure unregulated and unrestricted by my mother, she herself being the leading spendthrift, was the purchase or subscription for books, periodicals and newspapers. And whether we girls took down from the well-filled library shelves the *Confessions of St. Augustine* or those of Jean Jacques Rousseau, whether the parcel from Hatchett's contained the latest novels by Guy de

Maupassant and Emile Zola or the learned tomes of Auguste Comte or Ernest Renan; whether we ordered from the London Library or from Mudie's a pile of books on Eastern religions, or a heterogeneous selection of what I will call 'yellow' literature, was determined by our own choice or by the suggestion of any casual friend or acquaintance. When we complained to my father that a book we wanted to read was banned by the libraries: 'Buy it, my dear,' was his automatic answer. And if the whirl of society in which we lived undermined character by its amazing variety, it most assuredly disintegrated prejudices and destroyed dogma.

(b)

The Salvation Army, largely recruited from among the poorest of the poor, is often reproached by its enemies on account of the severity of its rule. It is the only religious body founded in our time that is based upon the principle of voluntary subjection to an absolute authority. No one is bound to remain in the Army a day longer than he pleases. While he remains there he is bound by the conditions of the Service. The first condition of that Service is implicit, unquestioning obedience. The Salvationist is taught to obey as is the soldier on the field of battle.

From the time when the Salvation Army began to acquire strength and to grow from the grain of mustard seed until now, when its branches overshadow the whole earth, we have been constantly warned against the evils which this autocratic system would entail. Especially were we told that in a democratic age the people would never stand the establishment of what was described as a spiritual despotism. It was contrary to the spirit of the times, it would be a stone of stumbling and a rock of offence to the masses to whom we appeal, and so forth and so forth.

But what has been the answer of accomplished facts to these predictions of theorists? Despite the alleged unpopularity of our discipline, perhaps because of the rigour of military authority upon which we have insisted, the Salvation Army has grown from year to year with a rapidity to which nothing in modern Christendom affords any parallel. It is only twenty-five years since it was born. It is now the largest Home and Foreign Missionary Society in the Protestant world. We have nearly

10,000 officers under our orders, a number increasing every day, every one of whom has taken service on the express condition that he or she will obey without questioning or gainsaying the orders from Headquarters. Of these, 4,600 are in Great Britain. The greatest number outside these islands, in any one country, are in the American Republic, where we have 1,018 officers, and democratic Australia, where we have 800.

Nor is the submission to our discipline a mere paper loyalty. These officers are in the field, constantly exposed to privation and ill-treatment of all kinds. A telegram from me will send any of them to the uttermost parts of the earth, will transfer them from the Slums of London to San Francisco, or despatch them to assist in opening missions in Holland, Zululand, Sweden, or South America. So far from resenting the exercise of authority, the Salvation Army rejoices to recognise it as one great secret of its success, a pillar of strength upon which all its soldiers can rely, a principle which stamps it as being different from all other religious organisations founded in our day.

With ten thousand officers, trained to obey, and trained equally to command, I do not feel that the organisation even of the disorganised, sweated, hopeless, drink-sodden denizens of darkest England is impossible. It is possible, because it has already been accomplished in the case of thousands who, before they were saved, were even such as those evil lot we are now attempting to deal with.

Our fifth credential is the extent and universality of the Army. What a mighty agency for working out the Scheme is found in the Army in this respect! This will be apparent when we consider that it has already stretched itself through over thirty different Countries and Colonies, with a permanent location in something like 4,000 different places, that it has either soldiers or friends sufficiently in sympathy with it to render assistance in almost every considerable population in the civilised world, and in much of the uncivilised, that it has nearly 10,000 separated officers whose training, and leisure, and history qualify them to become its enthusiastic and earnest co-workers. In fact, our whole people will hail it as the missing link in the great Scheme for the regeneration of mankind, enabling them to act out those impulses of their hearts which are ever prompting them to do good to the bodies as well as to the souls of men.

20. State Education

A. V. Dicey, *Law and Opinion in England* (London, Macmillan, 2nd ed., 1914), pp. 278–9.

The best document, perhaps, on the Education Act of 1870 is W. E. Forster's speech which introduced it to the House of Commons. Less well known, however, are the objections to it. Dicey, for example, saw State education as part of an insidious growth of collectivism.

In 1891 parents of children compelled to attend school were freed from the duty of paying school fees, and elementary education became what is called free.

This last change completely harmonises with the ideas of collectivism. It means, in the first place, that *A*, who educates his children at his own expense, or has no children to educate, is compelled to pay for the education of the children of *B*, who, though, it may be, having means to pay for it, prefers that the payment should come from the pockets of his neighbours. It tends, in the second place, as far as merely elementary education goes, to place the children of the rich and of the poor, of the provident and the improvident, on something like an equal footing. It aims, in short, at the equalisation of advantages. The establishment of free education is conclusive proof that, in one sphere of social life, the old arguments of individualism have lost their practical cogency. Here and there you may still hear it argued that a father is as much bound in duty to provide his own children at his own expense with necessary knowledge as with necessary food and clothing, whilst the duty of the tax-payers to pay for the education is no greater than the obligation to pay for the feeding of children whose parents are not paupers. But this line of reasoning meets with no response except, indeed, either from some rigid economist who adheres to doctrines which, whether true or false, are derided as obsolete shibboleths; or from philanthropists who, entertaining, whether consciously or not, ideas belonging to socialism, accept the premises pressed upon them by individuals, but draw the inference that the State is bound to give the children, for whose education it is responsible, the breakfasts or dinners which will enable them to profit by instruction. The State, moreover, which provides for the elementary education of the people, has now, in more directions than one, advanced far on the path towards the provision of teaching which can in no sense be called elementary. If a student once realises that the education of the English people was, during the earlier part of the nineteenth century, in no sense a national concern, he will see that our present system is a monument to the increasing predominance of collectivism. For elementary education is now controlled and guided by a central body directly representing the State; it is administered by representative local authorities, it is based on the compulsory attendance of children at school, it is supported partly by parliamentary grants and partly by local rates.

21. Ideas of Empire

(a) B. Disraeli, *The Selected Speeches of the late Rt. Hon. the Earl of Beaconsfield* (ed. T. E. Kebbel, London, Longmans, 1882), vol. II, pp. 528–31. Speech made at the Crystal Palace, 24 June 1872.

(b) J. A. Hobson, *Imperialism* (London, Allen and Unwin, 1901), pp. 286–7, 289–91.

This was the great age of imperialism, with the British Empire expanding dramatically. Disraeli has the reputation of an imperialist, but his ideas of empire are constitutional, and he has more in common with Lord Durham, who recommended self-government for Canada in 1837, than with Rudyard Kipling. J. A. Hobson was an early critic of imperialism, whose ideas much influenced Lenin; his thesis, briefly, was that 'the white man's burden' consisted mainly of loot.

(a)

Gentlemen, there is another and second great object of the Tory party. If the first is to maintain the institutions of the country, the second is, in my opinion, to uphold the Empire of England. If you look to the history of this country since the advent of Liberalism—forty years ago—you will find that there has been no effort so continuous, so subtle, supported by so much energy, and carried on with so much ability and acumen, as the attempts of Liberalism to effect the disintegration of the Empire of England.

And, gentleman, of all its efforts, this is the one which has been the nearest to success. Statesmen of the highest character, writers of the most distinguished ability, the most organised and efficient means, have been employed in this endeavour. It has been proved to all of us that we have lost money by our colonies. It has been shown with precise, with mathematical demonstration, that there never was a jewel in the Crown of England that was so truly costly as the possession of India. How often has it been suggested that we should at once emancipate ourselves from this incubus. Well, that result was nearly accomplished. When those subtle views were adopted by the country under the plausible plea of granting self-government to the Colonies, I confess that I myself thought that the tie was broken. Not that I for one object to self-government. I cannot conceive how our distant colonies can have their affairs administered except by self-government. But self-government, in my opinion, when it was conceded, ought to have been conceded as part of a great policy of Imperial consolidation. It ought to have been accompanied by an Imperial tariff, by securities for the people of England for the enjoyment of the unappropriated lands which belonged to the Sovereign as their trustee, and by a military code which should have precisely defined the means and the responsibilities by which the colonies should be defended, and by which, if necessary, this country should call for aid from the colonies themselves. It ought, further, to have been accompanied by the institution of some representative council in the metropolis, which would have brought the Colonies into constant and continuous relations with the Home Government. All this, however, was omitted because those who advised that policy—and I believe their convictions were sincere—looked upon the Colonies of England, looked even upon our connection with India, as a burden upon this country, viewing everything in a financial aspect, and totally passing by those moral and political considerations which make nations great, and by the influence of which alone men are distinguished from animals.

Well, what has been the result of this attempt during the reign of Liberalism for the disintegration of the Empire? It has entirely failed. But how has it failed? Through the sympathy of the Colonies with the Mother Country. They have decided that the Empire shall not be destroyed, and in my opinion no minister in this country will do his duty who neglects any opportunity of reconstructing as much as possible our Colonial Empire, and of responding to those distant sympathies which may become the source of incalculable strength and happiness to this land. Therefore, gentlemen, with respect to the second great object of the Tory party also—the maintenance of the Empire—public opinion appears to be in favour of our principles—that public opinion which, I am bound to say, thirty years ago, was not favourable to our principles, and which, during a long interval of controversy, in the interval had been doubtful.

(b)

The real questions we have to answer are these: 'Are we civilizing India?' and 'In what does that civilization consist?' To assist in answering there exists a tolerably large body of indisputable facts. We have established a wider and more permanent internal peace than India had ever known from the days of Alexander the Great. We have raised the standard of justice by fair and equal administration of laws; we have regulated and probably reduced the burden of taxation, checking the corruption and tyranny of native princes and their publicans. For the instruction of the people we have introduced a public system of schools and colleges, as well as a great quasi-public missionary establishment, teaching not only the Christian religion but many industrial arts. Roads, railways, and a network of canals have facilitated communication and transport, and an extensive system of scientific irrigation has improved the productiveness of the soil; the mining of coal, gold, and other minerals has been greatly developed; in Bombay and elsewhere cotton mills with modern machinery have been set up, and the organization of other machine industries is helping to find employment for the population of large cities. Tea, coffee, indigo, jute, tobacco, and other important crops have been introduced into Indian agriculture. We are gradually breaking down many of the religious and social superstitions which sin against humanity and retard progress, and even the deeply rooted caste system is modified wherever British influence is felt. There can be no question that much of this work of England in India is well done. No such intelligent, well-educated, and honourable body of men has even been employed by any State in the working of imperial government as is contained in the Civil Service of India. Nowhere else in our Empire has so much really disinterested and thoughtful energy been applied in the work of government. The same may be said of the line of great statesmen sent out from England to preside over our government in India. Our work there is the best record British Imperialism can show.

[Hobson then deploys the case against Britain]

A century of British rule, then, conducted with sound ability and goodwill, had not materially assisted to ward off the chronic enemy, starvation, from the mass of the people. Nor can it be maintained that the new industrialism of machinery and factories, which we have introduced, is civilizing India, or even adding much to her material prosperity. In fact, all who value the life and character of the East deplore the visible decadence of the arts of architecture, weaving, metal work and pottery, in which India had been famed from time immemorial. 'Architecture, engineering, literary skill are all perishing out, so perishing that Anglo-Indians doubt whether Indians have the capacity to be architects, though they built Benares; or engineers, though they dug the artificial lakes of Tanjore; or poets, though the people sit for hours or days listening to the rhapsodists as they recite poems, which move them as Tennyson certainly does not move our common people.' The decay or forcible supersession of the native industrial arts is still more deplorable, for these always constitute the poetry of common life, the free play of the imaginative faculty of a nation in the ordinary work of life.

* * *

Even from the low standpoint of the world-market this hasty destruction of the native arts for the sake of employing masses of cheap labour in mills is probably bad policy; for, as the world becomes more fully opened up and distant countries are set in closer communication with one another, a land whose industries had so unique and interesting a character as those of India would probably have found a more profitable market than by attempting to undersell Lancashire and New England in stock goods.

But far more important are the reactions of those changes on the character of the people. The industrial revolution in England and elsewhere has partaken more largely of the nature of a natural growth, proceeding from inner forces, than in India, and has been largely coincident with a liberation of great popular forces finding expression in scientific education and in political democracy: it has been an important phase of the great movement of popular liberty and self-government. In India, and elsewhere in the East, there is no such compensation.

An industrial system, far more strongly set and more closely interwoven in the religious and social system of the country than ever were the crafts and arts in Europe, has been subjected to forces operating from outside, and unchecked in their pace and direction by the will of the people whose life they so vitally affected. Industrial revolution is one thing when it is the natural movement of internal forces, making along the lines of the self interests of a nation and proceeding *pari passu* with advancing popular self-government; another thing when it is imposed by foreign conquerors looking primarily to present gains for themselves, and neglectful of the deeper interests of the people of the country. The story of the destruction of native weaving industry for the benefit of mills started by the Company will illustrate the selfish, short-sighted economic policy of the late eighteenth and early nineteenth centuries. 'Under the pretence of Free Trade, England has compelled the Hindus to receive the products of the steam-looms of Lancashire, Yorkshire, Glasgow, etc., at mere nominal duties; while the hand-wrought manufactures of Bengal and Behar, beautiful in fabric and durable in wear, have had heavy and almost prohibitive duties imposed on their importation to England.' The effect of this policy, rigorously maintained during the earlier decades of the nineteenth century, was the irreparable ruin of many of the most valuable and characteristic arts of Indian industry.

22. The Origins of Socialism

Beatrice Webb, *My Apprenticeship* (London, Longmans, 1926), pp. 178–80.

Socialism in Britain owed little to Continental example or theory; it was British in inspiration and development. Although the beginnings go back to the early trade unions and the radical writers of the early nineteenth century, the modern movement can be dated from the formation of the Fabian Society in 1884, and of the Democratic Federation and the Socialist League of 1881 and 1884. A Labour Representation Committee was founded in 1900, and by 1906 the parliamentary Labour Party was committed to an independent pursuit of working-class interests through political action.

Why this demand for State intervention from a generation reared amidst rapidly rising riches and disciplined in the school of philosophic radicalism and orthodox political economy? For it was not the sweated workers, massed in overcrowded city tenements or scattered, as agricultural labourers and home workers, in village hovels; it was not the so-called aristocracy of labour—the cotton operatives, engineers and miners who were, during this period, enrolling themselves in friendly societies, organising Trade Unions, and managing their own co-operative stores—it was, in truth, no section of the manual workers that was secreting what Mr. Asquith lived to denounce in the 1924 election as 'the poison of socialism'. The working-class revolt against the misery and humiliation brought about by the Industrial Revolution—a revolt, in spasmodic violence, aping revolution—had had its fling in the 'twenties and 'thirties and its apotheosis in the Chartist Movement of the 'forties. During the relative prosperity of the 'fifties and 'sixties the revolutionary tradition of the first decades of the nineteenth century faded away; and by 1880 it had become little more than a romantic memory among old men in their anecdotage. Born and bred in chronic destitution and enfeebling disease, the denizens of the slums had sunk into a brutalised apathy, whilst the more fortunate members of skilled occupations, entrenched in craft unionism, had been converted to the 'administrative nihilism' of Cobden, Bright and Bradlaugh.

The origin of the ferment is to be discovered in a new consciousness of sin among men of intellect and men of property; a consciousness at first philanthropic and practical—Oastler, Shaftesbury and Chadwick; then literary and artistic—Dickens, Carlyle, Ruskin and William Morris; and finally, analytic, historical and explanatory—in his latter days John Stuart Mill; Karl Marx and his English interpreters; Alfred Russel Wallace and Henry George; Arnold Toynbee and the Fabians. I might perhaps add a theological category—Charles Kingsley, F. D. Maurice, General Booth and Cardinal Manning. 'The sense of sin has been the starting-point of progress' was, during these years, the oft-repeated saying of Samuel Barnett, rector of St. Jude's, Whitechapel, and founder of Toynbee Hall.

When I say the consciousness of sin, I do not mean the consciousness of personal sin: the agricultural labourers on Lord Shaftesbury's estate were no better off than others in Dorsetshire; Ruskin and William Morris were surrounded in their homes with things which were costly as well as beautiful; John Stuart Mill did not alter his modest but comfortable way of life when he became a Socialist; and H. M. Hyndman gloried in the garments habitual to the members of exclusive West End clubs. The consciousness of sin was a collective or class consciousness; a growing uneasiness, amounting to conviction, that the industrial organisation, which had yielded rent, interest and profits on a stupendous scale, had failed to provide a decent livelihood and tolerable conditions for a majority of the inhabitants of Great Britain. 'England,' said Carlyle in the 'forties, 'is full of wealth, of multifarious produce, supply for human want in every kind; yet England is dying of inanition.'

23. Women's Rights

H. G. Wells, *Experiment in Autobiography* (London, Gollancz and the Cresset Press, 1934), pp. 469–70.

A general loosening of the Victorian moral code, which permitted married men moral lapses denied to their spouses, occurred under the new king. These changes were as important as the struggle of the suffragettes for political rights.

There was too a profounder, but connected, reaction against the mechanistic, conventional, and materialist views of the Victorian 'Establishment'. Writers like Shaw denounced bourgeois marriage as exploitation and Ibsen exposed some of its hypocrisy. H. G. Wells (1866–1946) was something of a rogue elephant among Edwardian intellectuals. Espousing 'socialism' he quarrelled with the Fabians, glorifying sexual freedom, he disliked the suffragette movement. His novels are an important source for this period, particularly for the lower middle class to which Wells had belonged.

The next book of mine in which unsolved sexual perplexities appear is *A Modern Utopia* (1905). Plato ruled over the making of that book, and in it I followed him in disposing of the sexual distraction, by minimizing the differences between men and women and ignoring the fact of personal fixation altogether. That is and always has been the intellectual's way out. My Samurai are of both sexes, a hardy bare-limbed race, free lovers among themselves—and mutually obliging. Like the people of the original Oneida community in New York State they constituted one comprehensive 'group marriage'. Possibly among such people fixations would not be serious; that is hypothetical psychology. I may have stressed the mutual civility of the order. The book was popular among the young of our universities; it launched many of them into cheerful adventures that speedily brought them up against the facts of fixation, jealousy and resentment. It played a considerable part in the general movement of release from the rigid technical chastity of women during the Victorian period. ...

* * *

A Modern Utopia was leading up to *Ann Veronica* (1909) in which the youthful heroine was allowed a frankness of desire and sexual enterprise, hitherto unknown in English popular fiction. That book created a scandal at the time, though it seems mild enough reading to the young of to-day. It is rather badly constructed, there is an excessive use of soliloquy, but Ann Veronica came as near to being a living character as anyone in my earlier love stories. This was so because in some particulars she was drawn from life. And for that and other reasons she made a great fuss in the world.

The particular offence was that Ann Veronica was a virgin who fell in love and showed it, instead of waiting as all popular heroines had hitherto done, for someone to make love to her. It was held to be an unspeakable offence that an adolescent female should be sex-conscious before the thing was forced upon her attention. But Ann Veronica wanted a particular man who excited her and she pursued him and got him. With gusto. It was only a slight reflection of anything that had actually occurred, but there was something convincing about the behaviour of the young woman in the story, something sufficiently convincing to impose the illusion of reality upon her; and from the outset Ann Veronica was assailed as though she was an actual living person.

It was a strenuous and long sustained fuss. The book was banned by libraries and preached against by earnest clergymen. The spirit of denunciation, latent in every human society, was aroused and let loose against me.

24. The Golden Age: Three Points of View

(a) C. F. G. Masterman, *The Condition of England* (London, Methuen, 1909), pp. 24–5.
(b) J. M. Keynes, *The Economic Consequences of the Peace* (London, Macmillan, 1924), pp. 6–7.
(c) C. W. Dilke, *Problems of Greater Britain* (London, Macmillan, 1890), p. 4.

(a)

Public penury, private ostentation—that, perhaps, is the heart of the complaint. A nation with the wealth of England can afford to spend, and spend royally. Only the end should be itself desirable, and the choice deliberate. The spectacle of a huge urban poverty confronts all this waste energy. That spectacle should not, indeed, forbid all luxuries and splendours: but it should condemn the less rewarding of them as things tawdry and mean. 'Money! money!' cries the hero—a second-grade Government clerk—of a recent novel—'the good that can be done with it in the world! Only a little more: a little more!' It is the passionate cry of unnumbered thousands. Expenditure multiplies its return in human happiness as it is scattered amongst widening areas of population. And the only justification for the present unnatural heaping up of great possessions in the control of the very few would be some return in leisure, and the cultivation of the arts, and the more reputable magnificence of the luxurious life. We have called into existence a whole new industry in motor cars and quick travelling, and established populous cities to minister to our increasing demands for speed. We have converted half the Highlands into deer forests for our sport; and the amount annually spent on shooting, racing, golf—on apparatus, and train journeys and service—exceeds the total revenue of many a European principality. We fling away in ugly white hotels, in uninspired dramatic entertainments, and in elaborate banquets of which every one is weary, the price of many poor men's yearly income. Yet we cannot build a new Cathedral. We cannot even preserve the Cathedrals bequeathed to us, and the finest of them are tumbling to pieces for lack of response to the demands for aid. We grumble freely at halfpenny increases in the rates for baths or libraries or pleasure-grounds. We assert—there are many of us who honestly believe it—that we cannot afford to set aside the necessary millions from our amazing revenues for the decent maintenance of our worn-out 'veterans of industry'.

To the poor, any increase of income may mean a day's excursion, a summer holiday for the children; often the bare necessities of food and clothes and shelter. To the classes just above the industrial populations, who with an expanding standard of comfort are most obviously fretting against the limitations of their income, it may mean the gift of some of life's lesser goods which is now denied; music, the theatre, books, flowers. Its absence may mean also a deprivation of life's greater goods; scamped sick-nursing, absence of leisure, abandonment of the hope of wife or child. All these deprivations may be endured by a nation—have been endured by nations—for the sake of definite ends: in wars at which existence is at stake, under the stress of national calamity, or as in the condition universal to Europe a few hundred years ago, when wealth and security were the heritage of the very few. But today that wealth is piling up into ever-increasing aggregation: is being scrutinized, as never before, by those who inquire with increasing insistence, where is the justice of these monstrous inequalities of fortune? Is the super-wealth of England expended in any adequate degree upon national service? Is the return today or to posterity a justification for this deflection of men and women's labour into ministering to the demands of a pleasure-loving society? Is it erecting works of permanent value, as the wealth of Florence in the fifteenth century? Is it, as in the England of Elizabeth, breeding men?

No honest inquirer could give a dogmatic reply. The present extravagance of England is associated with a strange mediocrity, a strange sterility of characters of supreme power in Church and State. It is accompanied, as all ages of security and luxury are accompanied, by a waning of the power of inspiration, a multiplying of the power of criticism. The more comfortable and opulent society becomes, the more cynicism proclaims the futility of it all, and the mind turns in despair from a vision of vanities.

(b)

That happy age lost sight of a view of the world which filled with deep-seated melancholy the founders of our Political Economy. Before the eighteenth century mankind entertained no false hopes. To lay the illusions which grew popular at that age's latter end, Malthus disclosed a Devil. For half a century all serious economical writings held that Devil in clear prospect. For the next half century he was chained up and out of sight. Now perhaps we have loosed him again.

What an extraordinary episode in the economic progress of man that age was which came to an end in August, 1914! The greater part of the population, it is true, worked hard and lived at a low standard of comfort, yet were, to all appearances, reasonably contented with this lot. But escape was possible, for any man of capacity or character at all exceeding the average, into the middle and upper classes, for whom life offered, at a low cost and with the least trouble, conveniences, comforts, and amenities beyond the compass of the richest and most powerful monarchs of other ages. The inhabitant of London could order by telephone, sipping his morning tea in bed, the various products of the whole earth, in such quantity as he might see fit, and reasonably expect their early delivery upon his doorstep; he could at the same moment and by the same means adventure his wealth in the natural resources and new enterprises of any quarter of the world, and share, without exertion or even trouble, in their prospective fruits and advantages; or he could decided to couple the security of his fortunes with the good faith of the townspeople of any substantial municipality in any continent that fancy or information might recommend. He could secure forthwith, if he wished it, cheap and comfortable means of transit to any country or climate without passport or other formality, could despatch his servant to the neighbouring office of a bank for such supply of the precious metals as might seem convenient, and could then proceed abroad to foreign quarters, without knowledge of their religion, language, or customs, bearing coined wealth upon his person, and would consider himself greatly aggrieved and much surprised at the least interference.

But, most important of all, he regarded this state of affairs as normal, certain, and permanent, except in the direction of further improvement, and any deviation from it as aberrant, scandalous, and avoidable. The projects and politics of militarism and imperialism, of racial and cultural rivalries, of monopolies, restrictions, and exclusion, which were to play the serpent to this paradise, were little more than the amusements of his daily newspaper, and appeared to exercise almost no influence at all on the ordinary course of social and economic life, the internationalization of which was nearly complete in practice.

(c)
While, however, we have so much of which to be proud in the development of our tongue, our trade, our literature, and our institutions, there is a corresponding present and temporary weakness to which it will be necessary in due place to call attention. The danger in our path is that the enormous forces of European militarism may crush the old country and destroy the integrity of our Empire before the growth of the newer communities that it contains has made it too strong for the attack. It is conceivable that within the next few years Great Britain might be drawn into war, and receive in that war, at the hands of a coalition, a blow from which she would not recover, and one of the consequences of which would be the loss of Canada and of India, and the proclamation of Australian independence. Enormous as are our military resources for a prolonged conflict, they are inadequate to meet the unprecedented necessities of a sudden war. We import half our food; we import the immense masses of raw material which are essential to our industry. The vulnerability of the United Kingdom has become greater with the extension of her trade, and, by the universal admission of the naval authorities, it would be either difficult or impossible to defend that trade against sudden attack by France, aided by another considerable naval power. Our enormous resources would be almost useless in the case of such a sudden attack, because we should not have time to call them forth.

Such is the one danger which threatens the fabric of that splendid Empire which I now attempt to describe.

THE ECONOMY

25. The Survival of Small-scale Industry after 1900: (*a*) General, (*b*) Sheffield

G. I. H. Lloyd, *The Cutlery Trades. An Historical Essay on the Economics of Small-Scale Production* (London, Longmans, 1913), pp. 425–7, 202–3.

The industrial revolution commenced in the eighteenth century; its main characteristic was the development of factory industry. It is important to realize, however, that domestic and small-scale industry persisted throughout the nineteenth century. Even in 1900 the 'average size' of factory was small, and there were still domestic or outside workers in many industries.

(*a*)

The factory system is so generally accepted nowadays as the normal type of industry that there is a tendency to overrate the extent to which it actually obtains. That industrial employment under the more primitive form is still very considerable in the manufacturing countries of Europe is beyond question, and in the less progressive countries it is probable that small industry still employs a larger total population than the more prominent factory system. In the older industrial nations, on the other hand, these proportions are of course reversed. For the United Kingdom we have no means of measuring, with any approach to accuracy, the extent of the employment afforded by small-scale industries. The returns of the Chief Inspector of Factories for the year 1904 give the number of persons employed in workshops in the United Kingdom as 653,912, and the number engaged in laundries as 104,477. Including workshops for which no returns are made, the total number of persons employed in workshops is estimated at about a million and a quarter persons. In comparison with the number returned as occupied in factories—more than four millions in all—even this total appears small, but the returns in question give no reliable indication of the real extent of small-scale enterprise. A rough comparison may be made between the employment figures of the United Kingdom, derived from the Census returns of 1901, and the above-mentioned factory returns for 1904. For this purpose certain broad and well-defined classes of employment may be selected, without, however, assuming that the basis of compilation makes them accurately comparable.

OCCUPATION RETURNS FOR THE UNITED KINGDOM

	Census 1901	Factory Returns 1904
Metals, machines, and implements	1,475,000	1,238,000
Textile fabrics	1,462,000	1,026,000
Clothing	1,396,000	307,000
Food, tobacco, etc.	1,301,000	293,000
Paper, books, stationery, etc.	334,000	266,000
Wood, furniture, etc.	308,000	206,000
Chemicals	150,000	113,000
Precious metals, jewellery, etc.	168,000	67,000
	6,594,000	3,516,000

This comparison, unreliable as it admittedly must be, does at all events serve to indicate the existence of a very large volume of employment beyond the purview of the Factory Acts, which must be accounted for in large measure by workers engaged in small workshops, domestic and otherwise. To any sociological investigator the small average size of industrial establishments is an arresting fact. In the majority of trades the number of persons employed per establishment commonly falls below twenty. Further, in trades suitable for small-scale operations there are found to be about as many individuals who work on their own account as there are of those who employ others. The immense number of small undertakings, apart from the railways, docks, and gasworks, and the great industries in which mechanical production has been highly perfected, is still the outstanding feature of modern industrialism. The experience of London in particular affords practical proof of the persistent vitality of small methods of business.

The most prominent examples of trades in which home-work is common are the various branches of the clothing industry—hand-weaving, knitting of hosiery, tailoring, dressmaking, millinery, bespoke boot-making, and so on; other instances are found in numerous trades which require little or no mechanical equipment, such as the manufacture of cardboard boxes, artificial flower-making, the chain and nail industry, and many others. It is a significant fact that many of these are cases in which 'sweating' is notoriously prevalent, and the group includes all those industries which, under recent legislation, have been scheduled for regulation by special wage-boards. The cutlery industry presents perhaps the best example of the persistence of the older forms, and in this instance the ancient organization manifests a continued vigour, although its vitality is now sadly impaired.

A return of the number of out-workers employed, in 1912, in trades scheduled under the Factory and Workshops Act of 1901 shows a total of 92,146 such out-workers in Great Britain, and it is admitted that the returns even in these cases are incomplete. The principal trades which provide employment of this kind are those connected with wearing apparel, in which 72,000 of the whole number are occupied.

(b)

In 1900 there were about 2,300 hand workers, comprising 1,250 men and 200 boys, as well as 600 women and girls in workshops, and 250 women home workers. Not counting those engaged at home, who were mostly solitary workers, the average number employed in one workshop was less than four persons. The shops in which the work of file-cutting is carried on are for the most part of a rude and primitive description, poor in structure and often dilapidated. The commonest form is a low lean-to shed in the yard at the rear of a dwelling-house. Before the trade was placed under special regulations there was usually no adequate provision for ventilation or cleanliness, the floors being of loose bricks or earth and the washing appliances being often inadequate or entirely absent. A marked improvement is now noticeable in these respects, and under more sanitary conditions the trade is taking a new lease of life, from five to six hundred shops being still in use—many being still found in the country round Sheffield, especially in the villages of Ecclesfield and Oughtibridge, as well as in the City of Sheffield itself.[1]

[1] In 1900 the number of hand file-cutting shops under inspection in Sheffield alone was 546. Report of Medical Officer for Sheffield, p. 70.

The organization is similar to that of other trades in which out-work prevails. The workers usually employ a little assistance, although many work alone; on the other hand, a few small masters take on a number of boys, the trade rule restricting each worker to one apprentice having broken down. The heavier classes of work are performed by men only, small articles and the edges of large files being cut by women.[2] There are regular lists of piecework prices, the scale being common to me and women workers alike. There is no reason to anticipate a rapid disappearance of the trade, since much of the work is required in quantities too small to make it worth while to employ a machine, while other kinds are too trifling and tiresome for the machine to undertake.

Looking now at the cutlery trades as a whole, we must remember that during the last two generations there has been no substantial expansion of employment. Under such circumstances the transition to machine methods, halting and gradual though it is, is accompanied by a redundant labour market, and is favourable to sweating. The small middleman is often able to secure trained workmen on terms which the respectable manufacturer could not imitate without provoking popular outcry and strenuous revolt. In short, the little master, by sweating his workers, can often produce common goods more cheaply than his reputable rival. On the whole, the condition of the man employed on his master's premises is uniformly superior to that of the out-worker. Employers naturally wish to attract the best and steadiest workmen as in-workers; hence, as a rule, the conditions as to deductions and supply of work are more favourable within the factory gates than outside. Not only is the employer virtually compelled to find work for his in-workers if they cannot otherwise secure it, but he also naturally dislikes seeing a rival manufacturer's work brought on to his premises for execution, though this the in-worker has, of course, a right to do if he is paying rent and is not supplied with work.

[2] The adoption of this employment by women is not a recent innovation. Cf. Mather's verse (*c.* 1785), ' 'Twas Jezebel's daughter I saw chopping files.'

26. The Organization of the Staple Industries: (*a*) Cottons, (*b*) Woollens

(*a*) S. J. Chapman, *The Cotton Industry and Trade* (London, Methuen, 1905), pp. 37-9.
(*b*) J. H. Clapham, *The Woollen and Worsted Industry* (London, Methuen, 1907), pp. 135-7.

Until 1914 Britain's economy depended on a narrow range of staple industries which provided the bulk of her exports: cotton and woollen textiles; iron and steel, including shipbuilding; and coal. These industries, moreover, were highly localized, forming the great industrial areas of the Midlands and North. Since all these industries depended on coal, the map of industrial Britain was the map of Britain's coalfields.

(*a*) THE FACTORY OF J. & P. COATS AND COMPANY, PAISLEY
We must now leave Dunfermline. There are no spinning mills in that town, and it is to a spinning mill that we must go to observe the furthest point of evolution reached. The one we shall visit presents a complete type of despecialisation under the factory system—5000 hands, 4000 of them women, absence of apprenticeship, large capital, engines of 12,000 horse-power, a world-wide custom—here we have every characteristic of the modern system of production. This factory is the Sewing Thread Factory belonging to Messrs. J. and P. Coats and Company, at Paisley, a short distance from Glasgow, and consequently in the heart of a region of extraordinary activity, in the neighbourhood of a great port, and in close proximity to a great city. This sum of conditions tends to make the evolution of working-class families more rapid and more complete.

There is nothing requiring special description in the processes employed. Machinery has simplified to the utmost the work of the persons engaged in this industry, so much so that in this immense factory I did not find a single individual who could properly be called a skilled workman. If we leave out the engineers engaged in superintending the engines, who really take no part in the operations carried on in the factory, and who would be equally good engineers in a foundry, the rest are only the servants of the machinery or foremen engaged in keeping order and general supervision.

The spinning looms are entirely in the hands of women. Girls of sixteen or eighteen, with a little practice, are quite capable of undertaking such a simple task. From time to time a broken thread needs to be joined or a defective layer of cotton wool must be taken away, but for the rest of the time they have nothing to do but to watch the spindles pursuing their incessant toil.

The reeling, labelling, and packing are also done by women, who, in a word, follow the cotton through its successive transformations, from the time the bales arrive at the factory to its final stage, when it is wound on a wooden reel and placed in a cardboard box.

Men are employed only in subsidiary tasks, such as superintendence, repairing the looms, and making reels.

The last operation, which is carried on in huge workshops, is a very simple one, and requires only a short apprenticeship. The wood is sawn by a steam saw into square lengths of uniform thickness, and these are then put into a machine and rounded. Another machine cuts them into little cylinders of the proper length for reels. Nothing then remains but to pierce the hole in the middle and to hollow out a place for the thread, and the reel is finished. Both these operations are done by machinery. It is also by the aid of machinery that the workshop is kept free from the debris which would otherwise accumulate very rapidly. A broad band, so arranged on the floor as to form an endless moving carpet, carries the sawdust and shavings to the end of the workshop.

As I have said, in Messrs. Coats's factory, both among the male and female hands, we find only unskilled labourers, individuals whose technical skill is reduced to a minimum. Therefore if the reflections upon the slavery of the worker under the modern industrial system, which we so constantly hear, were well founded, and if it were true that machinery, by depriving him of his professional skill, had at the same time deprived him of his dignity, and if by closing the little workshop in which he had a chance of becoming an employer, it had at the same time ruined his independence, it is among textile workers that we should find the most degraded condition and the least efficient organisation. It is there that energy would be most discouraged and confidence most dis-abused, and that we ought to find, not merely the greatest suffering, but also the most complete inability to devise any remedy.

(b) THE YORKSHIRE WOOLLEN INDUSTRY

When the woollen and worsted industries of the United Kingdom are compared with one another, two points of contrast are apparent in their general organisation—in the first place, the scale of operations in woollen is generally smaller than in worsted, and, secondly, specialisation has gone much further in worsted than in woollen. Taking the first point, one finds that in the year 1899 there were 1918 woollen and shoddy factories of all sorts in the United Kingdom, employing 153,232 persons; that is to say, as nearly as possible, 80 persons to a factory. At the same date there were 753 worsted factories, with 148,324 workpeople, which gives almost 200 to a factory. More recent and more exact figures yield similar results. For 1901 we have employment statistics not merely for factories 'in bulk', but for the various departments in the different classes of factories. From these figures it appears that in the average woollen spinning factory or woollen spinning department only 22 persons were employed in spinning and all incidental processes. In the case of worsted spinning the corresponding figure is 140. In weaving the difference is much less marked. The average woollen weaving shed, whether it is a separate business or a department, employs 50 workers in weaving and incidental processes; the average worsted weaving shed 106. It may be added that the employment figure for an average combing department or combing mill is 60.

The machinery statistics of 1904 bear out the employment statistics of 1901. The average number of woollen spindles working together in one place is 2354, of worsted spindles 8000. The average woollen weaving shed contains 49 looms, the average worsted weaving shed 125.

What is probably the commonest type of woollen mill, does not merely combine carding, spinning and weaving. It combines all processes, from opening the new wool or rag wool on a willey, to dyeing the cloth—when it is piece-dyed—and finishing it. In some cases, as we have seen, the rag wool itself is made on the premises. In others, wool dyeing also is done at home. At any point in the manufacturing process outside help may be required, either in emergency or as a regular thing. Rag wool is frequently bought dyed. If the shoddy or mungo maker owns 'engines', he will card the rag wool before he disposes of it to the manufacturer. Wool and yarn dyeing may be done by a firm of professional dyers. Woollen yarns, English or foreign, may be bought. At seasons of high pressure, a manufacturer who has difficulty in carrying out his orders to time, may give out spinning and weaving to be done by a neighbour whose machinery is less fully occupied than his own. Piece dyeing and finishing may also be done on commission, either for the manufacturer himself, or for a merchant to whom he has disposed of his unfinished cloth. But when all has been said, it remains true that combination of processes in the hands of a single firm is characteristic of the woollen industry.

The worsted industry, on the other hand, is marked by specialisation. At the bottom of the scale of processes comes the great specialised branch of wool combing, to which reference has already been made more than once. Combing was the last of the main processes in the worsted manufacture to be taken over by machinery. It was only between 1842 and 1853 that the inventions of Lister, Heilmann, Donisthorpe, Holden and Noble raised machine combing from the experimental stage to the stage of commercial success. At that time the combing of fine merino wool was in its infancy, and the task of organising the combing process, together with the associated processes of washing and carding, was heavy enough to occupy all the attention of a single firm. Moreover, in the early days, combing was an unusually profitable business, and the inventors, or some of them, were in a position to exploit their own patents, patents which were vigorously defended by much ligitation. So it came about that, although patented combs were sold at most remunerative prices, a large portion of the combing, both at home and abroad, was, from the outset, in the hands of firms which worked their own combs on commission for spinners. Now they work on commission for spinners and topmakers, more often the latter. Whenever possible, the combing mill runs day and night. Thanks to this and to the specialised knowledge of a very delicate and complicated series of processes that its managers and foremen acquire, it is able, if skilfully directed, to work more economically than a combing department in a spinning or general manufacturing mill can, as a rule, hope to do. The recent growth among spinners of the practice of buying tops instead of wool, has also tended to encourage the further organisation of combing as a distinct industry; while at the same time this practice owes its existence to the fact that combing was already in part so organised.

27. The Combination Movement

W. J. Ashley, *British Industries* (London, Longmans, 1907). Chapter by H. W. Macrosty, 'The Trust Movement in Great Britain', pp. 212–14.

American and German industry before 1914 was characterized by large-scale combinations (trusts or cartels), but British industry remained largely relatively small-scale and competitive. There was, nevertheless, some combination, a response to increasing competition, or to take advantage of the economies of scale.

If we seek to determine the extent to which British amalgamations dominate their respective industries, we will find ourselves much hampered by lack of information. The Wall Paper Manufacturers claimed 98 per cent of the trade in their prospectus; the Bradford Dyers' Association, the Bradford Coal Merchants' Association, the Aberdeen Comb Co., and the Textile Machinery Co., 90 per cent; the Calico Printers and the British Cotton and Wool Dyers, each 85 per cent; the Cement Manufacturers, 80 per cent; the United Velvet Cutters, 75 per cent; and W. Cory & Son and the British Oil and Cake Mills, each about 60 per cent. The Fine Cotton Spinners, the Yorkshire Indigo Dyers, the Rivet Bolt and Nut Co., the Linen Thread Co., and the English Velvet and Cord Dyers almost monopolize their respective trades; while the Bleachers' Association, the Imperial Tobacco Co., A. and J. Stewart and Lloyds, the United Turkey Red Co., the Metropolitan Amalgamated Waggon Co., and Rickett, Cockerell & Co., all occupy a dominant position.

The Fine Cotton Spinners have grown from 31 to 47 businesses; the Bradford Dyers from 22 to 33; the English Velvet and Cord Dyers from 11 to 22; the British Cotton and Wool Dyers from 46 to 51; the Yorkshire Woolcombers from 38 to 41; the Imperial Tobacco Co. from 14 to 19; the United Turkey Red Co. from 3 to 4; and the Yorkshire Indigo and Scarlet Dyers and the British Oil-cake Mills have also absorbed several businesses since they were started. Of these combines the Bradford Dyers and the British Cotton and Wool Dyers have had to meet fresh competition, though not to any important extent. The Imperial Tobacco Co. has caused some regrouping among its rivals—Messrs. Cope Bros. having amalgamated with a London firm; the National Provincial Co., including Messrs. J. & F. Bell, of Glasgow, and three English firms, was formed in August, 1902,* with a capital of £500,000, and about the same time three English wholesale firms joined hands. Further united action has been threatened by the Wholesale Dealers' Association.

When we come to consider the financial results of the combination movement, we find that the iron combinations, which are mostly all of the 'efficiency' class, did remarkably well in the three years, 1899–1901, when they enjoyed the effects of the 'boom'; but that the bad trade which followed its decline will try the merits of the new system is undoubted. Thus Richardsons, Westgarth & Co. paid a dividend of only 6 per cent in 1902, compared with 10 the previous year; and the South Durham Steel and Iron Co., which in 1901 showed profits of £105,680 compared with £125,834 in 1900, somewhat ominously announced in January of this year that while the result of the year's working might admit of the payment of the half-yearly preference dividend, they considered it judicious to wait until the accounts had been made up. Similarly, the shipping combines—such as the Shell Transport and Trading Co., the Union-Castle Line, Wilson's and Furness, F. Leyland & Co., France, Fenwick & Co., and the Ellerman lines—did well over the same period when the rates for chartering and freight were abnormally raised by the heavy Government demands for transport purposes during the war. But the fall in rates which has followed is undoubtedly affecting them adversely, as, in fact, is shown by the reports of the P. and O. Company and the Manchester Liners, among others. In the

* In the summer of 1903 it was announced that it had not been found possible to carry out this amalgamation.

wild anticipations which are formed of the powers and results of trusts, the experience of these two industries shows us that it is not possible to control deep trade movements. Cheap iron favours ship-building and low freights encourage trade, but neither alone can create a healthy demand. One industry recklessly conducted, or even one large financial house or manufacturing business, can cause a commercial crisis with ensuing depression; but the artificial production of good times is at present beyond our alchemy. The trusts must follow the course of trade; they cannot control it.

The 'horizontal' amalgamations naturally excite more interest, since their object is the far-reaching one of controlling production. In trying to arrange them according to their degree of success, all lines of division run criss-cross through the industries; we cannot say that any trade is unsuited or more suited than others for combination. The successful ones which have equalled or exceeded their prospectus anticipations are—

| J. & P. Coats
Fine Cotton Spinners' Association
Bradford Dyers' Association
Extract Wool and Merino Co.
United Turkey Red Co.
Linen Thread Co.
English Velvet and Cord Dyers' Association
Leeds and District Worsted Dyers' Association
Yorkshire Indigo and Scarlet Colour Dyers | } Textiles and Allied Industries | W. Cory and Sons
Rickett, Cockerell & Co. } Coal
Wallpaper Manufacturers' Association
Fairbairn, Lawson, Combe, Barbour (Machinery) |

28. International Competition: (*a*) Iron and Steel, (*b*) Cotton Textiles

(*a*, i) Lord Brassey, 'Introduction' to S. J. Chapman, *Work and Wages* (London, Methuen, 1904), pp. vii–ix.
(*a*, ii) E. W. Williams, *The Case for Protection* (London, G. Williams, 1899), pp. 146–8.
(*b*, i) Lord Brassey, op. cit., pp. xii–xiv.
(*b*, ii) Williams, op. cit., pp. 160–2.

Britain early in the nineteenth century had been 'the workshop of the world', dominating world industrial production and world trade. By 1914, however, Germany and U.S.A. were larger industrial producers, and Britain's share of world trade had much declined. There was from the 1880s increasing talk of foreign competition for British goods and even some challenging of Britain's economic policy of free trade.

(*a*, i)

In proportion to the natural resources at command, Great Britain fully holds her own. In quantity of production we cannot keep pace with the United States. While the output of ore from the mines in our own country is limited, we are advantageously placed for the importation of Spanish and Swedish ores. The cost of assembling the material for a ton of pig at Pittsburg has been reduced from 1*l*. 15*s*. 3*d*. to 1*l*. 3*s*. 9½*d*. At Middlesbrough it has been brought down to 16*s*. 5*d*. a ton.

As it has been shown by Sir Lowthian Bell, in able reports extending over many years, and embracing the state of the iron industry both in Germany and the United States, the costs of our Cleveland smelters are less than in any part of Germany. Sir Lowthian Bell found no smelting works in the Old World or the New to compare with those at Middlesbrough. He attributed our superior labour efficiency to our more liberal rates of wages.

Descending to a later date, in 1896 a British delegation, after a tour through Belgium and Germany, reported that the United Kingdom was at least as advanced as Germany. The delegation, while commending highly the physique of the German workmen and the value of their military training, were nevertheless confident that any difficulty we might have to encounter in competing with Germany would not be due to the greater cost of British labour; if more highly paid, our labour was efficient in proportion. The delegation were impressed with the perfect organisation of many works they visited in Germany. Care was bestowed on the smallest details. Germany had some advantage in having come late into the field. The metallurgical industries were up to date. The Bessemer and basic processes, and all the latest inventions, had been skilfully worked. In efficiency of labour Germany had no advantage. The production per man was less than in our British industries. If the German masters were in advance in technical attainments, their British competitors had more practical knowledge of the management of labour—knowledge often gained in early life while working in the ranks.

The advance of the United States to the supremacy they now hold as producers is due to the natural resources of that vast country and to a high level of efficiency. The chief improvements introduced in the United States are fully described in the chapter under notice. If wages are more liberal than with us, house rent is dearer. The cost of living is about the same. The rough work around blast furnaces is not attractive to the American. The workers are mostly foreigners, inferior in efficiency to the skilled men in British establishments.

The iron industry is extending in Russia, where large works have been established, but chiefly under foreign management. The demand for manufactured iron is, however, comparatively limited, and comes chiefly from the Government. The Russian worker is patient and obedient, receiving comparatively low wages as the reward of long hours of toil.

(a, ii)

'The situation is truly serious for British manufactures.' So writes a well-informed correspondent of the *Times* of April 18, 1899. This sentence was not written in a time of trade depression, when gloomy statements, even in well-informed quarters, need sometimes to be discounted, but in a period of exceptional prosperity, owing largely to the spurt in shipbuilding. The sentence therefore has special weight and significance. In no department of manufacturing industry was, until a very few years ago, England's supremacy greater and more unchallenged than in the production of iron and steel goods. That supremacy has gone. Not only do the combined nations of the world far exceed England's production, but one nation alone has beaten her hip and thigh—to wit, the United States. 'Tis essential to bear in mind that the use of iron and steel is growing very largely indeed. Railways are spreading themselves all over the world; nearly all ships of any size are now built of iron; the metal is coming into universal use for bridge-work and for building construction generally. Therefore England's iron and steel industries should be making rapid headway, in order to keep pace with the enlarged consumption. But, as matter of fact, they are, even during the present spurt, doing nothing of the kind. The spurt, which is proceeding as I write, renders it difficult to gauge the real permanent position of the industries; but a study of the available figures certainly indicates that the *Times* correspondent was right in describing the situation as 'truly serious'; for, apart from ship-building and armaments, a great deal of activity in which is obviously transient, signs are rather of retrogression than of progress. Just lately the output of pig-iron has advanced, but under the circumstances the advance is by no means remarkable. The production last year reached a total of 8,631,151 tons; but the production in the United States was 11,773,934 tons—greater than that of England by 3,142,783 tons. Germany's production rose from 6,881,466 metric tons in 1897 to 7,402,717 metric tons in 1898. Yet until recently the position was reversed, and more than reversed: England was well ahead.

(*b*, i)

The cotton industry of Great Britain has been long in a commanding position. In Continental Europe, and in the United States, the rate of progress has been more rapid than with us. We are still leading. No fewer than 500,000 operatives are employed in the British cotton industry—three and a half times as many as in Germany.

For any comparisons of British labour efficiency with that of our rivals in Continental Europe, Professor Schulze-Gaevernitz is the leading authority. In energy, skill, and watchfulness, English workers are unsurpassed. In England the number of workers employed to tend machinery is sensibly less than in Germany, and less supervision is required. Hence the cost of labour per pound of yarn spun is decidedly less in England than in Germany. We have approximately two looms to each operative. In Saxony the number of the looms and the number of operatives are about equal. Hence, though wages in Lancashire are considerably higher, the cost of spinning is less. In England the cotton industry is more specialised. Factory operatives in Germany are not, to the same extent as in England, a class trained from childhood. In Germany manufacturers are more ready to accept small orders for new designs.

In England we have an advantage over the Continent in the humidity of our climate. In an atmosphere less saturated with moisture it would hardly be possible to produce our goods of the finer counts. Our exports of cotton goods show some fall in quantity. The values are fully maintained, our exports consisting chiefly of the finest qualities.

The cotton industry of France is of long standing, and is mounted on a considerable scale. The success of the industry is due to the beauty and inexhaustible originality of French design.

In extent of production the cotton industry of the United States stands second only to that of England. American operatives are a migratory body, recruited from every nationality.

Bombay is becoming a powerful rival to Lancashire. Indian cotton is not suitable for fine spinning; though the range of Indian work is extending. In a volume chiefly devoted to the relative efficiency of labour it is of special interest to note that though the mills in India are run 350 days in the year, and for $11\frac{1}{2}$ hours a day, with swarms of hands in comparison with the numbers needed in English mills and with weekly wages incomparably less, the labour cost is far higher than in Lancashire, where the hours of labour are $55\frac{1}{2}$ per week, and 306 days only are worked in the year. The Indian operatives have less physical endurance and less power of continuous application. It is said that they are becoming more efficient.

(*b*, ii)

The statistics of the quantities of cotton goods shipped from this country do not show actual decrease; the trade is subject to ups and downs, but of late years a general view indicates stationariness, and might therefore be taken as negatively satisfactory, were it not that the world's consumption of cotton goods has been the while expanding very widely, partly in consequence of cotton edging out other materials, partly in consequence of the spread of civilization in barbarian and savage countries, partly owing to the world's increase in population and wealth. In this expansion England is having but little share. There has been a spurt of late in England, but other countries have 'spurted' so very much higher that it is doubtful if the supply is not beginning to outrun the demand; for we hear from various quarters ominous reports of accumulating stocks, only partially relieved by sales at unremunerative prices. This determination of other nations to become cotton manufacturers will undoubtedly increase England's difficulty in maintaining her position in the future, and will assuredly prevent any further expansion worth the name. You have on one side the constantly increasing output from American cotton-mills; the United States started out to-supply their home markets, and have achieved that purpose so well that they are now getting a surplus which will be thrown on the export market; a Consular report for 1898 points out that in the last eight years the number of spindles in the United States has increased 33 per cent, and the consumption of cotton nearly 37 per cent, though the population had only increased about 18 per cent. Experience in other trades must convince us of the serious character of American competition in the world market when that nation sets about the work in earnest; and the serious character in the case of cotton goods is heightened by the circumstances that, unlike England, the United States have their raw materials at their own doors. On the other side of the world you have India, Japan, and now China, threatening with their cheap labour, joined to European capital and enterprise, the most serious competition.

29. Growth of Welfare Expenditure

B. Mallet, *British Budgets, 1887–88 to 1912–13* (London, Macmillan, 1913), pp. 467–8, 470–1.

Despite the increases described here, the working classes as a whole probably contributed more in taxation and social payments than they received in benefits, until after the First World War. Nevertheless the period before 1914 saw the growth of more progressive taxation and of welfare services, for example old age pensions, which heralded the coming of 'the welfare state'. The war lifted both tolerable levels of taxation and expectations of welfare, thus giving 'displacement effect' to the growth of government expenditure.

While civil services proper have risen between 1888 and 1913 from something under £15,000,000 to £18,000,000, or by 19·61 per cent only, 'Social Services' have gone up from something under £5,000,000 to £35,500,000, or by 630·13 per cent, the increase since 1906 being 123·31 per cent. The great proportional increase in this case is, of course, due to the fact that the bulk of the expenditure is of recent origin, while the figures for the opening years were comparatively very small. The cost per head of the military services and social services is as follows:

	1887–8	1892–3	1898–9	1905–6	1912–13
Total Army and Navy	£0 16 8	£0 16 8	£1 1 9	£1 8 10	£1 11 8
Total Social Services	£0 2 8	£0 4 2	£0 5 5	£0 7 5	£0 15 7

But this is not the whole story as regards the latter group. A very large addition in the shape of local expenditure met mainly from the rates, but equally with Imperial expenditure a burden upon the community, must be added to the totals for Social Services if a true comparison of the burden per head is to be obtained. The two chief items of expenditure defrayed from the rates which thus fall to be added would be approximately as follows:

	1887–8	1911–12
Poor Law, England and Wales . . .	£7,000,000	£11,250,000
Poor Law, Scotland and Ireland . . .	£1,750,000	£2,100,000
Education, Great Britain	£2,900,000	£15,850,000
(Elementary and higher)		
Total	£11,650,000	£29,200,000
Per head	6s. 4d.	12s. 10d.

The increase is 150 per cent, and the total expenditure under the head of 'Social Services' *from taxes and rates combined* then becomes £16,500,000 for the first year of the series and £64,800,000 for the last, or per head 9s. and £1 8s. 5d. respectively.

Mr. Geoffrey Drage urged, in a letter to *The Times* on January 24th, 1913, and in subsequent letters, that an annual return should be given of the expenditure (central and local) on various forms of 'Public Assistance' which, as he defined it, would be fuller and more complete than, but on somewhat the same lines as, the above heading 'Social Services'. Mr. Pretyman accordingly asked on June 17th, 1913, for a figure shewing the 'direct beneficiary assistance' to individuals for the years 1890–1, 1900–1, and 1910–11, and the President of the Local Government Board, in reply, gave the following statement for England and Wales only. [not printed here.]

Mr. Drage, in a subsequent letter to *The Times* (June 30th, 1913), estimated that when figures for the current year and for the whole United Kingdom had been included the total would amount to not less than £66,000,000, and added 'when we have got these figures completed for direct beneficiary assistance we can then return to indirect assistance, such as that involved in cheap railway and tram fares, labour exchanges' (included, however, in the above head 'Social Services'), 'public baths, and workhouses, etc.; and even so, we shall leave out the vast expenditure on Public Health and Sanitation, Factory and Workshop and Mine Inspection, which appears to me to be more in the category of what I should term Sanitary Police'.

The growth which has thus occurred in this branch of expenditure is the outcome of what is nothing less than a revolution in public sentiment and political thought as to the limits of State action. In their extreme form these views have been represented in the parliamentary discussions of the last few years by those who advocate the appropriation of as large a proportion of the national income in taxation as possible, on the ground that its distribution by the State would lead to the needs of individuals being better supplied than at present, and the true burden on the nation as a whole thereby lightened. Many who would dissent from any such theory yet hold that a portion at all events of this outlay should not be considered a real increase of the national expenditure, such reforms, for instance, as Free Education and Old Age Pensions being rather of the nature of a transference of than an addition to expenditure, and that all such expenditure if it results in increased national efficiency may in the end prove reproductive, even from a revenue point of view. Giving the fullest weight to such considerations, however, it remains equally true that increased revenue has to be found to meet the increased charges thrown on the budget, and, when it is added that, as the experience of the last four years has shewn, this expenditure is liable to increase at a rate altogether beyond budget previsions, and that, owing to the variety of overlapping public authorities now responsible for it, its control is necessarily less effective than that of ordinary departmental expenditure, it is evident that this branch of public expenditure shares with those of expenditure on Defence and on Local Subventions an almost inevitable prospect of increasing growth. Together these three branches constitute over 60 per cent of the gross Imperial expenditure.

30.   Trades Unions: (*a*) Transport Strike (1907), (*b*) the Russian Revolution (1917)

(*a*) *Report of Proceedings at the Forty-fifth Annual Trades Union Congress*, Great Central Hall, Newport, 2–7 September, 1912 (London, Cooperative Printing Society, 1912), pp. 49–50.

(*b*) *Report of Proceedings at the Forty-ninth Annual Trades Union Congress*, Palace Hall, Blackpool, 3–8 September, 1917, p. 58.

Trade unions were a British invention, and after legislation in 1824, gradually strengthened their legal and political status until they were generally accepted in industry as necessary and desirable. The period after 1880 was one of great trade union growth in numbers, especially with the unionizing of unskilled workers, and of the establishment of collective bargaining for wages and conditions. The strike was an essential weapon for trade unions, even during the war. Generally there was little international co-operation between trade unions, but the Russian Revolution was seen, at least at first, as a working-class revolution.

(*a*)

In May last another strike of transport workers took place in the Port of London. We must admire the plucky and manly battle the men fought for more than ten weeks against a conspiracy of shipowners, wharfingers, and other employers, aided by the Port of London Authority, which is presided over by a Liberal peer, elevated to that position by the present Government. In this connection, it seems to me that the qualification most recognised by both Tory and Liberal Governments for a peerage is the payment of huge sums into the political fighting funds to maintain and uphold the privileges of the propertied classes. I have taken part in many industrial disputes and strikes during the past 23 years, but I cannot recall a strike where the forces of the Government were so obviously used to defeat the workers as was the case in London during the recent transport workers' dispute. In every way possible the police helped the blacklegs and owners.

On several occasions the question of the strike was before the notice of the House of Commons, but the chairman of the Port Authority, acting on behalf of the shipping interests, refused to make any settlement, and demanded unconditional surrender, with the deliberate intention of smashing the Transport Workers' Federation, and in this he received the staunch support of the Leader of the Opposition and many prominent members of his party.

The strike brought about a vast amount of misery and starvation in the dock districts, and when the general public and the organised workers became aware of the arbitrary and arrogant attitude taken up by Lord Davenport, very valuable assistance, both in finance and kind, was given to alleviate the sufferings of the people, and in the name of the transport workers I desire to express hearty thanks for such generosity. These strikes create a feeling of kindred interest among all classes of workers, which must tend to the solidarity of the whole working-class movement. The sacrifices which have been made by the wage-earning classes in defence of their rights have always been a prominent feature in the history of this country. In the early part of last century the Chartists were prominent in the political movement, and many lives were sacrificed, some not far from where we are now sitting. An attempt was made by the Chartists to seize the town of Newport, which failed, and 20 of them were shot dead. Three of the ringleaders were arrested, namely, Frost, Jones, and Williams, and were sentenced to death for high treason, which sentence was afterwards commuted to transportation for life. When the governing classes of the present day talk about the methods adopted now in carrying on trade disputes, we have to realise that the element of force is not so conspicuous as in bygone days. The present-day struggle is economic, for a better distribution of wealth and more political power for the wage-earning classes, which applies not only to this country but to all other countries. Labour unrest cannot cease, nor can the tide of industrial revolt be stemmed until remedial measures are brought about and the present social inequalities removed.

We want a better distribution of wealth. To-day the amount received by the workers is much less in proportion to the total wealth production than ever it was. At about the time when the first Trades Union Congress was held in Manchester in

1869 the annual wealth production was about 800 million sterling, and the amount received by the workers was about 400 millions, or just half. Last year the total wealth production was about 2,000 millions, and the amount received by the wage-earners was not more than about 800 millions, which is much less than one-half. Let us consider income tax statistics. The gross assessment for income tax in 1894–5 was 657 millions; in 1910–11 it was 1,046 millions. There we have a vast increase of wealth. To whom is it due? Not to the aristocracy or the capitalist classes, but to the workers, their labour, their skill, and their perseverance. When we are told that wages in some cases have doubled during the last 50 years, we answer that wealth has far more than trebled during that period in the country. There is sufficient proof to show that there cannot be any satisfactory solution to the wage system so long as a small class in the community own the means of producing wealth.

(b) The Russian Revolution

The Revolution in Russia, which occurred in the early months of this year, is an event of the utmost significance and importance to the people of all countries. Revolutionaries are not encouraged by Governments, and efforts have been made here and elsewhere to hide the truth regarding recent political events in Russia. To Britain, more than to any other country, Russia now looks for encouragement in her struggle to consolidate her new-found freedom. Russia remembers how in the days of despotism she looked with longing across the seas to England, and she remembers how thousands of her leaders and reformers were forced into exile, and on these shores of ours found that liberty for which they would freely have given their lives to have secured in their own land.

And now the people of Russia are in power!

Early in the war we gave whole-hearted support to the demands of the Czar and his traitorous Ministers. We broke our ideals of no annexation to grant the Czar's request that Constantinople should be added to his already too large dominions. At his request we sacrificed 200,000 of our best sons in Gallipoli and Greece, and with him we are responsible for the treachery to Roumania. Shall we now refuse the hand of friendship offered by the present rulers of Russia because that hand is hard with generations of toil? The present military situation of Russia is not the result of the Revolution. Had the Czar been left to carry through his treacherous designs, Russia would to-day have been an ally of Germany, and fighting with Germany against us. The present military situation in Russia is the inevitable result of the Czar's neglect of the Army and the people. The great wheat crops of Russia which should have come to this country have gone to Germany, and even their stores necessary for the people and the Army have also found their way there. Guns and shells manufactured here for Russia are also in German hands and some of the shells are now thrown on our sons on the Western front. Notwithstanding recent events, I have great faith in the new Government, and in the liberated peoples of Russia. The Revolution has not weakened, but strengthened, the Alliance.

PART II, 1918–1945

PART TWO: 1918–1945

The social history of the inter-war years is more of a piece than in the popular version which sets the roaring twenties, 'the jazz age', against the depressed thirties. The great social evil of both decades was undoubtedly unemployment. Once the short post-1918 boom was over, the country suffered endemic unemployment; after April 1921 there were always more than a million men without a job, a million concentrated particularly in those areas where the great basic industries were in decline. This irremedial and localized unemployment, in areas of past industrial achievement and prosperity, was the main social blight of the period, just as poverty because of low wages had been before 1914. The word 'dole' sets the tone of the period; the dole was the money received by an insured man after the benefits for which his weekly stamps had paid had expired. Yet it has been calculated that the unemployed married man on the dole in the thirties was, in real terms, better off than many a fully employed labourer in 1900! Nevertheless, the dole was a hand-out, and whatever it did to maintain the body, it failed to nourish the spirit; the effect of long-term unemployment was to destroy self-respect, to alienate from society, and to instil attitudes about work and society which persisted bitterly into the post-1945 world.[1] There was, however, an obvious increase in the standard of living for employed workers between the wars. The social surveys of the inter-war period—in Bristol, York, and London—show clearly how conditions had improved since the comparable surveys—of York and London—of the period before 1914. A Bristol worker in the new aircraft industry was approaching 'affluence' in the thirties,[2] and John Moore in his fictionalized portrait of Tewkesbury (Elmbury) showed how casual labourers in a small country town could live 'like lords'. But here is a paradox, and it is resolved only by realizing that there were two Englands between the wars, the England of 'the depressed areas', and the England of the industrializing and prosperous South and Midlands. Those in full employment not only received good wages, but they

[1] 37, 38.   [2] 39.

benefited from a world agricultural depression which brought down food prices. And all workers benefited from social services which were higher than before 1914. The social conscience of the day rightly stressed the miseries of unemployment, but it is against a background also of generally improving living standards that we should consider the history of the inter-war years.

We start with the legacy of war. Those millions who served abroad saw other societies, different habits and customs, different ways of life, with which to compare their own. If they returned with some new perspectives, their immediate wish was for peace, quietness, good food, social and family life; the normal characteristics of employment and freedom. When after 1921 the vagaries of the economy denied so many the modest requisites of reasonable living, there was a taking-up of old discontents and an intensification of social criticism and unrest. Abroad in Europe, revolutions in Russia, Germany, Austria, and Hungary, and closer home, in Ireland, suggested that a determined assault on existing and unsatisfactory political institutions could bring them down. And even during the war patriotism had not been enough to completely stop industrial strife in areas like Clydeside. But was there any revolutionary threat? Certainly genuine alarm was generated in official circles by the strike of London policemen in August 1918, and of Liverpool policemen in August 1919, which led quickly to improved conditions, coupled with harsh treatment for some of the strikers. There was also, in effect, a general strike in Belfast in 1919, where congested urban conditions were complicated by 'the Irish problem' and by rivalry between Catholics and Protestants. Most industrial action, however, aimed at winning better pay and hours, and was not politically motivated. There were exceptions; in May 1920 the London dockers refused to load *The Jolly George* with arms for Poland to be used against the U.S.S.R. But discontent rather than revolution motivated most British workers, and central to this discontent was unemployment. Even the suffragettes settled quietly into the post-war world. The women who retreated from the factories to the sink did so peacefully, assured that the age-old political discrimination against their sex was at last being reduced. There was some violence on the streets when the police clashed with strikers or the unem-

ployed on demonstrations;[3] but this was comparatively rare.
  Post-war industrial unrest centred in the coal fields. The miners, faced with attempts to cut wages after the short post-war boom, sought to revive the Triple Alliance with the railwaymen and transport workers, to fight the mine owners. On 'Black Friday', 14 April 1921, however, the miners, deserted by their allies, had to go it alone, denouncing the railwaymen and transport workers for their betrayal. They were defeated, as they were again in 1926, but this failure was as much the result of the decline of their industry as of any betrayal by their fellow workers. Miners gained concessions in 1924, a government subsidy to maintain wage rates whose expiry in 1926 led directly to the General Strike of 4–11 May.[4] The General Strike is an important event in British history. It unified and divided the people of Britain, unified the workers in common action, divided them from the rest of the community who saw the strike as a political revolution. It brought out reserves of quiet heroism in ordinary working people who were prepared to suffer hardship at a time of widespread unemployment. There was a romantic air about the events of May, the weak taking on the strong, middle-class undergraduates seeing the episode as an adventure as they manned the buses, conversions (like that of Hugh Gaitskell) to socialism, football between police and strikers. But the reality was the attempt by the unions to implement the old theory of the general strike, to coerce government with the weapon of industrial action. In the event the unions were not sufficiently well-prepared or organized to win, and it became evident that the population at large highly valued constitutional government. The unions seemed to drift into the strike almost by accident, and in one sense, the event can be considered as the last of a long line of coal stoppages rather than any attempt at revolution. The aftermath, however, was bitter, and the Trade Disputes Act of 1927 aimed conclusively at reducing the political power of the unions.
  Other features of the twenties had a more lasting effect. Addison's act, the Housing and Planning Act of 1919, which required local authorities to provide housing for the working class, resulted in over 170,000 houses being built, the Wheatley Act of 1924 in over 500,000, and various slum clearance acts

[3] 38. [4] 35.

in 265,500 houses before 1939. Various reports had laid bare the need for houses, and public opinion demanded that more should be spent on housing. Scotland after the war, for example, needed 236,000 houses to reach the modest spaciousness of three rooms per family, and over 80 per cent of all Scottish houses were of four rooms or fewer. In most parts of Britain, both in the twenties and thirties, great new housing estates were built on new sites, setting high standards of comfort and hygiene, but rarely of design. By the thirties, when private building was more important than local authority building, there was a real housing boom, at a time when the concept of a tolerable house had changed to include more rooms, more equipment, an inside toilet and a bathroom. Improved housing was evidence of a general rise in the standard of living, which was revealed by social surveys, like *The New Survey of London Life and Labour* of 1928 and Tout's *The Standard of Living in Bristol* of a decade later. John Boyd Orr's investigations into food, health, and income of 1936 also accepted that there had been a rise in the standard of living, but that still 10 per cent of the population (including a higher percentage of children) was under-nourished. The P.E.P. report of 1937 on Britain's health services, however, threw light on the inadequacy of hospitals and revealed a wide variety of standards between areas, while R. M. Titmuss, a formidable social critic, laid stress in 1938 on the continued inadequacy of housing, even after the building boom. J. B. Priestley, in his brilliant impression of the country in *English Journey* (1934), made a four-fold division of the country: Old England; nineteenth-century industrial England; post-war England of light industry and American influence; and England on the dole. W. H. Auden called the thirties 'a low dishonest decade'. Other writers were similarly pessimistic—T. S. Eliot, James Joyce, Virginia Woolf, Aldous Huxley, and George Orwell—and there emerged, also, another generation of intellectual social critics like L. T. Hobhouse and especially R. H. Tawney whose attitudes were well expressed in the title of his *The Acquisitive Society* (1923). P.E.P. (Political and Economic Planning) was founded in 1931 by a group of distinguished critics, including businessmen, to investigate social problems without party commitment. Disappointing in this context of self-critical zeal was the sad performance of the two Labour governments, neither of which was experienced enough administratively, or ruthless enough doctrinally, to pursue a self-interested policy on behalf of the working class.

What did the social critics and the social surveys reveal? Of major significance were population trends, where a marked decline in the birth-rate roused fears of a stagnant or declining population. In 1891 for every 1,000 women aged between 15 and 45, 121 children were born annually; by 1928, the figure was 83. At the same time the death-rate fell from 21·1 per thousand to 12·1, a tribute both to increasing standards of living and to better medical care. The excess of females over males increased from 11 to 16 per cent. Thus the population of Britain was becoming older, with obvious social implications. How much of this population was very poor? Real wages rose about a third between 1900 and 1930, with the unskilled gaining more than the skilled. Poverty, as defined by Booth, had decreased on average from over 10 per cent to less than 8 per cent of the total population; in Bristol only 4 per cent were in poverty while in the depressed north the figure was over 10 per cent. The causes of poverty, in order of importance, were unemployment, low wages, old age, absence of the male wage-earner, and sickness; each worsened according to the size of the family. George Orwell, writing in 1941, noted the emergence of classless sections of British society, a trend certainly brought about by increasing affluence.[5] The structure of incomes by 1940 allowed wage-earners often to do better than the lower-paid salary-earners. In Bristol 90 per cent of working-class families had more income than was 'barely necessary', and nearly 50 per cent had double the minimum standard. G. D. H. Cole in 1937 divided the British people conventionally into three main classes: the working class earning less than £4 a week now constituted 73·5 per cent of all families; the middle class, 21·3 per cent; and the upper class, 5·2 per cent, with much higher incomes. Wages added up to 40 per cent of total income, and salaries 25 per cent. Already a significant redistribution of income through the social services had occurred, in all, about 5 per cent of national income. It is clear that an immense change had come over British society since 1870, a change not measured just by the increasing standard of living. Even the attitude towards poverty had changed

[5] 40.

significantly. In 1900 a person in need had to apply for Poor Law relief, and, if able-bodied, had usually to go to the workhouse. In York in 1936, Rowntree noted eighteen times as much relief was given to people in their homes as in 1901, a relief which was paid without social stigma and as a right to which the workers themselves had contributed financially.[6]

In 1891, 50 per cent of people had had little or no formal education; by 1928, this was true of only 5 per cent. The number of teachers had increased over the same period from 16 per thousand children to 28 per thousand. The Fisher Education Act of 1918 proposed a school-leaving age of 15 which was not attained, however, until 1939. Compulsory continuation schools, also recommended in the act, were also not implemented except at Rugby. More important was the Hadow report of 1926 which proposed secondary education for all, starting at age 11, instead of the various provisions for secondary schooling which had grown up alongside the elementary schools established under the Balfour Act of 1901. Hadow also established the division of children into three streams—academic, technical, and modern—ostensibly classified according to ability. Here was the famous tripartite division, based on an 11-plus exam, which greatly favoured those children with privileged social backgrounds. This meant, in effect, reinforcing the educational advantages of the middle classes. The number of children at secondary school in Britain continued to rise, 394,105 in 1930 to 470,003 in 1938, but the number at primary school in 1938 was already nearly six million, of whom only 462,000 continued at school after the age of 15. Teachers' salaries were cut during the depression, but the Burnham Committee of 1938 established a national machinery for negotiating salary scales which still exists. At the tertiary level there was little expansion, with small university colleges being established at Swansea, Leicester, and Hull, and student numbers increasing from 40,000 to 50,000 between 1920 and 1940. Even in 1956 only 26 per cent of British university students were the children of manual workers, and the proportion was certainly less in the 1930s. Oxford and Cambridge enjoyed social and academic prestige above other universities, just as did the public schools, which survived a

[6] 41.

battery of grim revelations about their teaching methods and social customs by ungrateful alumni.

The more or less literate population produced by this complicated educational process was subjected to a battery of influences remote from much of the formal education with which it had been equipped. The popular press brightened drab lives with circulation races; advertisers sought custom even to the length of hiring aeroplanes to write messages in the sky; the B.B.C. under Lord Reith provided a rich programme of music, drama, and serious discussion as well as popular entertainment.[7] Aeroplanes, motor cars, new fabrics, jazz, cigarettes, more easily obtained contraceptives added some colour to life in an age which offered little in the way of richness or variety in experience except to the very rich. Good books, however, were plentiful and cheap; Penguins could equip a person with a first-rate library at sixpence per volume. Urbanization and increasing population densities made cities and towns larger, and city transport was improved to maintain mobility. Electric underground and buses gave Londoners access both to the central area, with its entertainment, and the countryside, with its open spaces. More and regular holidays took more working class people to the seaside or to London. Women completed their political victories by securing full electoral equality with men in 1928, but they played little part in politics. Margaret Bondfield, an unpopular Minister of Labour in MacDonald's second cabinet of 1929, was the first female cabinet minister. Most professions began slowly to admit women, except the Church, but during the long Depression women were often regarded as unfairly competing with men, and in many professions, like teaching, marriage meant automatic retirement. And nowhere did women receive equal pay for equal work, even where women dominated, as in teaching. The dutiful acceptance of inferiority by women, so that many, if not most, remained ill-paid drudges in inconvenient homes with poor equipment, is strange after their early militancy. Certainly women as servants began to disappear, domestic servants declining by one million between 1911 and 1931. There was, also, some loosening of Victorian and Edwardian standards, with greater freedom for women in both morals and clothes, and a greater tolerance of deviance in behaviour.

[7] 32.

The evacuation in 1939 of one and a quarter million children from the cities to the country and safer suburbs brought home to the well-to-do some of the degradation which was sometimes still found in the great cities. Of elementary schoolchildren under the London County Council authority 16·4 per cent were described as dirty and/or verminous; the East Sussex figure was 3·8 per cent, Northampton County Borough had 11·4 per cent, and Northamptonshire County 6·3 per cent. Skin disease, eneuresis, soiling were more common in the towns than in the country; no breakfast, inadequate sleep, too many cakes and sweets instead of a wholesome diet were also common. Urban society also had a high crime-rate, held by writers like Cyril Burt to be the price paid by society for squalid social conditions. For example, in 1936 of every thousand boys in Liverpool under seventeen, 24 had been found guilty of an indictable offence, most of them against property, and over a third of those had been found guilty on a previous occasion. Such children came mainly from large families with low incomes who lived in the slums. In Shoreditch, the most overcrowded of London's boroughs, 22 per cent of children had to share lavatories with other families and one in three houses had no indoor water supply; only 14 per cent had a bath and 66 per cent had no bathing facilities at all. Working class girls grew up without any basic sexual knowledge, except for taboos and nameless fears. Britain was certainly richer in 1939 than she had been in 1900, but there was still poverty and misery, and still a submerged, slum-dwelling, underprivileged, ignorant, and undernourished minority which improving economy and increasing social services had barely touched.

The intellectual history of the period can be written best in terms of science and, to a lesser extent, social science.[8] Fundamental scientific research, such as that which lead to radar and the atomic bomb, proceeded faster and further than ever before. Economics, similarly, developed beyond concepts of an economy in permanent equilibrium at full employment to an understanding of those processes which caused unemployment equilibrium and how they could be remedied. But what was the impact of science on society? How did science affect values? How was scientific advance translated into technical change? These are questions still unanswered. Undoubtedly science

[8] 50.

had a profound effect on the physical condition of man and on his state of mind. Increasingly he was subject to forces which he could not understand or even control. The era of the engineer was giving way to the era of the scientist, and when war came the scientists dominated in an enormous battle of move and countermove. But if science by its complex nature baffled the ordinary mind, political and economic beliefs seemed easier to grasp. It seemed obvious to many, for example, that socialism was superior as an economic system to the market economy of capitalism. Hence the almost uncritical adulation of Soviet Russia by such powerful minds as the Webbs and Shaw. The background of massive and apparently incurable unemployment, ineffective government, fascism, and, generally, the problems of the poor and underprivileged, stimulated a generation of young intellectuals not only to be critics of society, which was natural, but also to be socialist or even Communists, which was less rational. At home they failed to see the economic growth and structural change that had occurred; abroad, they saw clearly the political consequences of Fascism,[9] and fought it, but failed to see also the political authoritarianism of Communism. There was, in many creative and gifted minds, a despair which produced much fine but less clear social analysis. Yet J. M. Keynes was already providing an answer to economic ills by a clear understanding of how incomes are determined, by abandoning *laissez-faire*, and by accepting the mixed economy in which government plays an important and often decisive role in maintaining employment and fostering economic growth.

More than anything else, unemployment dominated the domestic scene in the inter-war years. Between 1925 and 1937 the percentage of insured persons unemployed in Britain never fell below 9·7 per cent, and rose in 1931 to 21·3 per cent. The average for 1891 to 1895 had been 6 per cent, for 1871 to 1875 1·6 per cent, the high and low of the period 1890 to 1914. The effects of long-term unemployment were extensively recorded, by newspapers, by social scientists, and by novelists. George Orwell's *Road to Wigan Pier*, for example, was a perceptive and moving account of the Depression. Ellen Wilkinson's *The Town that was Murdered* described how Jarrow's unemployment rate rose to over 70 per cent when shipbuilding

[9] 51.

ceased there in an effort to rationalize the industry. Indeed the literature of the Depression years, by sympathizers rather than by victims, can be compared with the literature of the First World War, a heart-felt reaction to man's follies and cruelty. Unemployment of the inter-war scale almost completely destroyed the long-held attitude that this social misfortune was the fault of the individual rather than of the system. Some of the middle classes protested that the dole encouraged idleness; some of the working classes argued that the means test added insult to the injury already done to them. Government was certainly not indifferent to the economy's formidable problems, nor to human suffering, but it was incompetent, timid, and conventional in its ideas and programmes, in spite of the more radical urgings of men like Oswald Mosley, David Lloyd George, and Harold Macmillan. All observers noticed the apathy induced by long-term unemployment. Fascists and Communists, with their cries for revolution, therefore, found little support among the workers. The Labour Party was also firmly non-Marxist, and was just as keen to balance the budget as the Conservatives. In any case, the Depression was seen as the result of world-wide conditions in which Britain was a common sufferer, conditions over which Britain had no control.

Britain's economy before 1914 had been highly specialized, with a narrow range of staple export industries whose origins date from the beginning of the industrial revolution. Britain's 'great specialization' had made her a highly dependent economy, dependent both on imported raw materials and foodstuffs, and on exports of manufactured goods to pay for those imports.[10] The First World War produced a discontinuity in world economic development which was disadvantageous to Britain. After a short post-war boom three major industries were never to recover their pre-war prosperity—cotton, shipbuilding, and coal—and others were to have a long and painful period of adjustment—for example, iron and steel. The main cause for decline was undoubtedly falling export markets, the result of both increasing production abroad and increasing tariffs.[11] But also important was a double failure in Britain: failure to modernize the staple industries which in many ways were less efficient than those of their competitors;

[10] 52.   [11] 54.

failure also to react quickly to the need for structural change in the economy, away from old into new industries. Quicker improvement in the staples would have helped exports but it would not have solved the basic problem of structural change. Only when the economy developed new industries in the thirties,[12] and during the Second World War, was the basic problem of unemployment solved. The so-called new industries were a complex of industries centring on motor vehicles, chemicals, electricity, and services. Most of them had their origins before 1914, had benefited from protection during and after the war (the McKenna duties), but had only come into prominence after 1931. The recovery of the 1930s was undoubtedly based on a housing boom and on the growth of the new industries. These industries, freed from the dominance of coal as the National Grid (established in 1926)[13] increasingly supplied them with necessary power, and avoiding the depressed markets of the north, emigrated to the south and midlands. It was here that a new industrial revolution was occurring. Nevertheless, even with recovery and structural change, the 10 per cent unemployed were not absorbed before rearmament and war finally solved this twenty-year-old problem.

Any chronology of the inter-war years must include trade cycles as well as the persistent structural unemployment. There were three main cycles between 1918 and 1939, with production falling away in 1920–1, 1928–31, and 1937–8. It was the central cycle, however, that dominated, with its depression and unemployment on an unprecedented scale. After the short post-war boom collapsed, Britain had uncertain prosperity during the twenties which did not compare with the real boom conditions enjoyed by the U.S.A. or even by Germany. At first high unemployment was thought to be temporary, and the return to gold in 1925[14] can be seen best as an attempt to get back to 'the normal conditions' of pre-1914. Similarly the abandoning of war-time controls, the returning of controlled industries to their owners, the cessation of bulk-purchases of raw materials by government, etc., were all moves to return to a *laissez-faire* economy. Indeed, little was learnt from the lessons of war economy. The problems of the peace-time economy which the government failed to solve were

[12] 53.   [13] 45.   [14] 55.

perhaps no more formidable than those of the war-time economy which, piecemeal it is true, the government had come gradually and successfully to solve. In 1918 Britain had had a command economy; by 1922, nearly all war-time controls were gone; therefore government had neither the will nor the power to solve the nation's economic problems. Conceptually it was difficult to realize that an economy which had functioned so successfully for so long under the principles of *laissez-faire* and free trade could not cope with the problems, formidable as they were, of these years. Even in the 1930s where there was more economic reality, government was still unwilling to intervene too much in the market. Recovery after the crisis of 1931 certainly owed something to government—protection and a policy of cheap money having some importance—but more to increasing investment and employment in the private sectors of housing and new industry.

The crisis of 1931[15] began with the collapse of the stock-exchange boom in the U.S.A. in October 1929. This created a cumulative downward spiral as each country scrambled for liquidity and repatriated foreign funds. London, a refuge for 'hot money' and now a short-term debtor rather than a short-term creditor as it had been before 1914, was weakened by continual withdrawals which threatened the gold reserves with extinction, and led to Britain abandoning the gold standard in September 1931. This was followed in 1932 by an Import Duties Act which imposed an average 25 per cent tariff on manufactured imports, and gave preferences to Empire economies. Thus in these two years *laissez-faire* policies which had been established in the 1840s by Peel disappeared; thereafter there was sterling diplomacy and protection. An era of British economic history had ended. It is a mistake, however, to concentrate only on the debits of the inter-war years. It *was* a period of structural change, technical advance, and rising living standards. In this period real wages on average rose by about 20 per cent; hours of work were reduced by 10 per cent; education and housing for the working classes were improved; the yearly week's holiday became a norm for the working man; health improved so that the recruits who went into battle in 1940 were superior in physique and health to those of 1914; there was an expansion of the professions and of the services

[15] 56.

generally. This was not an age of economic regression, only one in which economic advance was masked by an intolerable level of unemployment. It was a period, also, in which social thinking was changing significantly to accept government intervention in the economy as normal and desirable. This period saw the end of *laissez-faire*. It was, also, a time of remarkable technical progress: new metals (so that aluminium pots replaced the old heavy iron cooking utensils), new alloys (stainless steel), welding (to replace riveting), synthetic fibres, ball bearings, aerodynamics, radar, penicillin, were but a few important technical developments of the period. And, fundamentally important, the inter-war years saw the basic restructuring of the economy, a change in the location of industry, a change in the structure of production and exports, a change which enabled Britain to adjust quickly to the needs of war economy and which also enabled her to emerge from the Second World War with an economy capable of a new and sustained export performance.

GENERAL SOCIAL CONDITIONS

31.   The End of the War, 1918

Osbert Sitwell, *Laughter in the Next Room* (London, Macmillan, 1949), pp. 3–5.

Osbert Sitwell, man of letters, believed that the war was a great discontinuity which had broken up the old civilization of Europe. This was a view shared by many of his contemporaries who also believed that the war had robbed Britain of a generation of talent and leadership. Compare Keynes's memory of the pre-1914 era, above (**24** *b*).

So that night it was impossible to drive through Trafalgar Square: because the crowd danced under lights turned up for the first time for four years—danced so thickly that the heads, the faces, were like a field of golden corn moving in a dark wind. The last occasion I had seen the London crowd was when it had cheered for its own death outside Buckingham Palace on the evening of the 4th of August 1914; most of the men who had composed it were now dead. Their heirs were dancing because life had been given back to them. They revolved and whirled their partners round with rapture, almost with abandon, yet, too, with solemnity, with a kind of religious fervour, as if it were a duty. It was that moment which sometimes only occurs once in a hundred years, when strangers become the oldest friends, and the dread God of Herds takes charge. As a child I had first beheld his countenance, sweaty and alight with rage slaked and with pleasure, on Mafeking Night: but this evening he was in more benignant mood, chastened and by no means vainglorious. A long nightmare was over: and there were many soldiers, sailors and airmen in the crowd which, sometimes joining up, linking hands, dashed like the waves of the sea against the sides of the Square, against the railings of the National Gallery, sweeping up so far even as beyond the shallow stone steps of St. Martin-in-the-Fields. The succeeding waves flowed back, gathered impetus and broke again. The northern character of the revellers—if they may be described as that—was plain in the way they moved, in the manner, for example, in which the knees were lifted, as in a *kermesse* painted by Breughel the Elder, as well as in the flushed and intent faces. It was an honest, happy crowd, good-natured, possessed of a kind of wisdom or philosophy, as well as of a perseverance which few races knew: but it had nothing of Latin grace. ... The vision which rose before the eyes of the soldiers was very different from that which has lately opened for their sons: it was a deliverance from mud—and mud as one of the great engulfing terrors of mankind can easily be under-rated—from mud and poison-gas, from night patrols and No-Man's-Land, and going over the top, from tetanus, tanks and shell-fire, from frost and snow and sudden death. To their dung-coloured world of khaki in sodden trenches, it seemed until today as if the politicians had clamped them for ever.

No-one, then, who had not been a soldier, alive on the morning of the 11th of November 1918, can imagine the joy, the unexpected, startling joy of it; for in 1945 victory came with deliberate step, in 1918 at a gallop. It had flung itself on us. The news had been—or at any rate had seemed—beyond what could be believed: the only way of persuading oneself of its truth was by doing something one had never done before, such as dancing in Trafalgar Square. It was with this feeling, I think, that the units composing the crowd danced. ... When the news had first come with a ringing of bells and sounding of maroons, men and women who had never seen one another before, spoke, to ask if it were true. All day long the news pelted in, as the first of the two tidal waves that were to destroy Europe swept over it to the furthest end. Russia had already been submerged for a year: now the Kaiser was in flight, the routed German armies were returning home, as is their habit after laying a world waste, and were kicking their officers in front of them. In Bavaria, the Communists had seized power, and were torturing in a ten days' coven, and in Hungary, too, under their leader the infamous Bela Kun, persecution and death were rife, and anyone with a clean shirt had to hide. The Emperor Karl had been deposed in Austria, which was now little more than a derelict great city lost in the mountains. In remote turreted castles all over Germany and Austria, the princes cowered, had not even the heart to go hunting. The dark wind of destruction tore like a falling angel across the

European sky. In short, the popular reign of piracy, exalted to a creed, had begun.

32. Broadcasting, Programmes, and Audiences

(a) A. Briggs, *The Birth of Broadcasting* (Oxford University Press, 1961), pp. 258–9.

(b) *Survey of Listening to Sponsored Radio Programmes*, conducted in March 1938, for the Joint Committee of the Incorporated Society of British Advertisers and the Institute of Incorporated Practitioners in Advertising, under the chairmanship of Arnold Price, pp. 5–7.

The radio was the first expensive modern consumer durable to be purchased generally by the working classes. Its impact was important, both for challenging older forms of entertainment, and also as a medium for advertising and political propaganda. The survey of listening in 1938 showed how widespread was the use of radio on the eve of the Second World War.

(a)

Approximately one-quarter of the total BBC daily transmissions took the form of speech, but this figure included, of course, women's programmes, news bulletins, weather reports, and the Children's Hour. What were beginning to be called 'feature' programmes played a small part, although Corbett Smith as Artistic Director was particularly concerned with them between leaving Cardiff and his departure from the BBC in April 1925. Difficulties in execution and lack of expert advice about content held back progress in such programmes, although there were interesting experiments in 1925 in the combination of descriptive narrative, dialogues, sound effects, and music. A series of dramatized episodes in the history of famous British regiments certainly pointed the way 'to the possible development of the whole vast field—or jungle—of such programmes'.

The first regular daily weather forecast was broadcast as early as 26 March 1923. By then the most regular 'speech and music' programme of any length was the Children's Hour. It usually lasted for forty-five minutes, and it played a somewhat disproportionately large part in the early life of the broadcasting stations.

The first officers of the BBC both in London and the provinces had to reconcile themselves to becoming 'uncles' and 'aunts', with all that this meant, not only in the eyes of the children but to some extent at least in the eyes of the public as a whole. Uncle Arthur, Uncle Caractacus, Uncle Leslie, Uncle Jeff, Uncle Rex, Uncle Humpty Dumpty, Uncle Jack Frost, and Aunts Sophie and Phyllis became household names in the course of 1923. Burrows, who greatly enjoyed playing the part of one of the first radio Uncles, Uncle Arthur, wrote in 1924 that 'there is no section of our programme work upon which more time and thought is spent than that termed the Children's Hour'. C. A. Lewis, who perhaps enjoyed playing Uncle Caractacus rather less, was at least jolly enough to write, 'I wonder if there is any one in the world who has such a jolly mailbag as a broadcasting uncle.' Reith himself stressed the social value of the Children's Hour as a 'happy alternative to the squalor of streets and back yards': he scarcely mentioned 'uncles' and 'aunts' at all. Perhaps it was the fact that the Children's Hour was one of the earliest of programmes and in its origins one of the most informal which made many people cling to its fantasy world as long as they could: perhaps it was on more serious grounds that the young listeners of today would be the great wireless audience of the future or that character-building could be effected by radio that the BBC devoted so much attention to this programme.

(b)

The above figure [not printed here] for average listening at a specified period of the day are very different from the total audience which is reached through the various stations *at some time or other*. The total B.B.C. audience in terms of households must be something very close to the total number of set owners—8,442,050 according to the G.P.O. return of licences—and not the 2,462,000 households estimated (on certain assumptions) to be listening at the hourly period of maximum average listening.

The total audiences reached by Radio Luxembourg, Radio Lyons, Radio Normandy, the Paris Broadcasting Station, and Radio Toulouse respectively must also be greater (and possibly several times greater) than the figures quoted above for numbers of listeners at periods of maximum average listening. The present Survey does not provide any material to enable such total audiences to be estimated. Even a guess at them on the basis of the data contained in this report would be subject to such assumptions and such margins of error as to be practically useless.

Listening by Classes, Age, Sex and Territories

Separate tables analyse listening by social classes. The households covered by the Survey were divided into four groups—A. (wealthy and upper middle class), B. (lower middle class), C. (skilled working class), D. (unskilled working class and unemployed). The highest percentage of total listening was found in the B. class. Listening to commercial broadcasts was found to be most popular in the C. class, with the D. class returning only a slightly lower percentage of listening.

So far as sex was concerned a higher percentage of total listening was found amongst the females covered by the Survey (7·3 per cent.) than was discovered amongst the males (6·3 per cent.).

The individuals who were comprised in the Survey were divided into five age groups: 0–13; 14–20; 21–34; 35–54; 55 and over. Listening was found to be slightly more popular in the 35–54 age group than in any other.

So far as the different parts of the country are concerned, listening tends to decline as one proceeds north, with some recovery in the centres between the Trent and the Tweed. This applies to both total listening and listening to commercial broadcasts. Naturally the degree of listening to the different stations broadcasting commercial programmes varies considerably in the different parts of the country.

33. The Cinema

(a) *A New Survey of London Life and Labour* (London, P. S. King, 1928) Vol. I., pp. 292–5.
(b) J. B. Priestley, *Angel Pavement* (London, Heinemann, 1930), pp. 174–6.

The cinema replaced the music hall as the main form of entertainment, and its stars became popular heroes. On the silent screen, great stars like Rudolph Valentino, Mary Pickford, and Charlie Chaplin were world famous. The 'talkies' appeared first in the U.S.A. in 1928, fundamentally altering the cinema, and helping to make the Depression tolerable.

(a)

Although references to the invention of the 'bioscope' are found in London County Council reports as early as 1905, it was not until some years later that the 'Cinema' made itself felt as a force in the entertaining world. By 1911 however there was already 94 cinemas in the County of London, with an aggregate seating accommodation of 55,000, i.e. over three-quarters of that of the music halls. The early cinemas were relatively small, the average accommodation being about 600 seats, but with their lower costs and prices they soon became a serious challenge to the older form of entertainment. In the next ten years the number of cinemas increased from 94 to 266, while the theatres and music halls declined from 104 to 94.

The latest figures for 1929 show that the theatres and music halls have further fallen in number to 87, and their combined seating accommodation is now only 127,000 compared with 140,800 in 1911, and 115,500 in 1891.

In the past eight years (1921–29) there has been no increase in the total number of cinemas, but the present seating accommodation (268,000) shows that the average size of the halls has increased greatly in the last twenty years. There is no question that the cinema is at present the most popular form of London entertainment, and one by one several of the great theatres and music halls have been converted into 'picture palaces'. At the time of writing (1930) the problem of combining speech with moving pictures may be said to have been technically solved, but it is not yet possible to forecast the influence of the 'talkies' on the future popularity of the 'movies'.

The wide popular appeal of the cinema is of course largely due to the low level of admission prices. The application of 'mass-production' and mechanical methods to dramatic art enables a much cheaper, if inferior, article to be put upon the market. How low is the average price of the cheapest seats as compared with the theatre gallery is shown in another chapter. The cinema, whether silent or talking, may be but a poor substitute for good acting, but it enables thousands of persons who could never have afforded theatre prices to spend an enjoyable evening, and it is probably a more formidable competitor with the public-house and the streets than with the 'legitimate' drama.

(b)

At last the waitress came. She was a girl with a nose so long and so thickly powdered that a great deal of it looked as if it did not belong to her, and she was tired, exasperated, and ready at any moment to be snappy. She took the order—and it was for plaice and chips, tea, bread and butter, and cakes: the great tea of the whole fortnight—without any enthusiasm, but she returned in time to prevent Turgis from losing any more temper. For the next twenty minutes, happily engaged in grappling with this feast, he forgot all about girls, and when the food was done and he was lingering over his third cup of tea and a cigarette, though no possible girls came within sight, he felt dreamily content. His mind swayed vaguely to the tune the orchestra was playing. Adventure would come; and for the moment he was at ease, lingering on its threshold.

From this tropical plateau of tea and cakes, he descended into the street, where the harsh night air suddenly smote him. The pavements were all eyes and thick jostling bodies; at every corner, the newspaper sellers cried out their football editions in wailing voices of the doomed; cars went grinding and snarling and roaring past; and the illuminated signs glittered and rocketed beneath the forgotten faded stars. He arrived at his second destination, the Sovereign Picture Theatre, which towered at the corner like a vast spangled wedding-cake in stone. It might have been a twin of that great teashop he had just left; and indeed it was; another frontier outpost of the new age. Two Jews, born in Poland but now American citizens, had talked over cigars and coffee on the loggia of a crazy Spanish–Italian–American villa, within sight of the Pacific, and out of that talk (a very quiet talk, for one of the two men was in considerable pain and knew that he was dying inch by inch) there had sprouted this monster, together with other monsters that had suddenly appeared in New York, Paris, and Berlin. Across ten thousand miles, those two men had seen the one-and-sixpence in Turgis's pocket and, with a swift gesture, resolving itself magically into steel and concrete and carpets and velvet-covered seats and pay-boxes, had set it in motion and diverted it to themselves.

He waited now to pay his one-and-sixpence, standing in the queue at the Balcony entrance. It was only a little after six and the Saturday night rush had hardly begun, but soon there were at least a hundred of them standing there. Near Turgis, on either side, the sexes were neatly paired off. There were one or two middle-aged women but no unaccompanied girl in sight in the whole queue. The evening was not beginning too well.

When at last they were admitted, they first walked through an enormous entrance hall, richly tricked out in chocolate and gold, illuminated by a huge central candelabra, a vast bunch of russet gold globes. Footmen in chocolate and gold waved them towards the two great marble balustrades, the wide staircase lit with more russet gold globes, the prodigiously thick and opulent chocolate carpets, into which their feet sank as if they were the feet of archdukes and duchesses. Up they went, passing a chocolate and gold platoon or two and a portrait gallery of film stars, who eyelashes seemed to stand out from the walls like stout black wires, until they reached a door that led them to the dim summit of the Balcony, which fell dizzily away in a scree of little heads. It was an interval between pictures. Several searchlights were focussed on an organ-keyboard that looked like a tiny gilded box, far below, and the organ itself was shaking out cascades of treacly sound, so that the whole place trembled with sugary ecstasies. But while they waited in the gangway, the lights faded out, the gilded box dimmed and sank, the curtains parted to reveal the screen again, and an enormous voice, as inhuman as that of a genie, announced that it would bring the world's news not only to their eyes but to their ears.

34. The Lower-paid Worker in the Twenties

Wages Profits and Prices, A Report (London, Labour Research Dept., 1921), pp. 14–16.

Poverty was accentuated by the inflation of the war, when the lower-paid worker suffered disproportionately, and by the unemployment of the post-war period. This resulted in the revival of an old interest of the working classes and of the Labour Party: the establishment of a national minimum wage to safeguard the living standards of the lowest-paid workers.

The policy of focussing attention on the rise in the cost-of-living, by which all workers alike were affected, had indirectly the effect of promoting a policy of national awards and settlements. This tendency may be traced in the engineering and shipbuilding trades. The earlier awards of the Committee on Production related only to individual firms or to certain districts, and the amounts varied with the local circumstances, but the Committee decided in March, 1917, to consolidate all previous awards in a minimum national advance of 7s. a week, and upon this sum to base future advances. 'Those who had received less than 7s., and those receiving more than 7s.,' employers claimed 'should remain where they were,' but, 'any general advance on account of increased cost-of-living should be one sum.' The trade unions maintained on their side that 'special consideration should be given to districts which had not received 7s. advances in wages during the war period,' and this view was now adopted by the Committee. A supplementary agreement between the Engineering Employers' Federation and the engineering unions provided that 'the unions shall not be prevented from bringing forward for special consideration the case of any district in which they claim that the rates of wages are unduly low'. Similarly, in the mining industry wages varied before the war with the selling price of coal, and the miners in certain districts would at one time have been entitled to receive very large advances of wages, but these districts elected to come in with the rest and to promote a more equitable distribution of the surplus among the miners and consumers as a whole.

The attention of all sides was directed to the needs of lower-paid workers upon whom the cost-of-living pressed with the utmost severity. Representatives of the general labour unions pointed out to the Committee on Production the unfair operation of a mere percentage advantage. 'It gives,' they stated, 'a huge preference to the higher-paid men, whereas in the case of labourers with an average of 30s. it does not come out at all in their favour.' In the engineering and shipbuilding trades and other great industries, war advances took the form of a flat rate money amount, which had automatically the effect of levelling up wages for lower-paid workers. The same result was achieved elsewhere by a graduated scale of percentage advances. The various provisions of the cost-of-living sliding scales, which limited the amount of the basis wage to be subject to war bonus, have already been described. In the South Wales steel and tinplate trades, there was at one time a difference of fifty points between the highest and the lowest percentage increase received by different grades of workers, while the exchange from a 12-hour to an 8-hour shift entailed no loss of wages in respect of basic earnings under 50s. a week, but a graduated scale of reductions amounting to $33\frac{1}{3}$ per cent in the case of higher-paid workers. In the civil service there was not merely a substantially reduced bonus on the amount of salary above a certain figure, but badly-paid classes of women received special increases. Among commercial clerks it was usual under awards of the Committee on Production to fix an income limit above which the war bonus should not be paid. For example, where clerks in munition factories received the $12\frac{1}{2}$ per cent on gross earnings, this bonus was made 'payable on salaries up to and including £150 a year', but, 'to those who may be in receipt of more than £150 a year the bonus of $12\frac{1}{2}$ per cent is to be paid on that amount'. The income limit was raised to £250 in 1918. The policy sometimes adopted by the Committee on Production in regard to workers coming under the Trade Boards Act, showed a similar trend. The awards relating to the tailoring trade provided that 'the advances were independent of any increase in rates which may be necessitated by a determination of a trade board', but, where firms had given advances 'either voluntary or as a result of arbitration such advances shall merge in the advances given in the award'. In other cases where advances of wages were admittedly overdue in 1914, account of the fact was taken in the agreement of award. The Sankey award in the case of the miners, or the revised standard rates of railwaymen, were expressly granted with the object, not merely of meeting the rise in the cost-of-living, but of bringing up pre-war wages to a decent level of subsistence.

There was a corresponding movement to establish minimum rates of wages in hitherto unregulated trades. The Government was gravely concerned in the speeding up of output in munition factories and it was increasingly evident that good results could not be obtained from half-starved workers. 'The Government will see to it', Mr. Lloyd George declared in 1915, 'that there is no sweating in Government workshops'. The matter was taken up by the trade unions, and in response to their agitation national minimum rates for semi-skilled men and women munition workers were fixed by statutory orders under the Munitions of War Act, early in 1916. The Government next turned its attention to the improvement of food supplies; and, under the Corn Production Act, an agricultural wages board was set up in 1917, and minimum rates were agreed for agricultural labourers, who had hitherto received indefinitely low rates of wages. The new minima varied according to local conditions, but wages were levelled up so that the lowest general rate in England and Wales covered some thirty-five counties at the end of 1920. Again in 1919, when employers were eager to take advantage of the trade boom immediately following the armistice, the trade unions succeeded with comparatively little difficulty in establishing minimum rates in a number of important industries which were practically unregulated before the war. Under the typical national agreement of the printing trades, which covered all classes of workpeople, towns were classified into a London district and six provincial grades, and a minimum rate was fixed for each grade. Joint industrial councils, and other national machinery for fixing wages were set up in one industry after another. There was formed a National Industrial Conference Provisional Committee, representing employers and workers, and both sides were unanimous in recommending the enactment of a legal eight-hour day, and the fixing of a minimum wages for each industry—proposals, however, which the Government refused afterwards to carry into effect. Under the Industrial Courts Act, an Industrial Court was set up with special powers of enquiry into industrial conditions. Simultaneously, there was a rapid development under the Trade Boards Acts. Not only did advances of wages under the Act represent comparatively large amounts, but the Amending Act of 1918 provided for the setting up of boards in any trade where the Minister of Labour 'is of the opinion that no adequate machinery exists for the effective regulation of wages throughout the trade'. The amendment went substantially beyond the terms of the principal Act, whose operation was limited to trades 'where wages were exceptionally low as compared with that in other employments'. Some forty new boards were set up in 1919, and it was promised that between fifty and sixty others should follow immediately—a promise which unfortunately did not mature. Apart from mere compensation for increased cost-of-living, the worker's main achievement during the war was to redeem large tracts of industry from the area of unregulated and 'sweated' trades.

35. The General Strike, 1926

(a) John Simon, *Three Speeches on the General Strike* (London, Macmillan, 1926), pp. 57–9.
(b) Rival news-sheets, 7 May 1926: *The British Worker, The British Gazette*.
(c) H. Fyfe, *Behind the Scenes of the General Strike* (London, The Labour Publishing Co. Ltd., 1926), pp. 82–4.

The General Strike was probably the most important 'happening' of the 1920s, about which a large literature of condemnation and praise now exists. As a strike it failed; as a political move it failed; as a social revolution it failed; however, it left enduring attitudes which persisted into the post-1945 era.

(a)

The Trades Union Congress General Council issued the following 'proposals for co-ordinated action of Trade Unions' on Saturday May 1st, with a prefatory note saying, 'It should be understood that memoranda giving detailed instructions will be issued as required':

1. SCOPE

The Trades Union Congress General Council and the Miners' Federation of Great Britain having been unable to obtain a satisfactory settlement of the matters in dispute in the coal-mining industry, and the Government and the mine-owners having forced a lock-out, the General Council, in view of the need for co-ordinated action on the part of affiliated unions in defence of the policy laid down by the General Council of the Trades Union Congress, directs as follows:

TRADES AND UNDERTAKINGS TO CEASE WORK

Except as hereinafter provided, the following trades and undertakings shall cease work as and when required by the General Council:

TRANSPORT, including all affiliated unions connected with

transport, *i.e.* railways, sea transport, docks, wharves, harbours, canals, road transport, railway repair shops and contractors for railways, and all unions connected with the maintenance of, or equipment, manufacturing, repairs, and groundsmen employed in connection with air transport.

PRINTING TRADES, including the Press.

PRODUCTIVE INDUSTRIES

(*a*) IRON AND STEEL.

(*b*) METAL AND HEAVY CHEMICALS GROUP. Including all metal workers, and other workers who are engaged, or may be engaged, in installing alternative plant to take the place of coal.

BUILDING TRADE. All workers engaged on building, except such as are employed definitely on housing and hospital work, together with all workers engaged in the supply of equipment to the building industry, shall cease work.

ELECTRICITY AND GAS. The General Council recommend that the Trade Unions connected with the supply of electricity and gas shall co-operate with the object of ceasing to supply power. The Council request that the executives of the Trade Unions concerned shall meet at once with a view to formulating common policy.

SANITARY SERVICES. The General Council direct that sanitary services be continued.

HEALTH AND FOOD SERVICES. The General Council recommend that there should be no interference in regard to these, and that the Trade Unions concerned should do everything in their power to organise the distribution of milk and food to the whole of the population.

With regard to hospitals, clinics, convalescent homes, sanatoria, infant welfare centres, maternity homes, nursing homes, schools, the General Council direct that affiliated Unions take every opportunity to ensure that food, milk, medical and surgical supplies shall be efficiently provided.

(*b*)

(i) *The British Worker*

The General Council does not challenge the Constitution. It is not seeking to substitute un-constitutional government.

Nor is it desirous of undermining our Parliamentary Institutions.

The sole aim of the Council is to secure for the miners a decent standard of life.

The Council is engaged in an industrial dispute.

There is no Constitutional Crisis.

Appeals fail to get Blacklegs.

No Applicants for Jobs on the Railways.

Valleys Calm.

> From the special Commissioner of the British Worker, South Wales, Friday.

From all the valleys of the coalfield comes the same report—the workers are calm and steady. The efforts to organise blackleg services are a miserable failure.

For instance, an appeal was issued for 'intelligent and suitable men' for railway work. The wage offered was £2.6.0 a week. There was no response to the appeal. Then the offer was increased to £3 a week. But again there was not a single applicant.

The Amman Valley, in the centre of the anthracite coal field, is providing a wonderful example of solidarity and order. The official police report from this district says that the behaviour of the miners could not be better.

Sports programmes are in full swing at Swansea and in other large towns. Arthur Jenkins, the Eastern Monmouth Miners' agent, speaking at a mass meeting yesterday, said that if the military were sent into the industrial area the workers would play friendly football matches with them.

(ii) *The British Gazette*

The National Issue.

Constitution to be Vindicated.

Assault on Rights of the Nation.

Union Leaders Personal Responsibility.

Everyone must realise that, so far as the General Strike is concerned, there can be no [*sic*] of compromise of any kind. Either the country will break the General Strike or the General Strike will break the country.

Not only is the prosperity of the country gravely injured; not only is the immense and increasing loss and suffering inflicted upon the whole mass of the people; but the foundations of the lawful constitution of Great Britain are assailed. His Majesty's government will not flinch from the issue, and will use all the resources at their disposal and whatsoever measures may be necessary to secure in a decisive manner the authority of Parliamentary government.

The Prime Minister's message to the *British Gazette* expresses a decision from which there can be no withdrawal. All loyal citizens, without respect to party or class, should forthwith range themselves behind His Majesty's Government and Parliament in the task of defeating in an exemplary fashion the deliberate and organised assault upon the rights and freedom of the nation. The stronger the rally of loyal and faithful men to the cause of Parliamentary government, the sooner will victory be achieved and the shorter the period of waste and suffering.

(c)

THE AFTERMATH

May 13. What that big employer told MacDonald was true. I didn't, I confess, pay much attention to it at the time. Now it is clear that a plot was formed to hit back at the workers in every industry as soon as the strike ceased.

Their counter-attack began at once. They are paying no heed whatever to Baldwin's plea that 'all vindictiveness' should be put aside. They are determined, it seems, to act upon an earlier *dictum* of his—his statement last July that:

> 'All the workers of this country have got to take reductions of wages.'

The nation went to bed on Wednesday night immensely relieved and happy in the belief that work would everywhere be restarted this morning. It has suffered a bitter disappointment. What was a General Strike has become almost a General Lockout. Now it is the employers who are holding up industry and transport.

When railway men reported for work this morning they were told they must re-engage on the understanding that the companies 'reserved the rights they possessed in consequence of their breaking their contracts'. The railway unions thereupon ordered the men not to resume work.

A large number of other transport workers were also told they could only be re-employed on fresh terms. The same threat was made to workers engaged in the production of newspapers. Some employers declare that henceforward they will only employ non-union labour. Many say they are resolved to have 'open shops', where they can employ either union or non-union men.

Protesting against these attempts at victimisation, the General Council says it called off the strike believing that Baldwin meant what he said when he proclaimed his readiness to resume negotiations towards an honourable peace. Such peace, says the General Council, is not possible if employers seize the opportunity to humiliate and injure Trade Unionists. To this they cannot and will not submit.

At the same time they point out that Baldwin's personal honour is involved. He promised goodwill and co-operation. He said 'no recriminations'. Yet he allows employers in a spirit of mean revenge to endeavour to force important bodies of workers to surrender gains hardly won during many years.

The situation is now fraught with greater peril than before. During the strike every one was in a good temper; there was no malice or bitterness. Now vast numbers are indignant at what they call the dirty trick played on them.

Those who complained most loudly about the strikers' interference with the convenience of the public are now wantonly prolonging the dislocation of railways and street services. The plight of those working in city offices and shops was worse today than ever. Blackleg amateur drivers and conductors did not put in an appearance. (Many of them were recovering from the effects of celebrations last night which made parts of the West End of London what they are on Boat Race night); the regular employees were rebuffed and sent away. The public is angrily disappointed that the hope of the conflict being over is not yet realised. People are inclined to blame Baldwin as well as employers and to say that Trade Unions are justified in using any weapons against those who assail them in such a ruthless and rancorous way.

*   *   *

I went to the House this evening. Baldwin shuffled. Thomas made a good speech; he produced any amount of evidence that there is a concerted attack on the part of employers. He was very sore about the *Gazette's* heading, 'Total Surrender', but he stressed the fact that the General Council 'had got Baldwin's word'.

Not in writing, I fear, and you can't hold English gentlemen to anything but their signatures.

A friend of Samuel's, an M.P., says, by the way, that Baldwin almost fell on his neck when he announced that the General Council was agreeing to his terms of peace.

36. The Urban Environment, Merseyside 1934

D. Caradog Jones, *The Social Survey of Merseyside* (Univ. of Liverpool Press, 1934), pp. 54-7.

The social historian of England is fortunate in having a series of social surveys (Booth for London, Rowntree for York, Caradog Jones for Merseyside, Coates and Silburn for Nottingham) which give a realistic picture of living and working conditions in industrial towns. This survey of Merseyside between the wars is a typical example.

The site of the old mediæval town of Liverpool is the focus of the survey area. It is the administrative centre of the city and the commercial centre for the whole of Merseyside. Internal and external communications all radiate from here—the cross-river ferries, the rail and road tunnels under the Mersey, the Liverpool tramways, and, only short distances away, services by road and rail to all parts of Britain and shipping services to all parts of the world. Once shockingly overcrowded, this commercial and administrative centre has now a relatively insignificant resident population.

The warehouses and most of the factories are clustered close to the line of docks and around the commercial centre, but three smaller colonies of industrial plants are to be found along the canal banks in Bootle and Litherland, at Aintree close to the railway yards, and in Old Swan bordering on the

London and Manchester main road and Edge Hill station. Garston is a fifth, and quite separate, industrial area. It should be understood that the word 'industrial' is used in a broad sense; in a centre such as Liverpool a considerable proportion of the land is used for warehousing, railway depots, and shunting yards.

Inland, to a depth of two or three miles, is a vast expanse of dull streets in Kirkdale, Everton, Edge Hill, and Toxteth with scarcely an open space to rest the eye, a legacy of the rapid commercial expansion of the late eighteenth and early nineteenth centuries. Mostly working-class property from the beginning, those parts nearest the docks have sunk into slums. Only on the Great Heath, in what is now Abercromby Ward, are there dwellings of a different character. There many of the great houses of earlier days have been turned into flats, so preserving a sense of gentility though the density of population is increased, and they are now interspersed with institutions of all kinds, notably the two Cathedrals in process of building and the University. On the whole it is distinctly an overcrowded zone with few open spaces and few amenities.

Immediately beyond this again is a wide ring of parks setting a limit to the continuously built-up area. The parks themselves are not continuous: they are to some extent broken up by suburban housing, but they nevertheless mark the boundary between the unrelieved sea of streets and the patchwork of house and garden in the suburbs of the city and in the postwar municipal housing estates. This ring of parks serves as a lung to the thickly crowded area and it is of inestimable advantage to the children of the neighbourhood.

Between 1919 and 1931 some 22,000 houses have been built in Liverpool under municipal schemes, and the suburbs with these new housing estates form the outermost zone of the city. It contains also two small areas of an entirely different character built up previously—Garston, a little town by itself with its drab streets and industrial paraphernalia, and Much Woolton, a quarrying village distinctly rural in aspect, which is still separated from Liverpool by a green belt though it may not long remain so. The outermost zone is essentially residential: only in Garston and to a lesser extent in Old Swan is it industrial. It should be added, however, that the northern ends of Liverpool do not fit as readily as the southern and eastern fringes into this zonal division. Industrial and residential areas exist side by side in certain parts, like Walton—where there is an old village nucleus—Fazakerley, and Litherland, an urban district to be mentioned presently.

Bootle, continuous with Liverpool, contains somewhat comparable zones. Its riverside is given up entirely to docks: the northern wall of the Gladstone Dock is the northern boundary of the Bootle foreshore. That part of the borough nearest the docks resembles the corresponding part of Liverpool in being equally overcrowded. Inland the density of houses grows less, open spaces appear, and on the outskirts of the borough, in Orrell Ward, are to be seen again the familiar post-war suburbs and a municipal housing estate. Farther north along the coast passing from Bootle to Seaforth and Waterloo, and from Waterloo to Crosby, the social atmosphere becomes more and more middle-class. These are urban districts independent of Liverpool and in control, therefore, of their own development. Inland behind Seaforth is Litherland, for the most part an untidy medley of industry and working-class houses.

Birkenhead and Wallasey, on the west side of the river, are separated by the long inlet of Wallasey Pool, whose advantages as a site for docks have been fully utilised. Communication between them is difficult; it is accomplished only by floating bridges across the Pool, or by a circuitous rail or bus route: frequently it may be less troublesome to cross and re-cross the Mersey via Liverpool. The internal communication focus of each town, as of Liverpool, is the Mersey ferry. Thus, not only does the river dominate the whole economic life of Merseyside but it binds together the several parts. It is of especial significance on account of the daily movement of workers between Birkenhead, Wallasey, and Liverpool.

Wallasey, indeed, has the typical character of a dormitory town: over one-third of the total occupied population—according to the returns of the 1921 Census—spent their working hours in Liverpool. It is one vast suburb, almost entirely middle-class, which has spread around and between, without altogether destroying, the original village nuclei of the borough. On the extreme north-east of the peninsula it takes in New Brighton, which has enjoyed a fairly long though varied history as a seaside resort.

37. Unemployment in South Wales, 1934

Hilda Jennings, *Brynmawr* (London, Allenson, 1934) pp. 138–43.

South Wales, dependent on a traditional staple industry, suffered one of the most persistently high unemployment rates between the wars.

While some effects of unemployment are general, individual men and their families of course react to it in different ways, and out of some six hundred families normally dependent on unemployment benefit probably no two have precisely the same attitude to life and circumstances. Certain habits of thought and action are enforced by these circumstances. The unemployed man must register twice a week at the Exchange; he must draw his 'pay' there on Friday; if he has been out of work for some time, each Friday he will have a short period of sickening anxiety lest the clerk should single him out and tell him that he is to be sent to the 'Court of Referees'; then will follow a few days' consequent dread lest his benefit should be stopped and he be cast on to the Poor Law, have to do 'task work', for his maintenance, and take home less to his family in

POPULATION SHOWING PERIOD OF UNEMPLOYMENT
IN YEARS AND AGE-GROUPS

LENGTH OF UNEMPLOYMENT

Age	Unemployed under 2 Years	Unemployed 2–3 Years	Unemployed 3–4 Years	Unemployed 4–5 Years	Unemployed 5 or more Years	Totals
Over 65 Years	6	19	12	23	72	132
40–65 Years	77	99	107	104	143	530
30–40 Years	49	64	54	40	25	232
18–30 Years	77	51	52	24		204
14–18 Years	17					17
Never been Employed 14–18 Years	87					87
TOTALS	313	233	225	191	240	1,202

return for it. Having received his 'pay', duly contributed his 'penny' to the Unemployed Lodge of the Miners' Federation, and conversed with his fellow unemployed, he returns home. There, his wife awaits his return in order that she may do the weekly shopping, and in many cases almost all his unemployment pay, with the exception of a little pocket-money for 'fags' goes straight to her.

So far, there is similarity of practice, but beneath the surface this similarity does not reign. One man will approach the Exchange with impatience and bitterness at his dependence and impotency to help himself; one in a mood to find causes of complaint and irritation with the officials; one with growing apathy, and no conscious feeling except when his pay is threatened; one again with each visit feels the need for a change in the economic and social system; his political consciousness is inflamed, and he will fumble in his mind for an alternative, or shout the current formulae at the next 'unemployed' or 'party' meeting according to his mental outlook and capacity.

Visits to the Exchange at most take up part of two half-days in the week. For the rest, some men stand aimlessly on the Market Square or at the street corners, content apparently with a passive animal existence, or with the hour-long observation of passers-by, varied by an occasional whiff at a cigarette. Others work on allotment or garden, tend fowls or pigs, or do carpentry in their backyard or kitchen, making sideboards out of orange boxes stained brown with permanganate of potash, while their wives cook and tend the children in a restricted space around the fireplace, uncomplaining because they realise the necessity of providing some occupation for their husbands in order to keep them even moderately content. Some few attend classes on Political Economy or Industrial History; more talk or listen at frequent meetings of 'Unemployed Lodges' or the Trades and Labour Council, where grievances can be ventilated and pent-up bitterness find vent. Others again stroll 'down the valley' or sit on the banks in groups when the sun is warm. On wet days the Miners' Institute offers papers and a shelter, although shop doors and street corners satisfy many. At nights, there are the pictures, and the long queues outside the 'Picture House' probably account for more of the 'pocket-money' of the unemployed than do the public house, although the 'Clubs' are said to do good business all day long. In a drab and empty existence the 'pictures' provide some colour and excitement if little food for thought....

Varied though the individual and family habits and reactions to unemployment are, one general statement can be made of the unemployed homes in the mining areas of South Wales. With remarkably few exceptions, the bond of parental pride and affection has held firm. It expresses itself in diverse ways, not all of which are good, but which represent the various levels of family life and relationships. In some instances the old ambition of the collier to make his son a teacher holds good, and the whole family cheerfully goes short when necessary in order to give his boy a college career and to keep him decently clothed during it.

38. Violence in Belfast

Wal Hannington, *Unemployed Struggles 1919–1936. My Life and Struggles amongst the Unemployed* (London, Lawrence & Wishart, 1936), pp. 237–41.

Hannington belonged to the British Communist Party and was a bitter critic of the apathy of the Trades Unions and the Labour Party towards the plight of the unemployed. He organized some of the great Hunger Marches of the thirties (the first Hunger March in fact had taken place as early as 1922) which led to serious clashes with the police in the East End of London. Belfast suffered heavily from unemployment after 1921, depending as it did for work on shipbuilding, engineering, and linen manufacture. The Falls Road and Shankill area, mentioned in the extract, were to be the scene of bitter religious dispute in 1969 and after.

That same day information came to the National Unemployed headquarters in London concerning the struggles that had developed in Belfast, and within twenty-four hours the press were full of reports of severe fighting amongst the Irish unemployed against the police. It started over a strike of 2,000 Belfast unemployed who were being compelled to do relief work in exchange for a poor relief pittance. The maximum outdoor relief even for a man, wife and family had been fixed at 24*s*., and single men and women had been completely refused relief. Negotiations took place between the Irish Unemployed movement and the mayor of Belfast, who offered some concessions with the object of dividing the strikers and the mass movement. These were defeated by the unanimous vote of the strikers; then commenced mass marches, and demonstrations and collisions took place with the police on several occasions. The movement spread to Derry, a small border town, between north and south, where there were over one thousand families in starvation. The strike of the task workers in Belfast met with a wide response from the workers in employment. In a few days a relief fund of over £300 was raised and large supplies of food and clothing began to pour into the strike centre. The struggle was gaining momentum hour after hour. On Monday, 10th October, the Relief Workers' Strike Committee called for a rent strike and a school strike, and demanded from the authorities an increase in the scales of relief.

Bonfires were lit in the workers' quarters and round them gathered thousands of workers who were addressed by the unemployed leaders. The city of Belfast became an armed camp, thousands of police being imported. They vainly tried to intimidate the workers, and, heavily armed, they dashed through the streets in armoured cars. At several points the bonfires were extinguished, but the workers relit them. A special mass demonstration of women was held in St. Mary's Hall; they pledged themselves to stand shoulder to shoulder with their menfolk in the fight. On the following day, Tuesday, 11th October, workers were gathering everywhere in groups to discuss the events of the previous night and everywhere expressing the determination to stand solid against their starvation conditions. The police began to charge the crowds as they gathered, but after the first shock the workers met them with a hail of stones, and when the police got within striking distance of any body of workers a series of fierce battles broke out. Armoured cars were called out and drove into the crowds wherever they assembled. Squads of workers rushed to the sites of the relief work jobs and seized the tools with which they had been compelled to slave for a starvation pittance; armed with these, they returned to the demonstrations and fought desperately against the police.

In the Falls and Shankills districts very fierce hand-to-hand battles ensued; whilst the police used their batons, the workers used pickshafts and other weapons. Failing to intimidate and defeat the workers, the police then opened fire from rifles and revolvers. Five workers fell to the ground, so badly wounded that they had to be removed to the hospital, whilst others with lesser wounds were treated in the homes of comrades. The workers threw up barricades against the mounted police and the armoured cars; bravely fighting behind these barricades they repeatedly repulsed the attacks of the police. That night a police curfew was enforced in Belfast for the first time since the Irish struggles for independence in 1922....

During the night of Wednesday another worker who had been shot down by the police died in hospital. The fighting continued through the night and into Thursday. Then the 2nd Battalion of the King's Royal Rifles arrived from Tidworth. During the day conferences were resumed between the Belfast authorities and the Northern Ireland government. The Falls Road district was an extraordinary storm centre, barricades were everywhere in the streets. In an effort to smash the resistance of the workers in this district the police stopped the delivery of food by any tradesmen. By Thursday night there were hundreds of workers under arrest. Then came Friday, 14th October; Tom Mann, the national treasurer of the National Unemployed Workers' Movement, went over to Belfast to represent the British workers in the great funeral procession of those who had been murdered in the streets. Tens of thousands marched in the funeral cortège on Friday to Milltown Cemetery. Large forces of police were in attendance and the military were being held in readiness. Armoured cars moved along with the funeral procession and the whole funeral took place under the guns of those who had shot down the unarmed workers for protesting against their starvation conditions.

Tom Mann marched at the head of the cortège. At the gates of Milltown Cemetery the police placed him under arrest.

* * *

The same day that Tom Mann was deported from Ireland, the terrific struggles of the workers and the impressive funeral scenes compelled the Northern Ireland government to grant considerable concessions as follows:

The scale of relief for man and
wife was raised ... ... from 8s. a week to 20s. a week
Man, wife and 1 child ... from 12s. a week to 24s. a week
Man, wife and 2 children ... from 16s. a week to 24s. a week
Man, wife and 3 children ... from 20s. a week to 28s. a week
Beyond that number of children up to a maximum of 32s., a week, as against a previous maximum of 24s. a week.

Concessions also had to be made in the character of the relief work. Following these concessions the struggle gradually subsided, but thousands of meetings and demonstrations were held throughout Ireland and Great Britain to protest against the violence which had been used against the unemployed.

39. Working-class Standards of Living in Bristol, 1938

Herbert Tout, *The Standard of Living in Bristol, a Preliminary Report* (Bristol, P. Arrowsmith, 1938), pp. 16-20, 26.

Bristol was fortunate in having a variety of industries, including the new aeroplane manufacture, and was prosperous at the time of this survey in May–October 1937. The report was concerned with four-fifths of all Bristol families, the remainder excluded as being middle-class. Tout's basic conclusion was that 89.3 per cent of the 100,000 of the investigated families were above the line of 'standard needs', and only 10.7 per cent below it. Of the existing poverty, 32.1 per cent was due to unemployment, 21.3 per cent to low wages, 15.2 per cent to old age, and the rest mainly to the absence or sickness of the male wage-earner.

Food.—The British Medical Association set up a committee, which reported in 1933, 'to determine the minimum weekly expenditure on foodstuffs which must be incurred by families of varying size if health and working capacity are to be maintained'. It is the dietary recommended by that committee which is used. It is described as an adequate diet on which health and working capacity could be maintained over prolonged periods, and is said to be a typical diet as commonly used by the working classes in receipt of adequate wages. The diet has been criticised for not providing a sufficiently large allowance of milk, particularly for children over six. Adults are allowed a quarter of a pint, children aged 6–10 half a pint, and children aged 5 and under a pint daily. The Ministry of Health Advisory Committee on Nutrition and the League of Nations Technical Commission both suggested higher milk rations, and Mr. R. F. George has put forward a modification

B.M.A. Suggested Adult Ration For One Week.
(Male 14 years or over engaged in moderate work.)

	1937 Bristol Price	Cost
1 lb. Beef . . . . . at	6d. per lb.	6d.
½ lb. Minced Meat . . . ,,	6d. ,,	3d.
½ lb. Bacon . . . . ,,	10d. ,,	5d.
½ lb. Corned Beef . . . ,,	8d. ,,	4d.
¼ lb. Liver (Ox) . . . . ,,	8d. ,,	2d.
2 oz. Eggs . . . . . ,,	1s. 6d. per doz.	1½d.
½ lb. Cheese . . . . ,,	8d. per lb.	4d.
1¾ pints Milk . . . . ,,	3¼d. per pint	5¾d.
¼ lb. Fish . . . . . ,,	8d. per lb.	2d.
¼ lb. Butter . . . . ,,	1s. 1d. ,,	3¼d.
1 oz. Suet . . . . . ,,	8d. ,,	½d.
¼ lb. Lard . . . . . ,,	9d. ,,	2¼d.
7¼ lb. Bread . . . . ,,	4 lb. for 8½d.	15½d.
1 lb. Sugar . . . . ,,	2½d. per lb.	2½d.
¾ lb. Jam . . . . . ,,	5d. ,,	3¾d.
5 lb. Potatoes . . . . ,,	1·11d. ,,	5½d.
¼ lb. Peas (dried) . . . ,,	3d. ,,	¾d.
¼ lb. Tea . . . . . ,,	1s. 10d. ,,	5½d.
½ lb. Oatmeal . . . . ,,	3d. ,,	1½d.
¼ lb. Rice . . . . . ,,	3d. ,,	¾d.
½ lb. Syrup . . . . ,,	4½d. ,,	2¼d.
1 lb. Cabbage . . . . ,,	1½d. ,,	1½d.
¼ lb. Beans (Butter) . . . ,,	3½d. ,,	¾d.
½ lb. Barley . . . . ,,	3d. ,,	1½d.
Fresh Fruit and Green Vegetables		7d.
		88d.
	Cost . .	7s. 4d.

Daily intake:
- Calories 3,386
- Carbohydrates . . . 494 grams.
- Fat . . . . . 101 ,,
- Protein . . . . 99 ,,
- First Class Protein . . 50 ,,

of the B.M.A. diet to include them. It was decided, however, not to use this. Mr. Rowntree, on the other hand, has given a dietary which provides the same calories and proteins as the B.M.A. diet at considerably lower cost. It is even more vulnerable to the milk criticism and it, too, has been rejected.

An idea of the provision made can be obtained by studying the diet allowed a man (see table on opposite page).

Information on prices was collected from many sources in May and June, 1937, and care was taken to avoid pricing the foodstuffs extravagantly. There is no doubt that foodstuffs containing the necessary calories, proteins and the like could then be purchased at the prices named. The meat prices make allowance for the fact that a part of the week's supplies can be bought cheaply on Saturday nights.

Clothes.—The clothing allowances presented the greatest difficulty. Extensive inquiries were made, and it was found that it costs less to clothe a woman than a man. Very often second-hand clothes can be purchased quite cheaply in jumble sales, and although the allowances seem small, many persons in Bristol spend less than this amount on clothes in a year. A budget showing how a man might clothe himself, even with new clothes, for 73s. a year is printed below. He would have to wear his clothes with great care if they are to last the time ascribed to them, and he would have to repair his own boots.

The budget is only given as an illustration to make the

	s.	d.
1 Suit for 4 years at 36s.	9	0
3 Pairs of Trousers for 2 years at 6s.	9	0
2 Sports Coats for 3 years at 10s.	6	8
1 Pullover for 2 years at 4s.	2	0
1 Overcoat for 4 years at 36s.	9	0
3 Shirts at 3s.	9	0
3 Vests for 2 years at 2s. 6d.	3	9
3 Pairs of Pants for 2 years at 2s. 6d.	3	9
6 Pairs of Socks at 9d.	4	6
1 Pair of Braces		6
Handkerchiefs	1	0
Boots, including repairs .	13	10
say 2 pairs at 12s. 6d. every 3 years, and repairs at 5s. 6d. a year.		
1 Cap	1	0
	73	0

standard more vivid. In practice a man would probably buy some items second-hand. For clerks and black-coated workers an extra allowance is made, and a suggestion of Professor Bowley's that it is not reasonable to expect young women at work to restrict themselves to a bare minimum allowance has been adopted.

Fuel.—Previous surveys have usually made an allowance of the equivalent of $1\frac{1}{4}$ cwts. of coal a week for a family, but this has been slightly increased. The reason is that many poor families, particularly those in council houses, are unable to use coal for water-heating or cooking and must use gas. An average weekly consumption of coal throughout the year of $\frac{3}{4}$ cwt. is allowed for ($1s.$ $6\frac{3}{4}d.$) and $1s.$ $5\frac{1}{4}d.$ a week for gas, which (after allowing for the refund) will purchase 385 cubic feet of gas at the rate at which it was supplied through penny-in-the-slot meters. This consumption is slightly below the average ($1s.$ $7\frac{1}{2}d.$) in a certain test street on a council estate, details of which were kindly supplied by the Bristol Gas Company.

Light.—Following Mr. R. F. George's suggestion the allowance is an average of twenty hours a week burning a 60-watt light. At the rate at which electricity was sold through slot meters the average cost of this was $7.7d.$, and $0.3d.$ is allowed for lamp renewals (one a year). This estimate is approximately equal to average consumption in a test street on a council estate ($7.8d.$), and lower than the average of three test streets in Easton ($9.4d.$), details of which were kindly supplied by the Bristol Corporation Electricity Department.

Cleaning.—The allowance follows Mr. R. F. George's suggestion. It allows for a pound bar of soap and $2d.$ a week extra for scouring materials for the whole family.

It will be observed that the minimum standard makes no allowance whatever for sickness, savings, for old age or burial expenses, holidays, recreation, furniture, household equipment, tobacco, drink, newspapers or postage. In practice families whose income is below the standard do not forego all expenditure on these items, but everything they spend on them is at the expense of the meagre allowances made for the basic necessities. The 'needs' of each family are built up by adding the food and clothes allowances for each person, which depend upon the age, sex and occupation of the person, to the scaled family allowances for fuel, light and cleaning. Table I gives examples of the effect of the calculation : —

TABLE I
EXAMPLES OF STANDARD NEEDS

Family consisting of:	Family needs per week exclusive of rent
	s. d.
Man alone	12 9
Woman alone	11 4
Man and wife	20 3
Man, wife, child under 5	24 4
Man, wife, child 5–9	25 6
Man, wife, child 10–13	27 4
Man, wife, two children 5–9 and 10–13 . .	32 10
Man, wife, three children 0–4, 5–9 and 10–13 .	37 8
Man and wife 65 and over	14 8
Man alone 65 and over	9 5

* * *

Accepting the author's judgements as to the interpretation of the measurement technique, the study of incomes suggests that we have about 12 per cent. of all working-class families in the comfortable group, 56 per cent. with sufficient income for ordinary living, 21 per cent. with insufficient who are hard put to it to make a decent home, and 11 per cent. who are in poverty. If the last group is sub-divided, at the 25 per cent. below mark, we can say that about 4 per cent. of all families were living on the border-line of utter destitution.

40.   The Changing Social Structure
George Orwell, *The Lion and the Unicorn*
(London, Secker and Warburg, 1941), pp.
50–5.

Social class seems easy to recognize but difficult to define. Is it based on occupation, income, education, or birth? The Census of 1911 attempted to make meaningful class categories described as follows, in downward class order:
1. Professional and managerial;
2. Intermediate, e.g. pharmacists, teachers;
3. Skilled, e.g. coal miners, factory workers, office and shopmen;
4. Partly skilled, e.g. bus conductors;
5. Unskilled, e.g. labourers.

One of the most important developments in England during the past twenty years has been the upward and downward extension of the middle class. It has happened on such a scale as to make the old classification of society into capitalists, proletarians and petit-bourgeois (small property-owners) almost obsolete.

England is a country in which property and financial power are concentrated in very few hands. Few people in modern England *own* anything at all, except clothes, furniture and possibly a house. The peasantry have long since disappeared, the independent shopkeeper is being destroyed, the small businessman is diminishing in numbers. But at the same time modern industry is so complicated that it cannot get along without great numbers of managers, salesmen, engineers, chemists and technicians of all kinds, drawing fairly large salaries. And these in turn call into being a professional class of doctors, lawyers, teachers, artists, etc., etc. The tendency of advanced capitalism has therefore been to enlarge the middle class and not to wipe it out as it once seemed likely to do.

But much more important than this is the spread of middle-class ideas and habits among the working class. The British working class are now better off in almost all ways than they were thirty years ago. This is partly due to the efforts of the Trade Unions, but partly to the mere advance of physical science. It is not always realized that within rather narrow limits the standard of life of a country can rise without a corresponding rise in real-wages. Up to a point, civilization can lift itself up by its boot-tags. However unjustly society is organized, certain technical advances are bound to benefit the whole community, because certain kinds of goods are necessarily held in common. A millionaire cannot, for example, light the streets for himself while darkening them for other people. Nearly all citizens of civilized countries now enjoy the use of good roads, germ-free water, police protection, free libraries and probably free education of a kind. Public education in England has been meanly starved of money, but it has nevertheless improved, largely owing to the devoted efforts of the teachers, and the habit of reading has become enormously more widespread. To an increasing extent the rich and the poor read the same books, and they also see the same films and listen to the same radio programmes. And the differences in their way of life have been diminished by the mass-production of cheap clothes and improvements in housing. So far as outward appearance goes, the clothes of rich and poor, especially in the case of women, differ far less than they did thirty or even fifteen years ago. As to housing, England still has slums which are a blot on civilization, but much building has been done during the past ten years, largely by the local authorities. The modern Council house, with its bathroom and electric light, is smaller than the stockbroker's villa, but it is recognizably the same kind of house, which the farm labourer's cottage is not. A person who has grown up in a Council housing estate is likely to be—indeed, visibly *is*— more middle class in outlook than a person who has grown up in a slum.

The effect of all this is a general softening of manners. It is enhanced by the fact that modern industrial methods tend always to demand less muscular effort and therefore to leave people with more energy when their day's work is done. Many workers in the light industries are less truly manual labourers than is a doctor or a grocer. In tastes, habits, manners and outlook the working class and the middle class are drawing together. The unjust distinctions remain, but the real differences diminish. The old-style 'proletarian'—collarless, unshaven and with muscles warped by heavy labour—still exists, but he is constantly decreasing in numbers; he only predominates in the heavy-industry areas of the north of England.

After 1918 there began to appear something that had never existed in England before: people of indeterminate social class. In 1910 every human being in these islands could be 'placed' in an instant by his clothes, manners and accent. That is no longer the case. Above all, it is not the case in the new townships that have developed as a result of cheap motor cars and the southward shift of industry. The place to look for the germs of the future England is in the light-industry areas and along the arterial roads. In Slough, Dagenham, Barnet, Letchworth, Hayes—everywhere, indeed, on the outskirts of great towns—the old pattern is gradually changing into something new. In those vast new wildernesses of glass and brick the sharp distinctions of the older kind of town, with its slums and mansions, or of the country, with its manor-houses and squalid cottages, no longer exist. There are wide gradations of income, but it is the same kind of life that is being lived at different levels, in labour-saving flats or Council houses, along the concrete roads and in the naked democracy of the swimming-pools. It is a rather restless, cultureless life, centring round tinned food, *Picture Post*, the radio and the internal combustion engine. It is a civilization in which children grow up with an intimate knowledge of magnetoes and in complete ignorance of the Bible. To that civilization belong the people who are most at home in and most definitely *of* the modern world, the technicians and the higher-paid skilled workers, the airmen and their mechanics, the radio experts, film producers, popular journalists and industrial chemists. They are the indeterminate stratum at which the older class distinctions are beginning to break down.

This war, unless we are defeated, will wipe out most of the existing class privileges. There are every day fewer people who wish them to continue. Nor need we fear that as the pattern changes life in England will lose its peculiar flavour. The new red cities of Greater London are crude enough, but these things are only the rash that accompanies a change. In whatever shape England emerges from the war, it will be deeply tinged with the characteristics that I have spoken of earlier. The intellectuals who hope to see it Russianized or Germanized will be disappointed. The gentleness, the hypocrisy, the thoughtlessness, the reverence for law and the hatred of uni-

41. The Effect of the Social Services

B. Seebohm Rowntree, *Poverty and Progress* (London, Longmans, 1941), pp. 214–18.

Rowntree collected material for this book in 1935. He noted a marked improvement in living standards and a diminution in poverty compared with his earlier survey of 1901 (see p. 54) and since the end of the war. In 1913 the working class paid more in taxes than it received in social services, but by 1925 this position had been reversed. In 1937 over £200 million were redistributed in social welfare payments, over 5 per cent of the national income. Rowntree recorded part of this process in York.

SOCIAL SERVICES AND POOR RELIEF

In 1901 'Poor Relief' was the only source of financial assistance given from public funds to those who were in need, no matter from what cause. Those receiving it became 'paupers' and lost their rights as citizens. In that year the total sum paid in York in poor relief to people living in their own homes, was £5,950, equal to an average of $2\frac{1}{2}d.$ per working-class family per week, equivalent to $4\frac{1}{4}d.$ at 1936 prices.

Our schedules show that in 1936 the total sum distributed in York from public funds for the benefit of the sick, the unemployed, the aged, widows and orphans, and those who from any cause were destitute, was at the rate of over £275,000 a year, equivalent to an average of 6s. 6d. per working-class family per week.[1] This is over eighteen times as much as in 1901, after allowing for the difference in the value of money. The average amount received per family by those living below the minimum was 13s. 5d. per week.

Many references have been made in this book to these new

---

[1] The cost of the social services is met by contributions made by the workers, by employers, and by grants derived from State and Local taxation. Roughly speaking, one quarter of the benefits received are paid for by the workers.

social services and to outdoor poor relief or, as it is called, 'Public Assistance', but in order that the part which they play in contributing to the welfare of the workers may be fully realized, it will be worth while to bring a list of them together, and to show to whom they are rendered and upon what conditions. They fall under two heads:

(1) *Compulsory Insurance Schemes, to which all manual workers and other workers whose salaries do not exceed £250 a year must contribute.* The cost of these schemes is borne in approximately equal proportions by the workers, the employers and the State. The workers' contributions and the benefits payable under these schemes are as follows:

UNEMPLOYMENT INSURANCE

	Weekly Contribution	Weekly Benefits
Man	10d.	17s. for single man. 26s. for man and wife. 3s. for each child under 14.
Woman	8d.	15s.
Youths 18 to 21	8d.	14s.
Girls 18 to 21	7d.	12s.

HEALTH AND PENSIONS INSURANCE[1]

	Weekly Contributions	Weekly Benefits
Man	10d.	15s. for 26 weeks. 7s. 6d. for as long as disability lasts. Pension of 10s. a week for both man and wife at the age of 65. Maternity benefit of 40s. for wife. At death of man widow is given a pension of 10s. a week, with 5s. for first child and 3s. for other children under 14. Pension of 7s. 6d. a week for orphans.
Woman	7d.	12s. for 26 weeks. 6s. for as long as disability lasts. Pension of 10s. at the age of 65.

[1] Since 1936 the Health Insurance scheme has been extended to include working boys and girls from 14 to 16 years of age. They contribute 2d. a week and have the same medical attendance as adults but no sick pay.

(2) *Benefits towards the cost of which the workers make no direct contribution.* These comprise:
Non-Contributory Old Age Pensions. A pension of 10s. a week is payable at the age of 70 to all citizens whose resources do not amount to more than 25s. per week of which 15s. must be from means other than earnings. If their income is more than this the pension of 10s. a week is correspondingly reduced.
Public Assistance. The scale of relief under this head does not differ widely from that paid under the Unemployment Insurance scheme. It used to be administered by the Poor Law Guardians but is now administered by the Public Assistance Committee of the City Council. This relief corresponds to the old 'Out-door Poor Relief' but to-day recipients do not lose their rights of citizenship, and the receipt of Public Assistance is not regarded by most of them as carrying a stigma, though this is not true in all cases.

Previous to 1936 persons who had been unemployed for so long that their insurance benefits had run out were obliged to apply to the Public Assistance Committee of the City Council for relief, but during 1936 the Unemployment Assistance Board took over the responsibility for paying relief to all able-bodied unemployed. Two objects were attained by transferring the cost of relieving able-bodied men whose insurance had run out from the Public Assistance Committee to the Unemployment Assistance Board—it relieved the local authorities of a charge which in certain areas was unduly heavy, and it removed any stigma which unemployed persons might feel from being obliged to go for their relief to the Poor Law Offices, which are still associated in their minds with pauperism. Unemployment relief is now paid through the Labour Exchanges along with Unemployment Insurance Benefits. To-day Public Assistance is only paid to those not eligible for benefits from any of the social services, to some old people whose pensions are inadequate for their needs, and to sick people whose health insurance has run out or does not meet their needs. As the health insurance only provides for the needs of the sick person and makes no provision for those of his family, and as the benefit after 26 weeks falls to 7s. 6d. a week, a man with a family dependent on him is forced to apply for Public Assistance unless he has other sources of income which render this unnecessary. Non-contributory benefits are only given to those in real need, and after careful official investigation of the resources and means of those who claim them.

The sums received by the families included in our schedules during the week in which they were visited were as follows:

	£
Unemployment Insurance	1,802
Health Insurance	113
Contributory Old Age Pensions	636
Widows' and Orphans' Pensions	640
	£
Non-Contributory Old Age Pensions	1,347
Unemployment Assistance	—
Public Assistance	753

In view of the help now given in so many ways to people who are living at home, it is not surprising that the number of those who have to seek a shelter in the workhouse or 'Institution' as it is now called, is lower than in 1901, notwithstanding the great increase in the population of the city. The average number of persons in the workhouse in 1901 was 467 as compared with 453 in 1936. By 1939 the number had fallen to 403.

Almost all the occupants of the workhouse are either ill or old. In 1936, 149 of them were in the infirmary. In 1900 there were 84 children in the workhouse; in 1936 there were only 35, all of whom were either ill and in the infirmary or mentally defective, for no other children are kept in the workhouse but live in one or other of the scattered homes of which there are now four, sheltering 44 boys and 49 girls.

WORK

42. Office Work in the Twenties: the Cashier and his Children

J. B. Priestley, *Angel Pavement* (London, Heinemann, 1930), pp. 18–20, 70–1.

Angel Pavement was a 'typical City side-street, except that it was shorter, narrower, dingier than most'. Victorian novels, for example by Dickens, have long been recognized as important sources for social history. Priestley is equally illuminating for the twentieth century.

You could tell at once by the way in which Mr. Smeeth entered the office that his attitude towards Twigg and Dersingham was quite different from that of his young colleagues. They came because they had to come; even if they rushed in, there was still a faint air of reluctance about them; and there was something in their demeanour that suggested they knew quite well that they were shedding a part of themselves, and that the most valuable part, leaving it behind, somewhere near the street door, where it would wait for them to pick it up again when the day's work was done. In short, Messrs. Twigg and Dersingham had merely hired their services. But Mr. Smeeth obviously thought of himself as a real factor of the entity known as Twigg and Dersingham: he was their Mr. Smeeth. When he entered the office, he did not dwindle, he grew; he was more himself than he was in the street outside. Thus, he had a gratitude, a zest, an eagerness, that could not be found in the others, resenting as they did at heart the temporary loss of their larger and brighter selves. They merely came to earn their money, more or less. Mr. Smeeth came to work.

His appearance was deceptive. He looked what he ought to have been, in the opinion of a few thousand hasty and foolish observers of this life, and what he was not—a grey drudge. They could easily see him as a drab ageing fellow for ever toiling away at figures of no importance, as a creature of the little foggy City street, of crusted ink-pots and dusty ledgers and day books, as a typical troglodyte of this dingy and absurd civilisation. Angel Pavement and its kind, too hot and airless in summer, too raw in winter, too wet in spring, and too smoky and foggy in autumn, assisted by long hours of artificial light, by hasty breakfasts and illusory lunches, by walks in boots made of sodden cardboard and rides in germ-haunted buses, by fuss all day and worry at night, had blanched the whole man, had thinned his hair and turned it grey, wrinkled his forehead and the space at each side of his short grey moustache, put eyeglasses at one end of his nose and slightly sharpened and reddened the other end, and given him a prominent Adam's apple, drooping shoulders and a narrow chest, pains in his joints, a perpetual slight cough, and a hay-fevered look at least one week out of every ten. Nevertheless, he was not a grey drudge. He did not toil on hopelessly. On the contrary, his days at the office were filled with important and exciting events, all the more important and exciting because they were there in the light, for just beyond them, all round them, was the darkness in which lurked the one great fear, the fear that he might take part no longer in these events, that he might lose his job. Once he stopped being Twigg and Dersingham's cashier, what was he? He avoided the question by day, but sometimes at night, when he could not sleep, it came to him with all its force and dreadfully illuminated the darkness with little pictures of shabby and broken men, trudging round from office to office, haunting the Labour Exchanges and the newspaper rooms of Free Libraries, and gradually sinking into the workhouse and the gutter....

George had had opportunities that he himself had never had. But George had shown an inclination from the first, to go his own way, which seemed to Mr. Smeeth a very poor way. He had no desire to stick to anything, to serve somebody faithfully, to work himself steadily up to a good safe position. He simply tried one thing after another, selling wireless sets, helping some pal in a garage (he was in a garage now, and it was his fourth or fifth), and though he always contrived to earn something and appeared to work hard enough, he was not, in his father's opinion, getting anywhere. He was only twenty, of course, and there was time, but Mr. Smeeth, who knew very well that George would continue to go his own way without any reference to him, did not see any possibility of improvement. The point was, that to George, there was nothing wrong, and his father was well aware of the fact that he could not make him see there was anything wrong. That was the trouble with both his children. There was obviously nothing bad about either of them; they compared very favourably with other people's boys and girls; and he would have been quick to defend them; but nevertheless they were growing up to be men and women he could not understand, just as if they were foreigners. And it was all very perplexing and vaguely saddening.

The truth was, of course, that Mr. Smeeth's children *were* foreigners, not simply because they belonged to a younger generation but because they belonged to a younger generation that existed in a different world. Mr. Smeeth was perplexed because he applied to them standards they did not recognise. They were the product of a changing civilisation, creatures of the post-war world. They had grown up to the sound of the Ford car rattling down the street, and that Ford car had gone rattling away, to the communal rubbish heap, with a whole load of ideas that seemed still of supreme importance to Mr. Smeeth. They were the children of the Woolworth Stores and the moving pictures. Their world was at once larger and shallower than that of their parents. They were less English, more cosmopolitan. Mr. Smeeth could not understand George and Edna, but a host of youths and girls in New York, Paris and Berlin would have understood them at a glance. Edna's appearance, her grimaces and gestures, were temporarily based on those of an Americanised Polish Jewess, who, from her mint in Hollywood, had stamped them on these young girls all over the world.

43. Workers in the Ports, 1926

A. Creech Jones (National Secretary, Administrative, Clerical and Supervisory Group, Transport and General Workers' Union), in *Working Days*, ed. M. A. Pollock (London, Cape, 1926), pp. 190–2.

The difficulty of dock work was its lack of continuity, so that a casual labour system was characteristic of its organization. Such a system was fundamentally unsatisfactory for the labourer, and even with a strong union, dock labour remained an unpleasant and uncertain occupation until after 1950.

I am to describe the conditions of life of port and waterside workers, and shall make little reference to what is being done to find a solution to the problems I may mention. Out of the general background of port industrial life, it will be possible to build up the daily round of the docker.

Docks and poor social conditions seem to go together. Ugliness and squalor are the usual features of port neighbourhoods. Bad housing, indifferent health and poor conditions of life are the rule for the working population in these areas, even after allowing for the advance in recent years in general social amenities.

Dock employment is not an attractive form of employment and is popularly considered as undesirable and a 'last resort'. The ports have been thought of as places to which the unwanted and unemployable in our industrial towns may drift to try and eke out an existence by doing whatever casual work was offered. Certainly the docks have received in days gone by the men who were denied work in other industries—the men who have found all other openings of employment closed to them. The labour has been hard, conditions have been miserable, and the life has brought much wretchedness. But much of the work has called for little thought and next to no skill, and so the ports have been collecting grounds for those unfortunates thrust well below the poverty line. They have intensified the scramble for work and increased the precariousness of the livelihood which the ports offered. The districts of our ports have usually been seething areas of intense deprivation and poverty, of social horror and ugliness. These conditions have seldom failed to win from the public a ready sympathy for the dockers, and the subject of dockland has been a perennial theme for social investigators.

Dock labour is chronically under-employed, whether there are periods of trade prosperity or periods of depression. There is no definite relationship between the demand for labour and the supply available. The fluctuations in demand for workers can hardly be anticipated, and certainly cannot be brought under such control as to secure a regular volume of employment day by day. The forces of climate, tide and storm are beyond human control. The amount of work is influenced by the uncertainties and variations of trade and the effects abroad of seasons of abundance or seasons of drought. Fluctuations in demand for work are occasioned, too, by the variety of cargo to be handled. The method of handling, whether mechanical or by hand, and the amount of handling required by certain classes of cargo, are variable. Moreover, there can be no mechanical regularity in work where such influences as the diversion of a boat or the unexpected arrival of a ship, a traffic delay, the breakdown of machinery, a spell of wet weather, a rush order, affect the number of workers required. Even changes in freightage rates on competitive forms of transport have a bearing on the volume of work. Changes in national or international relationships have their reactions, as have been witnessed in the past few years in our east coast ports. It is really difficult for ports to attract sufficient seasonal work throughout the year in order to create a steady demand for labour. And these fluctuations, which not only occur as a result of the trade cycles but from season to season and from day to day, really mean that the labourers for the greater part only obtain work in accordance with the swelling or contracting of the traffic.

The work is therefore so precarious as to eliminate any certainty in the arrangement of the docker's life. Irregular income is not conducive to good housekeeping and careful expenditure. It tends to create a feeling of recklessness and irresponsibility. Men are thrown on their wits to eke out a livelihood as best they can. The anxious quest for jobs, the fretting resulting from baffled attempts to find work, the idleness in the streets and at home, the hanging about from one call-time to another, create far from healthful tendencies on character. The low standard of conditions that are produced as a consequence find expression in the unsatisfactory features of home life. Anxiety is never-ending. Hope and disappointment alternate in the mind. The family life is damaged and deranged, and the effects are pernicious to the individual, to the family and to social life generally.

There are probably 120,000 men engaged in dock work. Comparatively, only a few of them are in permanent employment. The vast majority are engaged day by day. A crowd of applicants assembles round the gate or office or at the calling-on places scattered often throughout the port; sometimes at the ship's side, at street corners, outside public-houses, wherever custom or the convenience of individual employers has established these places. When labour is wanted in the docks the foremen appear on call points, and this is a sign of struggle and confusion, pushing and elbowing, to attract his attention. Sometimes they take on those men they know or those with a preference card, for in certain ports there are preference men singled out from the rest. Speaking generally, however, the same competition for work described many years ago goes on day after day about the docks. Now in most ports only registered dockers can apply for work, and this practice is becoming increasingly the rule. Employment in the ports is being limited only to bona fide transport workers, and the influx from other industries and the drift of men from the towns is thus checked. This limitation of the volume of labour is doing something to average up the amount of employment to each individual man. Work is for a half-day engagement, and wages are paid at the end of each day, if the engagement lasts the day, except at Liverpool, where weekly wage payments have been established by a central clearance system. No longer can the employer engage a docker for an hour and throw him off when he has no further use for him. The men usually work in gangs, some at piece-work, where labour becomes intense, and the old have to keep pace with the young; some at day-work, where there is no special incentive to urge the men on.

The great competition for labour and the fluctuations in demand have led to the toleration of a much bigger army of workers about the docks than is really necessary. The docker, therefore, is lucky if he finds work on the average of more than eight half-days per week. It is difficult to obtain exact information, the average employment in the respective ports varies. A docker who is a genuine transport worker on the Hull docks during the past three years has been only able to average $2\frac{1}{3}$ days per week. At Swansea the average is said to be $3\frac{1}{2}$ days, at Bristol $3\frac{1}{2}$ days, at Glasgow 4 days; at seasonal ports the employment is more erratic.

44. Other Workers: (*a*) the Miner, (*b*) the Shirt-maker Machinist

M. A. Pollock, *Working Days* (London, Cape, 1926), pp. 62–5, 244–6.

Mine-workers have always been a class apart, living in their own communities, working in a dangerous industry, and bound together by a strong group consciousness. Their working and living conditions have fascinated both social investigators and novelists. Shirt-making, on the other hand, was a typical piece-work industry employing female labour, characterized by labour exploitation (sweating).

(*a*)

Down they plunge from daylight to darkness, with the swift motion of the engines paying out the cables that are attached to the cages. A short journey, though in cases of very deep pits, a swift journey, soon lands them at the bottom, for speed is essential in getting the men in in time.

Cages are quickly emptied, and away they go once more to the surface for another load of human beings. As soon as the men step out of the cage, they scurry off in different directions, each intent on getting to his work as quickly as possible, the young boys and youths jostling each other and joking and playing tricks on each other; the older men, more staid and sober, sometimes chiding the boys for their high spirits. But all is generally good-natured banter, and off they go, with hardly ever a thought of what lies ahead, except to be an occasional twinge that the task set may be too much to be accomplished in the time.

Arrived at the coal face, which may be two or three miles

in many cases from the bottom of the shaft, clothes are hurriedly torn off, until only a thin shirt and trousers are left (or in the case of deep, hot mines, nothing except a pair of bathing pants and boots are left), and the miner gets hold of his tools and prepares for his labour.

He carefully, though hurriedly, examines the roof and sides of his working place, tapping suspicious-looking places here and there with his pick or hammer, and being able to tell by the sound whether it is safe or not. If unsafe, timber must be 'sett' at once, so as to secure the roof and keep him safe from sudden falls of rock.

Then, if it is a thin seam—say two or three feet in thickness (in some cases seams are worked as low as eighteen inches)—he has to get on to his hands and knees and crawl into where the coal is, and, lying down in a curiously crumpled-up position upon his side, his legs drawn in beneath him, he begins to undercut the coal.

This is very skilled work, and only after a man has served through all the various grades, and has been for a period of some years in the pit, does he become a hewer. Brains as well as brawn are needed, and often a very small diminutive man will show a greater cutting capacity than another who is very much heavier built, just because he uses more skill in studying the nature of the strata, and knows how to use every ounce of his energy to the best advantage. Hour after hour passes, and the regular tapping of his pick continues, as he travels along the coal face, always lying in the same cramped position, and only changing over to his other side so that he can rest a bit, by changing his limbs in motion. A short respite is snatched for the eating of the frugal meal, generally of bread and butter or cheese, which his wife has prepared for him before he left in the morning, then with a drink of cold tea from his flagon, he is back at work again plying his pick, while the sweat pours from his body, and the work progresses.

When his undercutting is completed, he drills his shots, gets them charged, and withdraws to a safe distance till the explosion is past.

As soon as the shot has gone off, he is back in; for now he can tell whether his efforts will be successful for the day or not. If the shots have succeeded in bringing down the coal he has thus prepared, tubs must be got, the coal filled into them hurriedly, and away they go loaded, pushed sometimes by the boys and the young men, who will some day be hewers themselves, or drawn by ponies out to the main roads, when they are taken by haulage power to the pit bottom.

(b)

For the benefit of the reader, it is necessary to give a brief outline of the method usually adopted in the shirt-making industry. The makers are divided into four groups—toppers, seamers, sleeve-hands, and hemmers. There are also buttonholers, buttoners, examiners, and, finally, folders and packers; so that a shirt passes through a good many pairs of hands ere it is complete. In most factories these workers are paid by the piecework system. The topper lines the shirt, puts on the yoke, fly, collar and pocket, if any, and receives from 2s. 6d. to 1s. per dozen, according to the type of shirt. The sleeve hand receives about $7\frac{1}{2}d$. per dozen pairs of sleeves, the hemmer about 3d. per dozen shirts, and the seamer 1s. to 1s. 2d. The prices vary in different factories. The average working day is from 8 a.m. till 6 p.m. The piece-worker does not receive pay during holidays; in fact, she receives pay only for actual work done. If she is kept waiting for work (which often happens during slack periods), the loss is hers. Moreover, in many factories she has to buy the cotton and needles used in manufacturing her employer's goods. The piece-work system is one of the defects in our industrial machine. It is unfair because the slow worker is up against the difficulty of earning the time-rate, and is not wanted by the employer. It also encourages sweating because the quick worker is constantly trying to earn a little more, with the result that the price is often lowered on her account.

Here is an illustration. Millie had worked in this particular factory for eight years, and was one of the quickest toppers. When the prices were lowered a short time ago, the girls went in a body to the supervisor, and complained that they were unable to earn their money. Her reply was, 'Millie earns it, and if one girl can do it you'll have to; if not'—and she shrugged her shoulders.

What are girls to do in a case of this kind, especially in the East End of London where unemployment is rampant? I venture to think that the abolition of the piece-work system would in a measure alleviate this evil.

At one o'clock the shirt-workers were finished for the day, many of them having earned the small sum of 3s. Some went home to help the harassed and often irritable mother. Others went to an afternoon matinée at the cinema, and a few went for a walk, during which they discussed the possibility of trying to get another job.

45. New Industries: Cars and Electricity

(a) P. W. S. Andrews and E. Brunner, *The Life of Lord Nuffield* (Oxford, Basil Blackwell, 1955), pp. 197–8.

(b) The British Association, *Britain in Recovery* (London, Pitman, 1938), pp. 265–7.

The recovery of the 1930s owed much to the growth of the new industries. Lord Nuffield was well behind Ford in introducing the assembly line for the manufacture of cars, but he was the most successful of British manufacturers. The importance of electricity as a power for industry owed much to the rationalization of supply achieved through the National Grid.

(a)

The moving assembly line was the most substantial new installation. Until this time the chassis had been pushed up the line as each stage of work was completed, and the parts necessary for each stage were stacked available on the floor. Now the whole process of assembly was mechanized, with the chassis being slowly carried forward by a moving chain, and the parts were mostly delivered to the appropriate section by overhead conveyor direct from the stores, or from sub-assembly departments. The machine shops also were overhauled. In just over a year the works were put into a position to adopt the latest production methods, without which the subsequent expansion would have been impossible. The fact that 1933 was such a bad trading year gave some physical opportunity for the change-over, and the resources which Morris had husbanded enabled him to take the decision to carry it out even though sales and profits were falling so drastically. In this way he mitigated the effect of the falling output on local employment.

At the same time, Morris Motors also took advantage of the interference with production to introduce new models at the best-selling end of its range, where the cars were completely redesigned to suit the new production methods. The new Morris Eight came out at the end of 1934 and the Morris Ten in May 1935, the latter being complemented by a new Morris Twelve which was basically the same car except for its more powerful engine. There were changes in style—the old square lines giving place to rounded tops and sloping backs—and every design detail was thoroughly tested before the cars were put on the market. The new models were winners and production leaped ahead to meet expanding sales in 1935. The latest 'series' of the Eight and Ten were introduced in 1938–39 (and were, in fact, continued after the 1939–45 war while the business was developing its post-war designs). With more prosperous times, the range of models was reduced to suit production requirements, and the 16 and 18 h.p. cars were dropped altogether. There was also increased standardization of equipment generally.

(b)

The greatest single factor in the post-slump situation has been the entrance into operation of the Grid under the control of the Central Electricity Board. The Central Electricity Board was created by the Electricity (Supply) Act of 1926 and began the construction of the Grid in 1927. The last tower of the Grid was erected in September, 1933. Since then there have been additions to the transmission system and some expansion of transforming and switching equipment of the substations. At the end of 1937, however, the Grid comprised about 4180 miles of primary and secondary transmission lines, linking up selected generating stations with a capacity of over $7\frac{1}{2}$ million kilowatts by means of transforming and switching stations, with a capacity not far short of 10 million kVA.

The total cost of erecting the transmission system, exclusive of capitalized interest, was about £27,800,000, and the greater part of the expenditure involved was allocated during the slump years, 1930 to 1933, so that the Board was able to take advantage of falling prices, which made the cost of construction come very close to the estimates calculated in 1927, to carry out a slightly more ambitious scheme without additional cost and to keep the manufacturing workshops at a steady level of activity.

OPINION

46.   The End of Laissez-faire

J. M. Keynes, *The End of Laissez-Faire* (London, Hogarth, 1926), pp. 39–49.

Between the wars there was a marked shift in opinion towards more state intervention in the economy. The attack on *laissez-faire* came not only from the Fabians and great social critics like R. H. Tawney, but also from conservatives like Robert Boothby and Harold Macmillan. By 1939 there was general recognition that a free market economy had failed to solve the economic and social problems of the period.

We cannot therefore settle on abstract grounds, but must handle on its merits in detail what Burke termed 'one of the finest problems in legislation, namely, to determine what the State ought to take upon itself to direct by the public wisdom, and what it ought to leave, with as little interference as possible, to individual exertion'. We have to discriminate between what Bentham, in his forgotten but useful nomenclature, used to term *Agenda* and *Non-Agenda*, and to do this without Bentham's prior presumption that interference is, at the same time, 'generally needless' and 'generally pernicious'. Perhaps the chief task of Economists at this hour is to distinguish afresh the *Agenda* of Government from the *Non-Agenda*; and the companion task of Politics is to devise forms of Government within a Democracy which shall be capable of accomplishing the *Agenda*. I will illustrate what I have in mind by two examples.

(1) I believe that in many cases the ideal size for the unit of control and organisation lies somewhere between the individual and the modern State. I suggest, therefore, that progress lies in the growth and the recognition of semi-autonomous bodies within the State—bodies whose criterion of action within their own field is solely the public good as they understand it, and from whose deliberations motives of private advantage are excluded, though some place it may still be necessary to leave, until the ambit of men's altruism grows wider, to

the separate advantage of particular groups, classes, or faculties—bodies which in the ordinary course of affairs are mainly autonomous within their prescribed limitations, but are subject in the last resort to the sovereignty of the democracy expressed through Parliament.

* * *

I criticise doctrinaire State Socialism, not because it seeks to engage men's altruistic impulses in the service of Society, or because it departs from *laissez-faire*, or because it takes away from man's natural liberty to make a million, or because it has courage for bold experiments. All these things I applaud. I criticise it because it misses the significance of what is actually happening; because it is, in fact, little better than a dusty survival of a plan to meet the problems of fifty years ago, based on a misunderstanding of what someone said a hundred years ago. Nineteenth-century State Socialism sprang from Bentham, free competition, etc., and is in some respects a clearer, in some respects a more muddled version of just the same philosophy as under-lies nineteenth-century individualism. Both equally laid all their stress on freedom, the one negatively to avoid limitations on existing freedom, the other positively to destroy natural or acquired monopolies. They are different reactions to the same intellectual atmosphere.

(2) I come next to a criterion of *Agenda* which is particularly relevant to what it is urgent and desirable to do in the near future. We must aim at separating those services which are *technically social* from those which are *technically individual*. The most important *Agenda* of the State relate not to those activities which private individuals are already fulfilling, but to those functions which fall outside the sphere of the individual, to those decisions which are made by *no one* if the State does not make them. The important thing for Government is not to do things which individuals are doing already, and to do them a little better or a little worse; but to do those things which at present are not done at all.

It is not within the scope of my purpose on this occasion to develop practical policies. I limit myself, therefore, to naming some instances of what I mean from amongst those problems about which I happen to have thought most.

Many of the greatest economic evils of our times are the

fruits of risk, uncertainty, and ignorance. It is because particular individuals, fortunate in situation or in abilities, are able to take advantage of uncertainty and ignorance, and also because for the same reason big business is often a lottery, that great inequalities of wealth come about; and these same factors are also the cause of the Unemployment of Labour, or the disappointment of reasonable business expectations, and of the impairment of efficiency and production. Yet the cure lies outside the operations of individuals; it may even be to the interest of individuals to aggravate the disease. I believe that the cure for these things is partly to be sought in the deliberate control of the currency and of credit by a central institution, and partly in the collection and dissemination on a great scale of data relating to the business situation, including the full publicity, by law if necessary, of all business facts which it is useful to know. These measures would involve Society in exercising directive intelligence through some appropriate organ of action over many of the inner intricacies of private business, yet it would leave private initiative and enterprise unhindered. Even if these measures prove insufficient, nevertheless they will furnish us with better knowledge than we have now for taking the next step.

My second example relates to Savings and Investment. I believe that some co-ordinated act of intelligent judgment is required as to the scale on which it is desirable that the community as a whole should save, the scale on which these savings should go abroad in the form of foreign investments, and whether the present organisation of the investment market distributes savings along the most nationally productive channels. I do not think that these matters should be left entirely to the chances of private judgment and private profits, as they are at present.

My third example concerns Population. The time has already come when each country needs a considered national policy about what size of Population, whether larger or smaller than at present or the same, is most expedient. And having settled this policy, we must take steps to carry it into operation. The time may arrive a little later when the community as a whole must pay attention to the innate quality as well as to the mere numbers of its future members.

47. The British Fascists

John Strachey, *The Menace of Fascism*
(London, Gollancz, 1933), pp. 161–4.

John Strachey had a distinguished career as a left-wing intellectual. His brief association with Mosley (who left the Labour Party in despair at its ever achieving anything) indicates the political confusion created by the unprecedented Great Depression. The British Fascists, though interesting as the real manifestation of a small authoritarian minority, had no political importance in Britain, and failed conspicuously to gain mass support.

I recollect the figure of Mosley standing on the town hall steps at Ashton-under-Lyne, facing the enormous crowd which entirely filled the wide, cobbled market-square. The result of the election had just been announced, and it was seen that the intervention of the New party had defeated the Labour candidate and elected the Conservative. The crowd consisted of most of the keenest workers in the Labour party in all the neighbouring Lancashire towns. (Four or five million workers live within a tram-ride of Ashton.) The crowd was violently hostile to Mosley and the New party. It roared at him, and, as he stood facing it, he said to me, 'That is the crowd that has prevented anyone doing anything in England since the war.' At that moment British Fascism was born. At that moment of passion, and of some personal danger, Mosley found himself almost symbolically aligned against the workers. He had realized in action that his programme could only be carried out after the crushing of the workers and their organizations.

After Ashton, Mosley began more and more to use the word Fascism in private. Those members of the New party who, though they did not regret their break with the Labour leaders, were certainly not prepared to be Fascists, became more and more alarmed. Mr. Allan Young and myself were the two members of the Governing Council of the New party who were principally disturbed. We had been for some years past Mosley's closest lieutenants and we were willing to carry self-deception very far, much too far indeed, in order to avoid a break which was extremely painful to both of us. We pretended to ourselves that his talk about Fascism 'did not mean anything'. During the summer of 1931, however, new and definite signs of Mosley's determination to turn the New party into a Fascist type of organization became apparent.

For months the council wrangled over the question of what was discreetly called 'the Youth movement'. In fact, this organization represented Mosley's determination to create a private army. He had seen, with admirable realism, that the only thing that really matters in a Fascist party is the creation of a disciplined, military and, so far as possible, armed, force which can be used to break by terror 'that crowd which was prevented anyone doing anything in England since the war'.

The army of the Fascist terror was to be (and is now being) recruited ostensibly for such purposes as keeping order at meetings, engaging in athletic exercises, forming boxing and fencing clubs and the like. But Mosley naturally could not in the end maintain in argument with his colleagues that these trivial objects were the real purpose of 'the Youth movement'. He had to admit to us that this force was to be used in the revolutionary situation which we all agreed (and this is perhaps the one point on which we do still agree) must sooner or later come upon Great Britain. But on which side was our army to be used? asked Young and I. On neither side, was Mosley's answer. It was to come down on the supine Government and the disorderly workers and knock their heads together. And what next? we asked. Having obtained power in this way what should we do? Impose the Corporate State, declared Mosley.

Now Young and I had at that time no clear idea of what the Corporate State might be. But the more Mosley talked about it, the more it seemed to be remarkably like Capitalism: or rather it seemed to be Capitalism minus all the things which the workers had won during the last century of struggle. Our doubts grew and grew. Our personal loyalty to Mosley, with whom we had been through the long exacting struggle with the Labour party leadership, came into violent collision with our whole political and social outlook.

Certainly we had no one to blame but ourselves. We had got ourselves into an impossible position, partly by weakly allowing personal loyalties to blind our eyes and partly owing to defective comprehension of economic and social reality. The actual breaking-point came upon that touchstone of the modern world, our attitude to Soviet Russia. I was asked to write a memorandum defining the New party's attitude to Soviet Russia. I wrote an unequivocally pro-Russian document. Mosley equally unequivocally rejected it, and gave, quite frankly, his real reason for doing so. If the New party adopted a pro-Russian attitude, all hopes of support from the Conservatives and capitalists would be gone. Immediately after the Council meeting at which this discussion took place, Young and I resigned from the New party. This proved to be the end of the New party, and the hopeless attempt at obtaining the agreement of Capital and Labour for a quite impracticable policy of national reconstruction.

48. The Abdication of Edward VIII

Parliamentary Debates, 10 December, 1936 (5th Ser., vol. 318. vol. II of Session 1936–7, column 2179).

The abdication of Edward VIII was a dramatic event at the time, but, in retrospect, less important. There was some contemporary feeling that Edward's abdication was partly the result of his radicalism, but there is little evidence to support this view. Baldwin's opinions, here expressed, were widely held.

And then I reminded him of what I had often told him and his brothers in years past. The British Monarchy is a unique institution. The Crown in this country through the centuries has been deprived of many of its prerogatives, but to-day, while that is true, it stands for far more than it ever has done in its history. The importance of its integrity is, beyond all question, far greater than it has ever been, being as it is not only the last link of Empire that is left, but the guarantee in this country so long as it exists in that integrity, against many evils that have affected and afflicted other countries. There is no man in this country, to whatever party he may belong, who would not subscribe to that. But while this feeling largely depends on the respect that has grown up in the last three generations for the Monarchy, it might not take so long, in face of the kind of criticisms to which it was being exposed, to lose that power far more rapidly than it was built up, and once lost I doubt if anything could restore it.

49. Pacifism

G. Lansbury, *My Quest for Peace* (London, Michael Joseph, 1938), pp. 11–14.

The horrors of the First World War, and the socialist belief that working men everywhere had common problems and aims, made it difficult for the Labour Party to adjust its policy to the realities of European Fascism. Rearmament, therefore, created a great moral dilemma for the pacifists.

Nobody, however arrogant or powerful, in whatever land, openly or tacitly declares a love of war, no matter how he may extol the heroism of those who fight. Governments declare with a kind of deadly monotony their earnest desire for peace. At the same time all peoples are asked to supply an ever-increasing amount of money for armaments and war equipment of all kinds. 'Trust in God and make your poison gas deadlier' has taken the place of 'Keep your powder dry'. Nobody says a word in defence of war: in fact, all Government speakers make feeble and far-fetched apologies for the armament votes they demand. Most of them declare that the gases they will use are for self-defence. This is the most ancient of all excuses. It does not alter the fact that ever since the Armistice of 1918 the whole world has steadily and without rest been making ready for another great war. The present armaments race is not really rearmament, for no one ever disarmed: it is an accentuation of the armaments policy pursued everywhere except in Germany and other defeated nations since 1920. Germany started arming in dead earnest when the Allies refused to listen to her proposal that all nations should disarm, in which case she would be satisfied with an army of 300,000. This proposition was not even discussed. Neither were proposals for the international control of aviation and the complete abolition of aerial bombing. Great Britain shares a very large part of the responsibility for present-day aerial warfare, because of the refusal of her representatives in the first instance to accept without reservation the proposal to abolish all aerial warfare and bring aviation under international control. It has been stated that the Germans armed secretly and, without the knowledge of other Governments, created a huge air force and acquired other war equipment. Possibly; but it is equally certain that had any effort been made to treat the pre-Hitler Governments with the slightest recognition of their standing as heads of a great people, and efforts been made to bring them into European affairs on terms of equality, Europe would not now be faced with the fear of universal war. The late Arthur Henderson and Monsieur Briand worked hard towards this end, but neither of these men was in power sufficient time to enable him to win through.

I cannot forget my own part in this business of aerial warfare. I was a member of the Labour Government which not only retained the right to bomb people living within the Empire, but also defended this kind of warfare as being more merciful. I have since been quite ashamed that I was part of that Government. Whenever anyone joins a Government he must be prepared to accept the principle embodied in the words 'Cabinet Responsibility'. No person who holds the pacifist faith should take office in any Government other than one which is elected on a programme which will ensure the abandonment of imperialism, accompanied by disarmament. There is no possible compromise on this. I am, and always have been, a passive resister. If the masses in all lands would refuse to manufacture armaments there would be no war. Statesmen and prelates would soon find a better way of living in order to preserve the race from suicide. Gandhi has staked out a road which, if followed, would free mankind from the curse of war; but his policy must be accepted without compromise. The practice of non-violent resistance calls for patience, for courage, far in excess of what is needed in war, but in its result it is much more effective than meeting slaughter by slaughter. All through my life I have seen the wrong and dreadful results of war. The way out has never been so clear as to-day. It is not easy going, but it is sure: let the masses everywhere declare without reservation 'Away with the accursed thing', and peace would follow.

* * *

This quest for peace is not new. All through my lifetime it has been possible for people like myself to join with others in organised groups working for peace. The basis of appeal which has continually had my support is the Christian doctrine to be found in the life and teaching of Jesus. 'Forgive us our trespasses as we forgive them that trespass against us', 'Do unto others as you would they should do to you'—these and many similar injunctions seem to me to show what is needed from each of us to bring peace into the world. There are sayings of Christ which, men argue, conflict with these. This may be so: these I quote have the merit of realism; they are not mere idle words. Everybody knows that the practice of these principles of life and conduct by professing Christians in all lands would give peace in our time. We who are pacifists in this sense are neither dreamers nor self-righteous prigs. We contend that those are dreamers who declare that war is futile and criminal and at the same time declare their inability to do anything effective to prevent it happening, but on the contrary, use all scientific knowledge and inventions preparing for the very catastrophe they declare they wish to avoid.

50. Science and Society

(a) Sir E. Mellanby, *Recent Advances in Medical Science* (Cambridge University Press, 1939), pp. 18–27.

(b) J. C. Philip, 'The Chemist in the Service of the Community', in *What Science Stands For* (London, Allen & Unwin, 1937), pp. 45–9.

The First World War was called 'the engineer's war'; the Second, 'the scientists' war'. Between 1918 and 1939 notable advances in science had important effects on living and working conditions. As life became more science-based, so education also had to be more science-oriented. Research and development departments, also, were developed by large firms, and scientific research was increased in the universities.

(a)

Some changes in the mode of living become so ingrained in the people as social habits that their origin is often forgotten.

Probably the best example of this is seen in the modern habit of personal cleanliness, which has become a measure almost of aestheticism and social standing. But surely the primary stimulus to personal cleanliness of the present day is almost entirely the knowledge of bacteriology and its offspring infection. A good example of the effect of official action towards a greater cleanliness is seen in the case of the children of the London County Council schools, which is no doubt typical of most schools of the same kind in the country. In 1912, 39·5 per cent of these children had parasitic skin infections. Greater cleanliness was insisted upon by the authorities, with the result that by 1920 this percentage was reduced to 13·8, in 1934 to 4·5, and in 1937 to 2·6. Ringworm of the scalp was formerly one of the greatest scourges of school children. In these same schools in 1911 there were 6214 new cases of this disease, in 1920 there were 3983, in 1934 265 and in 1937 only 78. Many other instances of the same kind could be given, but these examples will probably suffice to show that personal cleanliness, as at present practised, is not an ingrained instinct, as many seem to think, but is largely the outcome of a public effort based on the knowledge acquired by medical investigation that many infective diseases of mankind are dirt diseases. Nor must we forget the part played in this crusade by those who manufacture and sell soap.

If there is still doubt as to the power of science in relation to disease of uncleanliness in the bacteriological sense, it is only necessary to study the story of the detection of the two persons who were the origins of the recent outbreaks of typhoid fever at Bournemouth and Croydon respectively. Neither of these men at the time of his detection had any idea that he harboured the typhoid micro-organism in his alimentary canal. Similarly, the modern method of tracking down individuals who have transmitted certain types of streptococci, often from their throats, to women in childbirth represents scientific detection at its best.

The scientist engaged in medical research will not complain of the basis assigned for public and private changes in habits of living which have proved so beneficial to health. What he does complain of is the great delay which often occurs before many of the teachings, which his investigations have elucidated, are adopted by public authorities and private citizens. In some cases the absence of application is due to administrative inertia or to lack of political and social interest, in some to ignorance or laziness on the part of the public, and more often it is due to such economic and social restrictions as prevent people from attaining the nutritional and hygienic conditions necessary for healthy existence. The medical scientist knows, for instance, that diphtheria could be cleared out of this country at once by the preventive inoculation of infants and children by diphtheria toxoid....

An even better example, although in this case social and economic conditions make the application of knowledge more difficult, is seen in the field of nutrition. Every expert in this field knows that the consumption of proper food from birth onwards would revolutionise the standards of physique and health. He sees enormous differences in the physique of the poor as compared with that of the well-to-do. For instance, the average height and weight of boys at the age of eleven attending a better-class school are 55·33 in. and 76·22 lb. respectively, while the corresponding figures for elementary school boys of the same age are 3 in. and 12 lb. less respectively. He sees also the great difference in the death-rate of children of the poor as compared with that of the children of the well-to-do. The following tables illustrate this point by giving the mortality rates of children up to one year and between one and two years, classified according to the social status of the father.

(In the tables, Class I represents the professional and generally well-to-do section of the population, Class III skilled

TABLE I

	Rates per 100,000 legitimate live births for deaths under 1 year					
	All	I	II	III	IV	V
Premature birth	1730	1050	1440	1680	1860	1960
Injury at birth	210	230	250	210	200	200
Congenital malformations	550	500	540	560	570	540
Congenital debility	300	140	220	290	330	380
Infantile convulsions	210	130	170	200	260	230
Whooping cough	180	30	100	160	210	270
Tuberculosis, all forms	100	30	60	90	110	130
Bronchitis and pneumonia	1270	280	610	1120	1450	1880
Diarrhoea and enteritis	520	200	260	460	540	790
Accident	80	60	80	70	90	100
Total	5150	2650	3730	4840	5620	6480

TABLE II

	Rates for deaths per 100,000 legitimate children at ages 1–2					
	All	I	II	III	IV	V
Measles	242	25	70	194	246	469
Whooping cough	127	28	52	109	140	209
Diphtheria	36	13	18	32	38	55
Tuberculosis, all forms	113	69	73	104	125	150
Influenza	28	16	23	26	24	39
Bronchitis and pneumonia	529	128	223	448	607	861
Diarrhoea and enteritis	72	28	40	60	76	118
Accident	53	19	39	48	56	70
Other causes	252	128	190	237	261	327
Total	1452	454	728	1258	1573	2298

artisan and analogous workers, and Class V labourers and other unskilled callings, while Classes II and IV are intermediates comprising occupations of mixed types or types not readily assignable to the classes on either side.)

(b)

Their work aims at the solution of some specific problem, concerned, it may be, with the improvement of an industrial process, the elimination of waste, the safeguarding of health, the utilization of by-products, the synthesis of antidotes. More definitely, and by way of example, the object may be to discover a fast blue dye, to purify a water supply, to find a rustless steel, to produce petrol from coal, to isolate a vitamin, to make a non-inflammable film or a creaseless cotton fabric. The general public, however dubious about pure research, would probably admit that the satisfactory solution of any one of these problems would be of service to the community; but it must be emphasized once more that the chemist can do these things only by virtue of his inheritance of knowledge and technique. The attack on such problems, to have a reasonable chance of success, must be organized on the basis of what is already known and what has already been achieved; nay, more, one has abundant ground for belief that the attack, so organized, is bound to succeed, even though it may be 'in the long run'.

In the last twenty years the amount of directed chemical research in this country has increased enormously. Industries of the most varied description have begun to realize the potential value of the trained chemist in solving their special problems and putting their manufacturing processes on a more rational basis. In this general movement the State, through the Department of Scientific and Industrial Research, has taken a prominent part by fostering Research Associations. The work of these organizations—such as those dealing with rubber; with paint, colour, and varnish; with cotton or wool; with non-ferrous metals; with sugar confectionery—is in many cases largely chemical or physico-chemical in character. The Research Associations have not only shown how general problems affecting an industry as a whole can be solved by joint research efforts, but their existence and activities have induced a notable degree of 'research-mindedness' in the individual associated firms. Financially, the work is based on co-operation between the State and industry, on the principle that the State helps those who help themselves. . . .

* * *

Again, the serious question of river pollution has been taken in hand with State help, and some years ago a chemical and biological survey of the river Tees was set on foot, the Tees being chosen for investigation because of the great variety of factory effluents discharged into it both in tidal and non-tidal reaches. Some of the newer industrial developments in Britain are presenting important problems in this direction. It has been estimated, for example, that if the waste waters from all the beet-sugar factories in this country were discharged into our streams they would cause as much pollution as untreated sewage from a population of four or five millions. The effluents from dairies and factories making milk products present a similar problem. Thanks, however, to research activity, largely at the instance of the Water Pollution Research Board, the disposal or purification of these and other trade effluents is being effectively achieved.

The question of river purification demands for satisfactory handling, as already indicated, the collaboration of other scientists with the chemist, and indeed the attack on many such problems, especially those affecting the health of the community, is likely to be successful only by the co-operation of teams of scientific workers from different fields. Smoke and fog, which not only present the scientist with interesting phenomena but constitute also a social and industrial problem of vital importance, concern the physicist, the physical chemist, the analyst, the fuel engineer, and the meteorologist, and it is only when the knowledge and experience of these workers are pooled that there is any hope of interpreting the phenomena and solving the problem. Again, recent developments in cancer research make it clear that apart from the pathologist, who is mainly concerned, the chemist has a very definite contribution to make to our knowledge of this baffling disease. Some of the most fruitful scientific investigation, indeed, is co-operative in character.

51. The War, 1939–1945

    (a) Leonard Woolf, *The Journey not the Arrival Matters* (London, Hogarth Press, 1969), pp. 28–9.
    (b) W. S. Churchill, a broadcast of 11 September 1940 in *Into Battle* (London, Cassell, 1940), pp. 274–5.

Leonard Woolf (1880–1969) lived through the period of this book, and his autobiography describes the intellectual pilgrimage of a moderate Labour man. Churchill used radio as a medium of inspiration unknown in previous wars and was able to appeal to the British population in a way not possible, for example, for Lloyd George. At the same time, of course, Goebbels was appealing to the German people, with similar effectiveness.

(a)

I am insisting on this [that victims of religious or political persecution, like Calas in the eighteenth century or Dreyfus in the nineteenth, though falsely accused, were at least treated as individuals] because it is essential to an understanding of the difference between the political and historical climate of 1939 and that of 1914; it also explains why people of my generation regarded with despair the world which Stalin, Mussolini, and Hitler had made, why so many people watched the war inevitably coming and entered it with a strange mixture of misery, calmness, and resignation. We knew that in Russia,

Italy, and Germany there were hundreds of Calases, thousands of Dreyfuses. The world had reverted to regarding human beings not as individuals but as pawns or pegs or puppets in the nasty process of silencing their own fears or satisfying their own hates. It was impossible even for that most savage of all animals, man, to torture and kill on a large scale peasants, fellow-socialists, capitalists, Jews, gipsies, Poles, etc. if they were regarded as individuals; they had to be regarded as members of an evil and malignant class—peasants, deviating socialists, capitalists, Jews, gipsies, Poles. The world was reverting or had reverted to barbarism.

Personally I had felt this deeply and bitterly all through the last two years before war broke out. There was the horrible ambivalence towards Chamberlain's shameful betrayal of Czechoslovakia. Chamberlain always seemed to me the most coldly incompetent, most ununderstanding, unsympathetic of the British statesmen who have mismanaged affairs during my lifetime. But when one stands on the very brink of war and suddenly, when one has practically abandoned hope, there is a shift in the kaleidoscope of events to peace instead of war, one cannot but feel an immense relief, release, and reprieve, even though at the same time one feels that the steps which have led to the avoidance of war ought not to have been taken, being shameful and morally and politically wrong. I suffered from this ambivalence all through the Munich crisis, for though the relief was extraordinary, I was convinced that by abandoning Czechoslovakia to Hitler we were only postponing war and that when it came we should have to fight it under conditions far more unfavourable to us than if we had Czechoslovakia and Russia as allies.

When the Polish crisis started, I felt that the end was coming. We went down to Monks House for the summer on July 26, but on August 17, just seventeen days before we were actually at war, we had to move from Tavistock Square to the house which we had taken in Mecklenburgh Square; so we had to be driving backwards and forwards between Rodmell and London. The air of doom and calm resignation both inside one and outside one is what I chiefly remember of those days. The appearance of sandbags, the men digging trenches, the man on the removal van taking our furniture from Tavistock Square to Mecklenburgh Square and as an ex-soldier, receiving his call-up notice (I shan't be here tomorrow, Sir)—all with this quiet, dull, depressed, resigned sense of doom.

I suppose that, from the beginning of human history, men and women, the nameless individuals, have always faced the great crises and disasters, the senseless and inexorable results of communal savagery and stupidity, with the calm, grim, fatalistic resignation of the furniture removal man and all of us in Rodmell and London in August and September 1939. It is a kind of sad consolation to think that it must have been almost exactly like this in Athens in August and September 480 B.C. [the second attempt of the Persian autocracy to overthrow Athens].

(b)

These cruel, wanton, indiscriminate bombings of London are, of course, a part of Hitler's invasion plans. He hopes, by killing large numbers of civilians, and women and children, that he will terrorise and cow the people of this mighty imperial city, and make them a burden and an anxiety to the Government and thus distract our attention unduly from the ferocious onslaught he is preparing. Little does he know the spirit of the British nation, or the tough fibre of the Londoners, whose forbears played a leading part in the establishment of Parliamentary institutions and who have been bred to value freedom far above their lives. This wicked man, the repository and embodiment of many forms of soul-destroying hatred, this monstrous product of former wrongs and shame, has now resolved to try to break our famous island race by a process of indiscriminate slaughter and destruction. What he has done is to kindle a fire in British hearts, here and all over the world, which will glow long after all traces of the conflagration he has caused in London have been removed. He has lighted a fire which will burn with a steady and consuming flame until the last vestiges of Nazi tyranny have been burnt out of Europe, and until the Old World—and the New—can join hands to rebuild the temples of man's freedom and man's honour, upon foundations which will not soon or easily be overthrown.

This is a time for everyone to stand together, and hold firm, as they are doing. I express my admiration for the exemplary manner in which all the Air Raid Precautions services of London are being discharged, especially the Fire Brigade, whose work has been so heavy and also dangerous. All the world that is still free marvels at the composure and fortitude with which the citizens of London are facing and surmounting the great ordeal to which they are subjected, the end of which or the severity of which cannot yet be foreseen.

It is a message of good cheer to our fighting Forces on the seas, in the air, and in our waiting Armies in all their posts and stations, that we send them from this capital city. They know that they have behind them a people who will not flinch or weary of the struggle—hard and protracted though it will be; but that we shall rather draw from the heart of suffering itself the means of inspiration and survival, and of a victory won not only for ourselves but for all; a victory won not only for our own time but for the long and better days that are to come.

THE ECONOMY

52.   An American View of the British Economy in 1927

The United Kingdom. An Industrial, Commercial and Financial Handbook. (United States Department of Commerce, 1930), pp. 1–3.

This quotation, from a commercial handbook, describes how Britain had developed a powerful economy on the basis of specialization and international trade.

The basic economic scheme of the United Kingdom has for decades been widely different from that of most other countries. To a degree almost unparalleled elsewhere, Britain is dependent upon the exchange of products of the mine and the factory for foreign foodstuffs and raw materials. Commerce is relatively more important to national life than in any other major country.

Coal had been king in nineteenth-century economic history. Great Britain was the first to mine and use the product in large quantities. With its rich deposits of high-grade coal lying close to the sea, the country held a sort of quasi monopoly of this all-important source of power during the greater part of the nineteenth century. For this reason, and also because of the strong spirit of initiative in the British people, Britain was the first to develop modern factory industry on a large scale and was long the principal supplier of manufactured goods in European markets and throughout the world. Moreover, in many parts of Europe as well as in South America and elsewhere coal was lacking, and that commodity itself became the greatest single export of the United Kingdom.

The development of mining and manufacturing brought with it expansion of British international commerce and international financial relations. The location of the country, practically in the center of the white population of the world, together with its excellent harbors, facilitated the distribution of the steadily rising volume of British products to foreign markets. From the opposite point of view, moreover, Britain was more and more under compulsion to import goods. Its population, increasingly devoted to mining and manufacturing, had grown greatly and had become more dense than that of any other major country. Consequently, the British people had become constantly more dependent upon imports of raw materials and foodstuffs to supply their needs, and the more so by reason of a relative decline in British agriculture resulting from the competition of new countries with abundant and cheap land.

The development of British commerce carried with it leadership in several related economic activities. The country became not only a great exporter and importer on its own account, but a great middleman for buying and selling the products of other nations, as well as a banker for the financing even of many transactions in goods not passing through the United Kingdom itself. Its shipping expanded enormously, and British vessels carried a major fraction of the commerce of the entire world. The savings accumulated from a prosperous industry, together with the relations established in a great international commerce, brought with them large investments of capital in foreign countries; British citizens before the war had probably larger sums placed abroad than the citizens of all other countries combined. London thus became the financial center of the world. The country was dominant also in the insurance business, especially marine insurance.

In the formation of Britain's basic scheme colonial enterprise played an important part. British colonial areas afforded favored markets for British goods and the development of these regions offered fruitful fields for capital investments. They aided British industry by furnishing sources of raw materials and by increased demand for manufactured goods used in development and required by the general advance in the purchasing power of the people in the areas. Colonial development, moreover, added enormously to British shipping trade and greatly stimulated the growth of middleman business in England.

In brief, the basic economic scheme of Great Britain became then the exchange of the country's coal, its manufactured goods representing the crystallized labor of a dense population, and its services of shipping, finance, insurance, and the like for

food and raw materials from overseas. While the United
States, for example, exports about 10 per cent of the movable
commodities which it produces, the corresponding proportion for the United Kingdom is around 30 per cent. So, too,
Britain has in proportion to population a larger trade than
most other countries. Its exports, on a per capita basis, were
$56 in 1913 and $76 in 1927, compared with $30 and $40,
respectively, for the United States.

53. The New Industrial Revolution

(a) H. D. Henderson, *The Interwar Years and Other Papers* (Oxford University Press, 1955), pp. 29–31.
(b) *Royal Commission on the Distribution of the Industrial Population* (Cmd. 6153, H.M.S.O. 1940), p. 37.

The industrial centre of gravity in Britain was shifting, geographically from the North to the Midlands and South, and struc-
turally from textiles and metals to engineering, chemicals, vehicles,
etc. These changes were reflected in a marked change in the
distribution of population.

We have repeatedly called attention to the transformation
that is coming over our economic life, amounting almost, as
Mr. Baldwin put it at the Guildhall, to a new Industrial Revolution. A fortnight ago we dwelt upon the regional aspect of
this process, pointing out that 'the old industrial North is
yielding place to the Midlands and the South'. We may perhaps convey our meaning more exactly and more vividly as
follows. Draw a line across the map of England, starting from
the mouth of the Severn, and proceeding straight to Stafford,
from there to the High Peak, and from there to Scarborough.
This line divides Great Britain into two parts of almost equal
economic importance, but presenting an extraordinary contrast of fortune. The numbers of insured work-people on either
side of the line are approximately the same; but to the left of
it (i.e. to the north and west) unemployment is almost exactly
double what it is on the right (i.e. to the south and east); or

rather it was double just before the coal stoppage began; the contrast must be much sharper at the present moment. To the left of the line lie all the 'black spots', the regions of the Tyne, the Clyde and South Wales, the Potteries, the textile districts of Lancashire and Yorkshire. Almost every area of present development or future promise, for example, Doncaster, Coventry, Kent, lies to the right.

* * *

Let us now turn to the occupational changes which are taking place. The November issue of the MINISTRY OF LABOUR GAZETTE contains some very important tables, to which we referred briefly last week, which throw fresh light upon the matter. They show, industry by industry, the changes over the last three years in the numbers of work-people insured. In one of the tables, industries are arranged in two groups; according as their personnel has increased or diminished. We give below the figures for the latter group, simplifying the classification a little:

Industry	Numbers of insured work-people				Decrease over the three years
	July 1923	July 1924	July 1925	July 1926	
Coal mining . .	1,256,000	1,260,350	1,240,450	1,227,870	28,130
Woollen and worsted .	271,000	260,890	257,700	254,750	16,250
Bread-, biscuit-, cake-, &c., making . .	157,700	144,540	141,790	145,830	11,870
General engineering .	669,000	627,380	627,280	615,920	53,080
Marine engineering .	66,300	66,110	61,720	58,370	7,930
Shipbuilding . .	270,200	255,090	241,700	224,120	46,080
Iron and steel . .	242,000	237,460	225,910	218,340	23,660
Railway service (non-permanent workers) .	191,100	173,210	168,610	160,650	30,450
National government .	179,600	160,970	156,490	151,470	28,130
Vehicle building . .	27,700	24,550	24,630	21,700	6,000
Total of above industries	3,330,600	3,210,550	3,146,280	3,079,020	251,580

There are many features of this table which deserve attention. In the first place, it should be observed that the figures cover all insured persons attached to the industry, whether in or out of work; so that to obtain the decline in the volume of employment, we need to add the figures of the last column to any increase which there may have been in unemployment. All our basic metallurgical industries appear in this list. It is hardly less noteworthy that, while the woollen industry also appears, the cotton industry, which has fared materially worse, does not. The cotton industry has actually increased its personnel by about 8,000, an interesting sidelight on the working of the short-time system. The table of industries which have increased their personnel is too long for us to reproduce in full. The distributive trades alone show an increase of 260,000, or more than the aggregate decrease in the declining industries. Building and public works have increased by 128,000; the motor-car industry and road transport by 75,000; furniture, brick-making, printing, silk, the electrical trades, all show important increases.

Now figures such as these convey a very inadequate impression of the real magnitude of the changes that are taking place. The changes are partly occupational, and partly regional; and figures relating only to one or to the other tendency do not do justice to the combined effects of the two. Engineering, for example, as a whole, presents an unsatisfactory picture; but engineering is a large category comprising many essentially different trades, with very diverse fortunes and localized to a large extent in different centres. General figures for engineering thus tend to cloak the plight of engineering on the Tyne, say, or in South Wales, as can be seen from the following figures from the same issue of the LABOUR GAZETTE:

ENGINEERING
PERCENTAGES UNEMPLOYED AT 25 OCTOBER 1926

London	6·9		North-eastern	26·9
South-eastern	6·1		North-western	19·0
South-western	6·3		Scotland	23·6
Midlands	13·2		Wales	33·6
			Northern Ireland	29·0

Such are the diversities concealed beneath the average unemployment percentage for engineering of 16·4.

(b)
RECENT POPULATION CHANGES

82. Between 1921 and 1937 the population of Great Britain increased from about 42¾ to 46 millions, that is to say, by 7½

per cent., or rather less than one-half of 1 per cent. per annum. But the increase was very unequally distributed. Thus, for example, in London and the Home Counties the increase was about 18 per cent., while in the Midland group of counties, viz., Staffs., Warwicks., Worc., Leic., and Northants, it was about 11 per cent. No other area shows a comparable rate of increase. In the West Riding, Notts and Derby group the increase was of the order of 6 per cent., and in Mid-Scotland of 4 per cent. In Lancashire the increase was less than 1 per cent., while in Glamorgan and Monmouth there was a decrease of about 9 per cent., and in Northumberland and Durham of 1 per cent. In the 'Rest of Great Britain' there was an increase of 6 per cent. The rate of increase of population in London and the Home Counties was nearly $2\frac{1}{2}$ times that of the population of the country as a whole; although it contains only slightly more than 25 per cent. of the total population it includes about 55 per cent. of the population added during the sixteen years in question. Although London and the Home Counties and the Midland group of counties contain between them only about 35 per cent. of the total population of Great Britain, they contain nearly 70 per cent. of the population added during the period.

83. Between 1923 and 1937 the estimated number of persons insured against unemployment increased by 22·3 per cent. in Great Britain as a whole, by 42·7 per cent. in London and the Home Counties, by 28·2 per cent. in the Midland group of counties, by 15 per cent. in the West Riding, Notts and Derby, by 9·5 per cent. in Mid-Scotland, by 7·6 per cent. in Lancashire and by 4·7 per cent. in Northumberland and Durham. In Glamorgan and Monmouth there was a decrease of 4·3 per cent., while in the 'Rest of Great Britain', i.e., excluding the areas already specified, there was an increase of 27·8 per cent.

54. The Decline of Overseas Trade: Cottons and Coal

(a) *Report of Committee on Industry and Trade, Survey of Overseas Markets* (H.M.S.O., 1925), pp. 5–6.

(b) *The United Kingdom. An Industrial, Commercial and Financial Handbook* (United States Department of Commerce, 1930), pp. 4–5.

The massive and persistent unemployment of the inter-war years stemmed from one basic cause, the decline of foreign demand for the goods of Britain's old staple industries. This decline was the result, partly of increasing foreign competition, partly of a general decline in world trade, and partly of technical change (for example, the substitution of oil for coal in ships).

(a)

It is evident that if statistics disclose a serious falling off in British exports to a particular market (other than a merely casual decline due to transitory and superficial conditions), the observed falling off must be due to one or more of three main groups of causes:

I. decline of purchasing power of the local population;
II. growth of local manufacture;
III. displacement of British imports by imports from other sources.

Frequently an observed decline is due to a combination of all these causes, and they are sometimes so inter-connected that it is difficult to estimate precisely the relative importance of each. An illustration may, however, be given, in which the statistical data happen to be sufficiently complete to enable the different factors to be measured separately.

The exports of British cotton piece goods to India declined by 57 per cent. between 1913 and 1923. The statistics [not printed here] show the movement of Indian imports and home production over the same period, together with the percentages of imports derived from the United Kingdom at the two dates. These data are sufficient to enable a rough calculation to be made of the relative extent to which the decline of consumption, increase of home production, and keener foreign competition have contributed to the aggregate decrease of British exports of cotton piece goods to India. The result is to show that, in this particular case, about three-fifths of the total decline is to be attributed to diminished consumption, about a quarter to increased local production, and about one-seventh to increased foreign competition. It is, of course, only in exceptional cases that all the requisite data for such an estimate are available, information as to the magnitude of local manufacture being frequently wanting, or out of date. Another illustrative example, may, however, be taken from the export of pig iron. In 1923, our exports of pig iron to France, Italy and Japan showed decreases, compared with 1913, of 63, 38, and 87 per cent., respectively. The relative importance of the various factors in the decline was, however, totally different in the three markets. In the case of France, the predominant cause was increase of local production (owing to the recovery of Alsace-Lorraine); in Italy, decline of local consumption; and in Japan, the competition of imports from other sources (*i.e.*, from China and India).

(b)

The British coal industry, which had so long played the major rôle in national prosperity, has suffered particularly since the war. The causes of its difficulties are more or less common throughout the world and have brought about depression in the American coal industry as well. Coal is no longer so supreme as a source of power. Oil and water power have come rapidly to the fore, while invention has made it possible to obtain much more power from a given quantity of fuel. Everywhere the growth in demand for coal has thus been checked. The failure of British coal mining to maintain its prosperity has figured prominently in the post-war difficulties of the nation. British industry has been described as an inverted cone resting on a coal point. This carbon point has written much of the country's economic history for the past 10 years.

Here, again, pre-war events cast their post-war shadows before them. The coal stoppage of 1912 was a forerunner of the serious stoppages of 1921 and 1926. The latter, together with the 8-day general strike which it brought on, involved a loss in working time to the extent of more than 162,000,000 days and a loss to the nation's income estimated as high as £400,000,000.

The coal problem continues to be outstanding. The presence of some 200,000 miners in excess of the absorption capacity of industry is at the heart of the nation's unemployment problem. It is true that pit-head coal prices, which for some years were so high as to make competition in world markets difficult, have been brought down until in 1928 they are low compared with general price levels. This feat, however, was accomplished for the most part by reducing the standard of living for miners, which in turn has meant lower buying power for a large section of the population, with consequent injury to domestic markets.

The United Kingdom since the war has lost one-third of its former oversea coal market—a fact which has had a detrimental effect on the important shipping trade, since the outbound movement of coal constitutes probably four-fifths of the bulk of British exports. Fortunately, however, the reduction in coal exports has been for the most part in consignments to European countries, and, therefore, the synchronization between coal movements to Latin America and other distant destinations and the imports of foodstuffs and raw materials has not been so greatly disturbed.

55. The Difficulties of the Gold Standard

Report of the Committee on Finance and Industry (Cmd. 3897, H.M.S.O., 1931), pp. 106–7.

The return to gold in 1925 was not successful and after five difficult years Britain was forced off gold in 1931. The main difficulties were, first, the over-valuation of the pound (making exports dearer), second, the competitive devaluations of other countries, especially France, and, third, flows of 'hot money' (speculative funds) which made London's short-term debtor position highly unstable.

245. The first set of difficulties has been caused by the fact that the various gold-parities established by the countries returning to the gold standard did not bear by any means the same relation in each case to the existing levels of incomes and costs in terms of the national currency. For example, Great Britain established a gold-parity which meant that her existing level of sterling incomes and costs was relatively too high in terms of gold, so that, failing a downward adjustment, those of her industries which are subject to foreign competition were put at an artificial disadvantage. France and Belgium, on the other hand, somewhat later established a gold parity which, pending an upward adjustment of their wages and other costs in terms of francs, gave an artificial advantage to their export industries. Other countries provide examples of an intermediate character. Thus the distribution of foreign trade, which would correspond to the relative efficiencies of different countries for different purposes, has been seriously disturbed from the equilibrium position corresponding to the normal relations between their costs in terms of gold. This, however, has been a consequence of the manner in which the post-war world groped its way back to gold, rather than of the permanent characteristics of the gold standard itself when once the equilibrium of relative costs has been re-established, though, even after six years, this is not yet the case.

246. The second set of difficulties has resulted from the international lending power of the creditor countries being

redistributed, favourably to two countries, France and the United States, which have used this power only spasmodically, and adversely to the country, Great Britain, which was formerly, the leader in this field and has the most highly developed organisation for the purpose. This re-distribution of lending power has been largely due to the character of the final settlement of the War debts in which this country has acquiesced. For although Great Britain suffered during the War a diminution of her foreign assets of some hundreds of millions, she has agreed to a post-war settlement by which she has resigned her own net creditor claims, with the result, that on a balance of transactions, virtually the whole of the large annual sums due from Germany accrues to the credit of France and of the United States. This has naturally had the effect of greatly increasing the surplus of these two countries, both absolutely and relatively to the surplus of Great Britain. The diminution in Great Britain's international surplus, due to her war sacrifices remaining uncompensated by post-war advantages, has, however, been further aggravated recently by the adverse effect on her visible balance of trade of the first set of difficulties just mentioned, namely, the differing relationships between gold and domestic money-costs on which different countries returned to the gold standard.

56. The Financial Crisis, 1931

P. Snowden, *An Autobiography* (London, Ivor Nicholson and Watson, 1934), pp. 894–6.

The Labour Party, in office at the beginning of the Great Depression, had an orthodox monetary policy which was inappropriate as remedy for the country's problems. Snowden's attitude towards the Budget was little different from that of Gladstone.

'I believe, if I may put it so bluntly as this, that an increase of taxation in present conditions which fell on industry would be the last straw. Schemes involving heavy expenditure, however desirable they may be, will have to wait until prosperity returns. This is necessary—I say this more particularly to my hon. friends behind—to uphold the present standard of living, and no class will ultimately benefit more by present economy than the wage-earners. I have been in active political life for forty years, and my only object has been to improve the lot of the toiling millions. That is still my aim and my object, and if I ask for some temporary suspension, some temporary sacrifice, it is because I believe that that is necessary in order to make future progress possible.

'The Budget position, as the right hon. gentleman said, is serious. It is no secret that I shall have a heavy deficit at the end of this year. No Budget in the world could stand such an excessive strain as that which has been placed upon it by the increase of unemployment during the last twelve months. The depression has affected both sides of the Budget. Expenditure has increased, revenue has declined. There is this fact which I think we sometimes ignore. Productive capacity has now fallen off by 20 per cent. That means 20 per cent. less in those resources from which the Exchequer must draw its revenue. Capital values have fallen, except in the case of gilt-edged stocks. And may I say, in reply to what the right hon. gentleman stated about British credit, that, in spite of the depression, British credit is standing higher today than it has done during the last five years. Of course, I do not mean exactly at this precise moment, but taken over a few weeks.

'We have the burden of War Debt. I do not want to give offence to anybody when I make this statement, that when the history of the War in which that Debt was incurred, its recklessness, its extravagance, commitments being made which were altogether unnecessary in the circumstances at the time, when that comes to be known, I am afraid posterity will curse those who were responsible. Though the industrial slump has affected this country so seriously, we have suffered less than others of the great industrial countries of the world. Their budgetary positions are worse than ours. I am quite familiar with what the right hon. gentleman has said about the talk which is going on in certain quarters, but I am sorry to hear that the right hon. gentleman associated that with the responsibility of a Socialist Government. This is not a situation and this is not an occasion when people should talk of taking action which might ruin the country in order to gain a Party advantage.

'There is, as the right hon. gentleman said, one vulnerable spot in our position, and that arises from the fact that we are the world's great financial centre. It is quite true, as he said, that if there were well-grounded fears that this country's budgeting was not sound, then it might have disastrous consequences, which would have their repercussions abroad. It is quite true that other countries are watching, and we must maintain our financial reputation. That we can do. Our position is fundamentally sound, sounder than that of any other country in the world, and all that is required is an effort to get over the present temporary crisis, and that can be done without any very great efforts. It will involve some temporary sacrifices from all, and those best able to bear them will have to make the largest sacrifices. In the general sacrifice, the Members of the Cabinet are prepared to make their substantial contribution. As I have said before, this is a problem which no Party can solve, but the country and the House of Commons must realise the gravity of the position. Instead of Party bickering, which we can resume later, we must unite in a common effort to take effective measures to overcome our temporary difficulties and to restore our former prosperity.'

It is difficult to convey an impression of the effect of this statement. Members turned deadly serious, and listened with strained attention to this unexpected development of the debate. It was felt that a House which a few minutes before had been cheering the familiar reproaches on an ordinary Party occasion now realised that it was faced with a situation which would demand the co-operation of all Parties. The task was too big for one Party, and a united national effort would be needed to deal with the crisis.

After my statement, Sir Donald Maclean moved the Liberal Amendment in a brief speech, in which he said that my grave warning was not unworthy of some of my great predecessors. Then the House emptied, and members congregated in the Lobbies, the library and the smoke-room to discuss the statement. Its effect on the Labour members was stunning. They regarded this as the end of their hopes that the Labour Government would proceed with a policy of spending public money on extravagant schemes of social reform. I was very sorry for the Labour members. I had so phrased my statement as to prepare them gradually for the unpleasant truth, and the abandonment for the time being of schemes upon which they had set their hearts. The Left Wing members of the Labour Party at once began to express their dissatisfaction and disgust with the statement. One of them declared: 'It's bigger, not smaller, Budgets we want!' They shewed not the least appreciation of the national situation, nor of the fact that the decline in revenue and in trade made it impossible to carry out a policy of increased expenditure which might have been possible when trade was booming and revenue was expanding.

57. International Influences on Commercial Policy

Commercial Policy in the Interwar Period: International Proposals and National Policies (League of Nations, Geneva, 1942), pp. 52, 70–1.

The inter-war years were characterized by increasing economic nationalism, in spite of much talk to the contrary. Increasing protection undoubtedly contributed to the decline of international trade.

The state of apparent, if precarious, economic equilibrium broke down in the summer of 1929. Before the end of the year measures of intensified agricultural protectionism had been introduced in Germany, France and Italy; upward tariff revisions had occurred in Roumania, Norway, Hungary and Finland and in many other countries higher schedules were in preparation. The movement, which was accompanied by deconsolidation of duties and denunciations of existing treaties, was accelerated and extended as the economic depression spread and deepened. The final adoption of the Hawlay-Smoot tariff in the United States in June 1930 was shortly followed by higher tariffs in Canada, Cuba, France, Mexico, Italy, Spain, Australia, and New Zealand. The United Kingdom abandoned her traditional free-trade policy with the imposition of emergency duties in the autumn of 1931 and the first general tariff in February 1932.

A new and far more critical phase in the development of

restrictions on trade opened with the financial crises in Austria and Germany in the early summer of 1931, followed by the widespread abandonment of the gold standard some months later. The upward trend of duties was accelerated and affected almost all countries. Moreover, tariffs were supplemented—and before long overshadowed—by direct quantitative restrictions and the control of foreign exchange transactions. At the close of 1931, foreign exchange controls were in force in Austria, Bulgaria, Czechoslovakia, Denmark, Estonia, Germany, Greece, Hungary, Latvia, Portugal, Spain, Yugoslavia, Argentine, Brazil, Bolivia, Colombia, Chile, Uruguay, Turkey, Iran; customs quotas in Czechoslovakia, France, Italy, Latvia, the Netherlands, Turkey.

* * *

The system of multilateral trade, already seriously affected, broke down with the collapse of the world monetary system. There ensued a general movement towards bilateralism—the endeavour by each country to achieve reciprocity in trade by reducing imports from countries with which its trade balance was passive.

The same general factor provoked attempts on the part of many countries to develop their exchanges of goods and realize a system of settlements within restricted areas. Thus, the United Kingdom and France expanded their imperial trade. Germany sought new outlets and sources of supply in Central and South Eastern Europe and in Latin-America. The members of the 'gold bloc' endeavoured to expand their mutual trade (Brussels Protocol, 1934), while several of the smaller European countries concluded—or adumbrated—regional trade agreements for the same purpose.

With certain notable exceptions (for example, the efforts of the 'Oslo Group', the above developments were accompanied by the creation of new or the extension of existing preferential systems and the emergence of new forms of commercial discrimination. By the Ottawa Agreements of 1932 and the Import Duties Act introduced in the United Kingdom the same year, a general preferential system within the British Commonwealth and the Colonial Empire was established. The German trading methods were frankly and flagrantly discriminatory. Through the use of exchange control and quantitative restrictions, the M.F.N. clause lost much of its value in European commercial relationships. Those relationships were as complex and disparate as they were unstable. Each bilateral agreement was *sui generis*, designed to meet the special trade requirements of, and to afford effective reciprocal advantages to, the signatories. Commercial agreements, in truth, became instruments of commercial warfare. The degree of instability in commercial relationships may be illustrated by the fact that the Economic Committee, when requested by the League Council in 1935 to examine the feasibility of an international agreement providing for notification one month in advance of proposed changes in tariffs and other restrictions, reported that there was 'no chance at present of achieving such an agreement'.

PART III, 1945-1970

PART THREE: 1945–1970

The twenty years after the Second World War divide, by sound popular repute, into two main periods, of austerity and affluence. 'Work or Want', from the posters of the Cripps era, and Harold Macmillan's 'You've never had it so good' of July 1957, describe attitudes, if not the whole truth, succinctly. 1951 marks the beginning of the end of austerity; 1955 marks the beginning of affluence. Labour fell from power, having governed a full term for the first time in the history of the Labour Party, in 1951; the Tories governed for the next thirteen years. Rationing, the symbol of equality of sacrifice during the war, was already coming to an end in 1950. In the period between 1950 and 1957 the nation turned away from its long haul towards solvency, to enjoy some of the fruits of effort. At first, however, there was no dramatic change, and the contrast with the period before is a muted one until the mid-fifties. Nor is 1951 a decisive date in other respects. The granting of independence to India in 1947 and the Suez affair of 1956 were surely more important dates? As a background to the history of the period the debates over Britain's Imperial, European, and world position were surely more relevant to an understanding of modern Britain than the end of post-war rationing? Nor are austerity and affluence clear divisions in the history of the welfare state which has been continuously extended by all governments since the war. Thus the people of Britain had both increasing affluence and increasing social welfare, and the questioning of either affluence as a necessary aim of economic policy, or the paradox of an increasingly wealthy people getting more and larger government hand-outs, came only at the end of the period, in the sixties.

What were the economic bases of this affluence? Immediately after 1945, economic conditions were largely determined by the war. The war had resulted in a running down of domestic capital, through lack of replacement or repairs, as well as through physical damage from bombing and shipping losses, by some £3,000 millions. About £1,000 millions of overseas assets had been sold, external debts of £3,000 millions had accumulated, and exports had fallen to less than one-third of their pre-war level. It was necessary, therefore, to continue war-time restrictions to encourage exports and to keep imports down to a minimum, as well as to repay debts. Rationing and other war-time controls, in contrast to the period after the First World War, were relaxed only gradually and terminated finally almost ten years after the war, in 1954. The abolition of restrictions on imports, except from the dollar area, was achieved only in 1955. Exchange controls between the sterling area and the rest of the world were also only gradually eased, and non-residents were not allowed to convert sterling on current account freely until 1958. Partly as a result of these restrictions the period after 1945 was one of increasing levels of production and employment, of rapidly expanding exports, and of rising living standards. Partly also as a result of government policies, this growth was accompanied by persistent economic problems, especially balance of payments disequilibria as there was a continuing tendency for the growth of imports to be faster than the growth of exports, including invisibles.[1] This was so especially in periods of rapid economic growth and inflation. Briefly this meant that Britain's current surplus on visible and invisible trade was not large enough to finance her overseas expenditure (on goods and services) and capital exports, and also to uphold the position of the City of London as banker to the sterling area. All postwar governments, whether Labour or Conservative, have had the problem of recurring balance of payments deficits, which they have tried to solve with deflationary policies aimed at creating spare capacity that could be devoted to exports. Domestic demand has also been reduced, by taxation and credit restrictions, and by high interest rates. The result has been a policy of 'stop–go', the government alternately restraining and encouraging growth to produce unwanted fluctuations in economic activity, and obviously making it impossible for the economy to realize its full growth potential. On two occasions, also, in 1949 and 1967, there were devaluations, reducing the value of sterling in terms of other currencies so as to make exports cheaper and imports dearer, thus encouraging the former and discouraging the later. Thus the £1 sterling which had emerged from the war at a value of $4·03, was devalued to $2.80 in 1949 and to $2.40 in 1967, on both occasions with beneficial results for exports.

Although relatively small in area and with less than 2 per cent of the world's population, Britain by the mid-sixties was still the world's third largest trading nation, being responsible for 11 per cent of world international trade in manufactured goods. For over a century international trade had been the basis of British growth, and in the period after the war trade in commodities, since trade in service had become relatively less important, was more important than it had ever been before, Britain relying on imports for about half her food needs, and, with few exceptions, needing to import nearly all the raw materials used by industry. Exports, including goods and services, amounted in the sixties to about 20 per cent of British national income; Britain was the largest single food importer and among the largest importers of metal ores, textile raw materials, and petroleum; Britain was a major world exporter of vehicles and machinery, metal manufacturers and textiles, electrical goods and chemicals. In the twenty years after the war there were two significant changes in the composition of this trade: agricultural imports as a percentage of total imports were reduced from 40 to 25, as domestic agricultural production[2] rose; there was, at the same time, a marked increase in the proportion of imports of finished and semi-finished manufactures, including capital goods, from less than 10 to over 20 per cent of total imports. As regards the direction of trade, the most notable post-war trend was the growth of trade between Britain and Western Europe; by 1962, for the first time, exports to Europe exceeded those to the sterling area. At the same time Britain's protective tariff was considerably modified, as a result of the General Agreement on Tariffs and Trade. The European Free Trade Association (EFTA) was set up in 1960 by Austria, Denmark, Norway, Portugal, Sweden, Switzerland, and Britain. In 1961 Britain applied to join the European Economic Community, the Common Market set up in 1956 by Italy, France, Germany, Belgium, the Netherlands, and Luxemburg, but negotiations were suspended in 1963.[3] Negotiations were resumed in 1970.

The British economy in the twenty years after 1945 undoubtedly performed remarkably. New industries flourished,

[2] 88.   [3] 95.

productivity increased, living standards improved. Britain was able to maintain a heavy defence burden and to resume massive investments abroad. Nevertheless the rate of growth of national income was slow compared with those of many foreign countries, especially in Europe. And, as has been indicated, growth was punctuated by balance of payments crises: in 1949, 1951, 1955, 1957, 1961, and 1964. Much economic thinking in Britain has been devoted, not to the obvious successes of the economy, but to its relatively slow rate of growth,[4] and the explanations for this latter phenomenon have been legion: too much spent on defence; too much invested abroad; poor industrial relations and too high wages; too much government interference in the economy; too little government interference; an inappropriate educational system; the failure to get into the Common Market with its growth economies; the decline of the Commonwealth; an obsolete social structure which starves industry of managerial talent; 'stop–go' economic policies which inhibit growth; too high rates of interest; insufficient research and development programmes. There is something to be said for placing the blame on each of these alleged weaknesses, but aggregated they make a formidable list of disqualifications for growth. How is it, then, that Britain performed as well as she did? Are these characteristics unique to Britain? Industrial production in 1969 was 30 per cent higher than in 1959, and double the prewar level. Between 1958 and 1968 manufacturing in the chemical industries rose 82 per cent, in engineering and electrical industries 57 per cent, in printing and publishing 52 per cent, in gas and electricity 64 per cent, and in construction 49 per cent. These figures, in the context of rapidly rising exports, are impressive, and even if other countries achieved higher rates of growth, most of the countries of the world did not. The problem of explaining Britain's 'slower' rate of overall growth is an important one, but it should not disguise the problem also of explaining Britain's remarkable rate of growth of exports after 1945.

The Second World War provoked an upheaval in British society more profound than that of the First, to judge by the social consequences. Destroyed or damaged houses, worn-down industrial equipment and rolling stock, neglected schools and

[4] 85.

hospitals, indeed the whole apparatus of an industrial society was in need of urgent repair and replacement at the end of the war, and needed as great a collective effort by society as had the war. The direction of the war had been achieved by a total control of the economy previously unknown; such control had been ruthless and effective, as convincing in its efficiency as in its equity, and pointed morals for the post-war world. Wartime planning for post-war reconstruction aimed primarily at full employment; if the First War had as its slogan 'Homes for Heroes', the Second had its unsung slogan as 'Full Employment'. As events turned out, full employment was easily attained, not so much because of Keynesian budgetary techniques for boosting the economy, but because of a buoyant world demand for both raw materials and manufactured goods. Whereas the pre-war average unemployment rarely fell below 10 per cent, the post-war figure rarely rose above 3 per cent. Thus was removed the specific social evil of the inter-war years. At the same time, real incomes began to increase, at first slowly, and by the mid-fifties quite rapidly. There was, also, compared with before the war, a more equitable distribution of income, and a much greater redistribution of income by government. By the 1960s there were two broad bands of income: $10\frac{1}{2}$ million people had incomes between £275 and £750, and $9\frac{1}{2}$ millions with incomes between £750 and £2,000. Those with incomes above £2,000 numbered only half a million. With such incomes the working classes began to buy desirable consumer goods in greater quantities until in the 1960s they were regularly purchasing cars. In 1945 there were $1\frac{1}{2}$ million cars on the road; in 1955, over 6 millions; in 1965, over 9 millions. The result was increasing congestion on the roads, but greater pleasure for an increasing number of people as their mobility was increased. Prosperity was also obvious in the expansion of shopping facilities and advertising. A retail revolution, which had commenced in the thirties, exploded in the fifties and sixties, with chain stores and supermarkets, and intensive advertising in the press and on television. There was a corresponding expansion of hire-purchase, especially for consumer durables.[5]

The welfare state, enlarged from its pre-war base by the Labour government of 1945, and subsequently accepted and

[5] 81.

enlarged by the Conservatives, made prosperity secure because it offered a safety net in case of sickness, accident, or unemployment. Acclaim of the welfare state, however, was a great change from the days when working men and duchesses had joined in denouncing Lloyd George's pioneer insurance of 1911. The foundations of British post-war thinking on welfare were laid by William Beveridge who was asked during the war to produce a plan for social welfare. Beveridge proposed a single insurance stamp to cover all the disasters that cause poverty, a national health scheme for medical treatment, and children's allowances. Beveridge's report was received with general enthusiasm and was implemented after the war. The real novelty of Beveridge was that he envisaged community action to provide not only full employment but also to defeat poverty and sickness. The National Health Service, as well as family allowances, were to be provided for all, without a means test, and out of general taxes. The Poor Law officially disappeared in 1948, but, of course, not poverty. Still the very poor, mainly the old, unable to subsist on the old non-contributory pensions, had to fall back on the National Assistance Board for out-door relief. By the end of the Labour administration the range of State services available was truly impressive: allowances for children after the first; free medical and dental treatment in or out of hospital;[6] grants for births and deaths; residential homes for old people; national assistance without a family means test; free primary and secondary education and, when qualified, university education. There was then, and afterwards, some criticism, but generally this vast welfare programme, and its costs, were accepted by both political parties, and by the British public. The National Health Service attracted the most attention and most criticism, and marginal private charges were added (for spectacles and medicine). Nevertheless all classes continued to use the free services, although private treatment, and insurance to cover that treatment, grew with increasing prosperity.

The Education Act of 1944 provided for universal compulsory secondary education to the age of 15, following universal primary education to age 11. As the Hadow Report of 1926 had foreshadowed, three types of secondary school emerged: the grammar, the technical (though few actually appeared),

[6] 58.

and for the bulk of children, about 75 per cent, the secondary modern. Theoretically they were different but equal; in practice, they reflected social and academic differences. A new examination, the Certificate of Secondary Education, provided a test of achievement for the secondary moderns, and, particularly in the south, large numbers of children began to stay on beyond the age of 15 to gain this educational qualification. Many of the secondary moderns were unsuccessful; the best offered a lively and relevant education. The grammar schools pursued academic excellence and their pupils were selected by an eleven-plus examination that came in for much criticism. In the sixties, especially under the Labour government, the comprehensive schools sought to avoid the selection problem, with its social implications, by taking in all secondary pupils in a given area and streaming them within the schools. This move roused bitter antagonism from the grammar schools and their defenders who argued that what was obviously academically good should not be destroyed in the cause of doctrine, even a doctrine of social equality. What is most interesting is that education became, by the sixties, a lively electoral issue and increasing numbers of parents became concerned about the amount and the quality of the education provided for their children.

Higher education also expanded, but rather to meet the need for skilled and trained people, for professional training, rather than to provide a liberal and literary education for a leisured ruling class. Technical Colleges expanded; Teachers' Training Colleges became Colleges of Education with longer courses; Colleges of Advanced Technology, of university status, were founded along with a crop of new universities. The Robbins Report of 1962 recommended a vast increase in full-time tertiary education, from 216,000 to 560,000 places in twenty years. It argued that there was an untapped 'pool of talent' which would easily fill these places, and provide the increasing stream of teachers and technical professionals necessary for the expanding economy. In education, perhaps more than in any other sphere of social endeavour, attitudes had changed most since 1870. They reflected a general growth of democratic sentiment combined with an acute appreciation of the practical needs of a growing economy, stimulated by the threat of being outdone by overseas competitors. The cost was alarming, the more so because it impinged directly through the local rates. The cost of public education in 1964–5 was £1,449.5 millions, which absorbed nearly 5 per cent of the national income.

A deliberate attempt, also, was made to house the population decently, and increasingly to protect the national environment from spoiling and pollution. The environment for most of Britain's population was urban, 80 per cent of the population living in towns, with the greatest concentrations in the great conurbations: London, the West Midlands, South East Lancashire, Merseyside, Tyneside, Clydeside, and the West Riding. Within the cities, districts changed status as the larger houses of a previous age were subdivided and deteriorated into poorer living quarters, or even into slums. Much elegant Georgian and Regency building disappeared, to be replaced by office buildings or blocks of flats. The worst urban areas tended to be inhabited by coloured immigrants—the Asians and West Indians—who were also among the lowest-paid workers.[7] Such areas, in proximity to low-wage and poor-housing districts inhabited by the white British, were the scenes of some but relatively rare racial tension and violence. Standards of housing were still low in 1966 in the most crowded areas: the 1966 sample Census showed that there were in England and Wales $2\frac{1}{4}$ million houses (14·9 per cent of all houses) without a fixed bath, and 180,000 that were overcrowded. But since the percentage of houses without a bath in 1951 had been 22 per cent, already much had been achieved. The number of houses, also, was increasing. In 1938, 364,000 houses had been built, and this figure was not exceeded until 1964. The Macmillan housing drive took the figure of new permanent dwellings over the 300,000 mark in 1954, for the first time since the war. Thereafter annual house construction figures remained high. Rents remained controlled until 1957. Prices of houses went up continuously, and the Building Societies emerged as one of the largest institutional investors. Slum clearance proceeded, with half a million houses demolished by 1963, and new housing estates were criticized as impersonal and lonely compared with the cosy friendliness of the replaced slums. Urban development was directed by the Ministry of Town and Country Planning which exercised con-

[7] 67.

trol over local authorities. Twelve new towns were provided for by the New Towns Act of 1946. With all this development, the modern traveller in Britain is impressed with the countryside, and with how much of the age-old pattern of fields, hedges, and stone walls remains.

The grappling with the problems of poverty, education, health, and housing certainly helped to maintain and to improve the quality of life in Britain. Better housing, for example, partly transformed the long-established working-class habit of the husband spending his leisure time outside the home; he could now watch television with his family and could share more sympathetically in the bringing-up of his children. More middle-class marriages, also, changed, from the bad economic arrangement that the Fabians had denounced, into partnerships. Children were better able to do their homework; parents to entertain their friends. Only sentimental refugees to the professional middle classes like Richard Hoggart regretted the changes. This rapid social change was also reflected at work, where perhaps the most noticeable change was the growth of the service industries. The less than half a million in the armed services were professionals, led by a highly trained élite, with a pattern of life not unlike civilians; the Civil Service rapidly increased in numbers; personal services—not in the form of domestics, but in the form, for example of hairdressers, garage mechanics, and photographers, as well as in the form of the old learned professions—also expanded absolutely and proportionately. By the end of the sixties Britain had 'a service economy' in which almost half of the employed population was not producing tangible goods.

What do these changes in employment, spending, standard of living, education, environment amount to? Does the welfare state and the redistribution of income that goes with it stand for something quite new in Britain? The argument that the actions of the post-war Labour government amounted to a social revolution, must be mainly discarded.[8] It was widely believed at the time, when middle-class newspapers like *The Daily Telegraph* complained bitterly that the middle-class contribution to national well-being was ignored. Yet Labour left the public schools alone, and accepted the hierarchical plan for State education embodied in the 1944 Act. Even

[8] 77.

when the coal-owners, traditional symbols of capitalism, lost their property, they received fair compensation. By 1951, far from planning the whole economy, the Labour government had abandoned most wartime controls, and proclaimed its belief in a mixed economy. The more powerful agent of social change came later, in the fifties, namely the affluence brought about by a changing industrial society, in which its resources for the first time were being fully used to create working-class wealth. Rising wages even brought the new welfare society under critical scrutiny. Was there the need for such a structure if private affluence brought the services it provided into the reach of most individuals?

What has affluence done to the British class structure? Carr-Saunders and his collaborators wrote, in 1957, just before the full onset of the affluent society, that the threefold division of society into upper middle class, lower middle class, and working class was meaningful but that occupation was the best guide to the class to which a person belonged. When asked, nearly half the population thought of themselves as middle class. Certainly in our period there has been a great working-class move towards middle-class status, sometimes accompanied, especially in the educated, by a guilty conscience. Compared with the 1870s there is much less social rigidity between classes, and being 'lower class' does not now carry any necessary implication of being inferior. New élites, widely recruited, have appeared, especially top managers and scientists, but there have also been levelling influences, for example the expansion of consumer spending and of the mass media. The nation clothes itself at Burtons and at Marks and Spencers. All classes watch the television. Accents are less important than they were once, and a classless semi-transatlantic accent has generally influenced speech. In the universities, in the newer services like advertising, in the eating houses of the cities, the classes have merged increasingly. It is now quite different from the twenties of P. G. Wodehouse's Jeeves and Wooster, or from the thirties of Oxford and of the rigid grammar and public school division. Education nevertheless remains one of the methods of achieving class distinction, hence the attraction of the private schools, even while it remains also one of the main vehicles of class mobility.

Class identification is an important social factor, and certainly in the sixties there was a strong feeling in favour of classlessness. It came to be accepted that a man was not inferior for working in a factory instead of a bank. The bank clerk, indeed, awakening to the fact that of the two he was often the worse paid, began to turn to the possibilities of industrial action. In this general classless euphoria, of course, a small number of people clung ostentatiously to their distinctions of school, accent, friends, and life styles. The Englishman continued to love a lord, and any serious sentiment for republicanism was conspicuously absent. However, the general thesis that there has been a significant reduction in inequality in Britain since 1945, and a significant move towards a classless society, has not been unchallenged. R. M. Titmuss in particular has argued that income statistics derived from the Board of Inland Revenue Reports are misleading. He claimed that the well-to-do succeeded in disguising their true income in a number of ways, while at the other end of the social scale, national insurance payments should be seen as a 'regressive poll tax' which took no note of personal circumstances. Titmuss also suggested that wealth as distinct from income was perhaps becoming more concentrated than before. Brian Jackson argued that Seebohm Rowntree's poverty remained at least in towns like Huddersfield. There a working man without savings enjoyed affluence only at times, for example when his children were working and living at home before marriage. In old age he might well sink to subsistence.[9] His life style and values still differed from those of the middle classes. He did not seek promotion at work, as in many mills this was impossible. He eschewed abstract conceptual speech for the descriptive and anecdotal. He did not seek the privacy so dear to the middle classes. Perhaps the deepest division in British society was the distinction between those with intellectual interests and those without. This was more fundamental than the division C. P. Snow saw in his concept of the two cultures, the scientific and the literary.

Whatever its defects, this post-war society has a good deal to be said for it at an everyday level. The typical modern Briton worked (1968) five days a week for a wage of £22 and lived with his wife and 2·8 children in a clean, light house with 'mod. cons.'. Saturday he went shopping, or watched a

[9] 65.

football match, and on Sundays he drove round the countryside in his family car. He might have managed a short holiday abroad and was open to a host of world influences through the liberating power of the T.V. His children enjoyed full-time education to age 15 and were more likely to 'get on' than he was. He had no religion except in the burial rites of his family but he might well have had an informed view of the major political and social questions of the day. He was tolerant, probably treated his wife as an equal and his children as friends. It was accepted, but not thought about, that this pleasant life might end in a nuclear holocaust. Some publicists vigorously denounced this kind of life, as lacking depth and commitment; as being rootless, materialist, and callous towards the old. But the thinkers who made the charge that this was a decadent society, could do so only by ignoring its vitality, change, and growth.

Certainly British society in the sixties had serious defects, including an increasing crime rate. In 1939 of 283,220 indictable offences brought to the notice of the police, over 50 per cent resulted in conviction; of 1,133,882 in 1965 less than 40 per cent were solved. The statistics are difficult to interpret, but there was evidently an increase in crime over the period, especially crimes of violence in the age-group 17 to 21. The State improved conditions of work and pay for the police and continued to liberalize the treatment of convicted offenders; the death sentence was finally repealed. Nevertheless, crime rates crept upwards, although there was no break-down of law and order, and no serious loss of faith in the British policeman. As crime increased, so did other indicators of social malaise: illegitimacy was high; venereal disease on the increase; student unrest, though far less than abroad, was widespread; industrial unrest and unofficial strikes threatened productivity; the aged poor suffered under inflation; there was an increase in drug taking; the reputation of Parliament and the credibility of the major political parties declined. Britain, the oldest of the industrialized countries, the oldest democracy, was suffering, perhaps, from a degree of obsolescence in her social, economic, and political institutions. It would be absurd, however to be pessimistic about her post-war achievements or about her future promise.

GENERAL SOCIAL CONDITIONS

58. The Crisis of 1947

Annual Register (London, 1947), pp. 6–7.

The period 1945–50 included the great constructive achievements of 'the ablest government of modern times' (Harold Macmillan), but as a result of the war, its social flavour was of austerity. The overriding economic need was to close 'the dollar gap', and although the American loan of 1946 and Marshall Aid of 1948 'saved' Britain, there was still considerable hardship. Between these came the crisis of 1947. The weather was the worst since 1880–1, three months of extreme cold followed by spring flooding.

A partial thaw in the South added floods to the weather complications, while the North-East had more snow. On 5 February the Ministry of Fuel and Power announced 'a most serious situation ... in regard to the supply of coal', in consequence of 'exceptionally cold weather'. Railways were blocked. Industrial coal could not be moved by sea or land. On the 6th it was forbidden to load coal into bunkers—an order immediately withdrawn—gas pressures were further reduced, and the Central Electricity Board described the stocks at power stations as 'critically low'. In Birmingham alone 60,000 men were idle or on short time. And the snow continued to fall!

On 7 February Mr. Shinwell told an astonished House of Commons that from the following Monday (10 February) there would be no electricity for industrial consumers in London, the South-East, the Midlands, or the North-West. Domestic heating by electricity would be forbidden between 9 a.m. and noon, and from 2 p.m. to 4 p.m. The proposal was that a number of power stations should close, so that the authorities could concentrate on supplying coal to the remainder. Yorkshire, and perhaps London, would be exempt from the industrial ban. The shortage of stocks, said the Minister, was 'entirely attributable to the weather'. Before Mr. Shinwell spoke Captain Prescott, Conservative, delivered the first of many 'crisis' attacks on the Government, accusing Mr. Shinwell of too much optimism and too little planning. Mr. Eden said that the Cripps Plan had come too late, and that a bad distribution of coal had led to the worst economic crisis for twenty years. Major Lloyd George accused his successor of inconsistency, and of ignoring the warnings given by the Opposition in the previous summer. Whatever Mr. Shinwell did now, stocks would continue to fall until the end of April. It was already time to plan for the next winter. The Liberal leader, Mr. Clement Davies, demanded the establishment of a Council of State. On the other hand, Mr. Ellis Smith, while recalling that the Government had only just weathered the previous winter and contending that there should have been a plan and more delegation of authority, blamed the crisis on the carelessness of the Tories before the war and the miscalculations of the Coalition. The following day Mr. Shinwell told a press conference that unless consumption were reduced there would be complete disaster within ten days. A rapid thaw would increase the difficulty. There were 3 million tons of coal on the ground or in transit. The Ministry was trying to build up a week's stock at the power stations. He had increased the stocks by 4 million tons in the summer; the trouble was the amazing consumption. The Cripps Plan had never had a chance, owing to the weather.

The operation of the new orders put 1,800,000 men out of work in the first three days of the week beginning 10 February. Sheffield, though not within the banned area, was idle because fuel failed to arrive. In a dimly-lit House of Commons the Prime Minister reported that the first day's response to the cut had been excellent, and announced a ban on greyhound racing, the strict curtailment of transport services in London—where snow ploughs were at work in the streets—and the suspension of the B.B.C.'s Third Programme and television services. Mr. Attlee broadcast to the nation an appeal for cooperation and economy in the use of fuel. It became clear, however, that domestic users were not complying with the Prime Minister's request, backed up though it was by frequent warnings from the Central Electricity Board. The first day's saving of coal, 22,550 tons, had not sufficed to improve the stock position. There were 125 colliers stormbound in the Tyne. Coal was therefore given complete priority on the frozen railways.

The prospects of an early thaw had by now receded. The British crisis was a subject of world-wide discussion, and President Truman asked whether coal bound for Europe should be diverted to this country. This offer the Prime Minister declined, on the ground that it was unfair to Europe. Mr. Shinwell, by now involved in frequent storms in the Commons, said that there were no negotiations in progress for the purchase of American coal. A message from Downing Street declared 'the same speed and urgency as a major military operation during the war' would be applied to the fuel problem. On 12 February the Prime Minister presided at the first meeting of the Joint Committee of Ministers and representatives of the Coal Board, the Railway Executive Committee, and the Central Electricity Board, to make emergency decisions. Restrictions on domestic consumption of power were applied to the whole country and reinforced with heavy penalties under the Defence Regulations. Power was taken to requisition coal.

59. The National Health Service

Aneurin Bevan, *In Place of Fear* (London, Heinemann, 1952), pp. 75–7.

At the end of the war Britain had widespread but not comprehensive health services, which had grown up piecemeal. Lloyd George had introduced insurance against sickness for workers earning under £250 p.a. in 1911 and this was extended by stages until perhaps half the population was covered by a basic family doctor scheme. Hospitals had in some cases grown out of the sick-bays of the workhouses, in others been set up by the local authorities, or as voluntary hospitals depended largely on charity days. Schoolchildren had been medically inspected and treated since 1907 and many firms had their own schemes like that described by Lady Bell at Middlesbrough. Since 1875 local authorities had Medical Officers of Health and the better ones provided extensive services for mother and child care. The Beveridge plan proposed a comprehensive health service to ensure that 'for every citizen there is available whatever medical treatment he requires'. The principle was accepted by the war-time government and implemented by the post-war government which gave the responsibility for working out the scheme in detail to Aneurin Bevan, the Welsh socialist (1897–1960).

... The collective principle asserts that the resources of medical skill and the apparatus of healing shall be placed at the disposal of the patient, without charge, when he or she needs them; that medical treatment and care should be a communal responsibility; that they should be made available to rich and poor alike in accordance with medical need and by no other criteria. It claims that financial anxiety in time of sickness is a serious hindrance to recovery, apart from its unnecessary cruelty. It insists that no society can legitimately call itself civilised if a sick person is denied medical aid because of lack of means.

Preventable pain is a blot on any society. Much sickness and often permanent disability arise from failure to take early action, and this in its turn is due to high costs and the fear of the effects of heavy bills on the family. The records show that it is the mother in the average family who suffers most from the absence of a free health service. In trying to balance her domestic budget she puts her own needs last.

Society becomes more wholesome, more serene, and spiritually healthier, if it knows that its citizens have at the back of their consciousness the knowledge that not only themselves, but all their fellows, have access, when ill, to the best that medical skill can provide. But private charity and endowment, although inescapably essential at one time, cannot meet the cost of all this. If the job is to be done, the State must accept financial responsibility.

When I was engaged in formulating the main principles of the British Health Service, I had to give careful study to various proposals for financing it, and as this aspect of the scheme is a matter of anxious discussion in many other parts of the world, it may be useful if I set down the main considerations that guided my choice. In the first place, what was to be its financial relationship with National Insurance; should the Health Service be on an insurance basis? I decided against this. It had always seemed to me that a personal contributory basis was peculiarly inappropriate to a National Health Service. There is, for example, the question of the qualifying period. That is to say, so many contributions for this benefit, and so many more for additional benefits, until enough contributions are eventually paid to qualify the contributor for the full range of benefits.

In the case of health treatment this would give rise to endless anomalies, quite apart from the administrative jungle which would be created. This is already the case in countries where people insure privately for operations as distinct from hospital or vice versa. Whatever may be said for it in private insurance, it would be out of place in a national scheme. Imagine a patient lying in hospital after an operation and ruefully reflecting that if the operation had been delayed another month he would have qualified for the operation benefit. Limited benefits for limited contributions ignore the over-riding consideration that the full range of health machinery must be there in any case, independent of the patient's right of free access to it.

Where a patient claimed he could not afford treatment, an investigation would have to be made into his means, with all the personal humiliation and vexation involved. This scarcely provides the relaxed mental condition needed for a quick and full recovery. Of course there is always the right to refuse treatment to a person who cannot afford it. You can always 'pass by on the other side'. That may be sound economics. It could not be worse morals.

Some American friends tried hard to persuade me that one way out of the alleged dilemma of providing free health treatment for people able to afford to pay for it, would be to fix an income limit below which treatment would be free whilst those above must pay. This makes the worst of all worlds. It still involves proof, with disadvantages I have already described. In addition it is exposed to lying and cheating and all sorts of insidious nepotism.

And these are the least of its shortcomings. The really objectionable feature is the creation of a two standard health service, one below and one above the salt. It is merely the old British Poor Law system over again. Even if the service given is the same in both categories there will always be the suspicion in the mind of the patient that it is not so, and this again is not a healthy mental state.

The essence of a satisfactory health service is that the rich and the poor are treated alike, that poverty is not a disability, and wealth is not advantaged.

60. Working-class Standards of Living, 1950

B. Seebohm Rowntree and G. R. Lavers, *Poverty and the Welfare State* (London, Longmans, Green, 1951), pp. 76-7.

The indefatigable Rowntree collected data on York for the third time in 1950. The proportion of the working class suffering from primary poverty (for definition, see above, p. 54) had dropped from 6·82 per cent in 1936 to 2·77 per cent in 1950. In 1899 it had been 15·6 per cent (the figures are not exactly comparable). The causes of poverty had also changed: none was due to unemployment, against 28·6 per cent in 1936; only 1 per cent due to low wages, against 32·8 per cent in 1936; old age accounted for 68·1 per cent against 14·7 per cent in 1936. The proportion of the working class now in the top category E was now 58·8 per cent, and the following schedules give a picture of 'life above the minimum'.

CLASS 'E'

528 Five rooms, no bathroom. Rent and rates 9s. 11d. Man 56, wife 56. Man is bill-poster and earns £5 6s. which is total family income, giving surplus of 19s. 5d. per head. Wife says they live quietly but have no difficulties. Are able to go away for an annual holiday, spend a reasonable amount on pleasure and still save a bit for their old age.

551 Four rooms, no bathroom. Rent and rates 12s. per week. Man 47, wife 42. One child aged 7. Man is a machinist at a factory power plant and earns £7 6s. 1d. Value of school milk 9d. Value of home-grown vegetables 5s. Total income £7 11s. 10d. Surplus £1 2s. 3d. per head. Wife very cheerful and happy. No financial problems.

15 Five rooms and bathroom. Rent and rates £1 3s. 4d. per week. Man 44, wife 51. Two children aged 16 and 13. One male lodger. Man is railway worker and earns £9 5s. 3d. Family allowance 5s. Value of school milk 1s. 6d. Value of home-grown vegetables 2s. 6d. Lodger pays £2 2s. Total income £11 16s. 3d. Surplus £1 8s. 2d.

per head. Family circumstances very good. House is fitted with telephone. Elder son goes to a boarding school. Parents hope it may be possible for both children to go to a University.

2018 Five rooms and bathroom. Rent and rates 16*s.* 5*d.* per week. Man 44, wife 39. Three children 16, 12, and 4. Man is railway worker and earns £7 2*s.* 8*d.* Wife works part-time in factory for £2. Eldest child is an apprentice and contributes 15*s.* to family income. Family allowance 5*s.* Value of school milk and cheap milk 2*s.* 6*d.* Total income £10 5*s.* 2*d.* Surplus 17*s.* 10*d.* per head. Wife says she has a hard struggle to make ends meet, and they could not do so without her earnings, though her absence at work means that the child aged 4 has to be sent to a day nursery.

61. The Quality of Working-class Life

(a) The Milk-bar
R. Hoggart, *The Uses of Literacy* (London, Penguin, 1957), pp. 203-5.
(b) The Club
B. Jackson, *Working Class Community* (London, Routledge, 1968), pp. 45-7.

Richard Hoggart's famous book is a personal view rather than a sociological study of the effects of literacy on working-class ways of life. He feared that the commercial exploitation of literacy was leading to a trivialization of life, to a substitution of a 'candy floss world' for a real one. Such a world was classless, culturally, but gutless as well. The extract pinpoints one of the most vulnerable, small, working-class groups. Brian Jackson presents a more optimistic view of working-class life in the working men's clubs. Both extracts emphasize relative well-being, and the first underlines the economic importance of the teenager. The teenage consumer, whose real spending power increased rapidly after 1945, has been a potent force in several industries; for example, clothing and entertainment. Teenagers no longer copied their parents; they created their own life style, although, it is important to remember, 90 per cent of their spending has been conditioned by working-class values.

(a)

Like the cafés I described in an earlier chapter, the milk-bars indicate at once, in the nastiness of their modernistic knick-knacks, their glaring showiness, an æsthetic breakdown so complete that, in comparison with them, the layout of the living-rooms in some of the poor homes from which the customers come seems to speak of a tradition as balanced and civilised as an eighteenth-century town-house. I am not thinking of those milk-bars which are really quick-service cafés where one may have a meal more quickly than in a café with table-service. I have in mind rather the kind of milk-bar—there is one in almost every northern town with more than, say, fifteen thousand inhabitants—which has become the regular evening rendezvous for some of the young men. Girls go to some, but most of the customers are boys aged between fifteen and twenty, with drape-suits, picture ties and an American slouch. Most of them cannot afford a succession of milk-shakes, and make cups of tea serve for an hour or two whilst—and this is their main reason for coming—they put copper after copper into the mechanical record-player. About a dozen records are available at any time; a numbered button is pressed for the one wanted, which is selected from a key to titles. The records seem to be changed about once a fortnight by the hiring firm; almost all are American; almost all are 'vocals' and the styles of singing much advanced beyond what is normally heard on the Light Programme of the B.B.C. Some of the tunes are catchy; all have been doctored for presentation so that they have the kind of beat which is currently popular; much use is made of the 'hollow-cosmos' effect which echo-chamber recording gives. They are delivered with great precision and competence, and the 'nickelodeon' is allowed to blare out so that the noise would be sufficient to fill a good-sized ballroom, rather than a converted shop in the main street. The young men waggle one shoulder or stare, as desperately as Humphrey Bogart, across the tubular chairs.

(b)

Working needs not only dictate lunchtime opening. Some clubs re-open their bars as early as 4 p.m. for early evening shift workers to enjoy a quiet drink, along with a read of the papers. During the lunch hours the clubs are seldom full. The old men may be there, and perhaps someone is off work through illness. There may be a handful of local workers and perhaps a soldier or sailor home on leave. Very little drinking is done; mostly it is talking, reading the paper, playing cribbage.

A club may be dominated at lunchtime by a group of pensioners, meeting there as they do every day. But on Saturdays it is different. Many men will be off work and the place will be quite crowded, and on Sundays at this time it will be packed. Regular members attend their own club for a drink before Sunday dinner, even members who have now moved some distance away. Women are not usually allowed in the club at this time, and a gathering is created—kin, workmates, old schoolmates, neighbours—different from that on any other occasion. It cannot be like this in a pub, and in the evenings they may have their wives with them or be visiting other places, or they may have more visitors in their own club. But Sunday lunchtime is a cherished occasion. It doesn't last long. But while it does the notice may go up above the beer pumps 'No ladies will be served at this bar'. Men gamble, make wagers, drink, talk. It's common to see a small girl sitting by her father, who called in from his Sunday morning walk.

Tuesday and Wednesday evenings are quiet ones. Pay day is a long way behind. And Wednesday, end of the financial week, is called 'common sense night'. Thursday night is much busier as some trades get paid then, and Friday and Monday lead in and out of the very busy weekends. At Almondbury, a very flourishing club, we were told: 'Tuesday's our slackest night —there's about forty people in the club.'

* * *

Washing facilities matter a great deal in some clubs and attract people in, for just that purpose, on a Thursday or Friday evening. In any club a workman can have a dowse down, of course, in a way he never could in a public house. Some clubs go further. At Oakes the notice said: 'Baths available Thursday onwards' and about a dozen members took advantage each week. At Lindley Working Men's Club there are four baths. These are used on a Thursday and Friday night. Thursday night for women only—which means members' closest feminine relatives, wives, mothers, daughters, sisters. There is a charge of fourpence.

At the weekends, the clubs are at their liveliest. But it is holiday spirit rather than holiday clothes. The lounge of the Friendly and Trades Club on a Saturday evening is much noisier than the lounge of a pub would normally be. Some of the men and women are dressed up for Saturday night, but most are in their daily clothes and had taken no special trouble before coming out. One or two in flat caps, but not best flat caps; one man with a flower in his buttonhole; and upstairs three hundred people in the Concert Room.

Saturday night is Concert Night and Sunday night is Bingo Night, though one or two clubs disapprove of either one or the other, or both. Bingo accounts largely for the flourishing financial state of the clubs. They have never been richer. But drinking, and concerts, and gambling, each deserve treatment of their own.

62. Women After the War

R. M. Titmuss, *Essays on the Welfare State* (London, Allen and Unwin, 1958), pp. 90-3.

Professor Titmuss illuminates the personal reality behind social statistics on the role of women. The extract also emphasizes the characteristic socialist theme that social and, in this case, sexual equality, or nearly equality, can be so much more important than political equality. Other relevant facts about marriage are: the percentage of women who marry has been increasing since 1911; marriages take place earlier in life (52 per cent of women aged 20-4 in 1954 were married against 24 per cent in 1911); more married women work (48 per cent of all women workers were married, 3.75 million in 1955).

It would seem that the typical working-class mother of the 1890's, married in her teens or early twenties and experiencing ten pregnancies, spent about fifteen years in a state of pregnancy and in nursing a child for the first year of its life. She was tied, for this period of time, to the wheel of childbearing. Today, for the typical mother, the time so spent would be about four years. A reduction of such magnitude in only two generations in the time devoted to childbearing represents nothing less than a revolutionary enlargement of freedom for women brought about by the power to control their own fertility. This private power, what Bernard Shaw once described as the ultimate freedom, can hardly have been exercised without the consent—if not the approval—of the husband. The amount and rapidity of the change together support such a proposition. We are thus led to interpret this development as a desired change within the working-class family rather than as a revolt by women against the authority of men on the analogy of the campaign for political emancipation.

What do these changes signify in terms of 'the forward view' —the vision that mothers now have and have had about their functions in the family and in the wider society? At the beginning of this century, the expectation of life of a woman aged twenty was forty-six years. Approximately one-third of this life expectancy was to be devoted to the physiological and emotional experiences of childbearing and maternal care in infancy. Today, the expectation of life of a woman aged twenty is fifty-five years. Of this longer expectation only about 7 per cent of the years to be lived will be concerned with childbearing and maternal care in infancy.

That the children of the large working-class families of fifty years ago helped to bring each other up must have been true; no single-handed mother of seven could have hoped to give to each child the standard of care, the quantity of time, the diffusion and concentration of thought that most children receive today. In this context, it is difficult to understand what is meant by those who generalize about the 'lost' functions of parents in the rearing of children. Certainly the children themselves, and especially girls, have lost some of these functions. But despite the help that the mother had from older children she could not expect to finish with the affairs of child care until she was in the middle-fifties. Only then would the youngest child have left school. By that time too her practical help and advice would be increasingly in demand as she presided over, as the embodiment of maternal wisdom, a growing number of grandchildren. In other words, by the time the full cycle of child care had run its course the mother had only a few more years to live—an analogous situation to the biological sequence for many species in the animal world. The typical working-class mother in the industrial towns in 1900 could expect, if she survived to fifty-five, to live not much more than another twelve years by the time she reached the comparative ease, the reproductive grazing field, of the middle fifties.

The situation today is remarkably different. Even though we have extended the number of years that a child spends at school and added to the psychological and social responsibilities of motherhood by raising the cultural norms of child upbringing, most mothers have largely concluded their maternal role by the age of forty. At this age, a woman can now expect to live thirty-six years. And if we accept the verdict of Parsons and Bales, Margaret Mead and others, she has also been largely divested of her role as a grandmother by the professional experts in child care.

63. Influences on Family Size

Royal Commission on Population. *Report* (Cmd. 7695, 1949), pp. 56–8.

After a long period in which the size of family was decreasing, 'the bulge' in births during and after the war caused considerable speculation as to its causes. It is possible that whereas before 1939 babies competed with consumer goods, increasing affluence allowed couples to have more of both. The Royal Commission on Population was set up in 1946 'to examine the facts relating to the present population trends in Great Britain', 'to investigate the causes of those trends', and to make recommendations about any measures that should be taken 'in the national interest to influence the future trend of population'.

137. There is some evidence that not only during the war but during the last two decades, a change of attitude to size of family has been taking place. The extremely small family is no longer as 'fashionable' as it was in the early 1920's. Much has been written in recent years of the disadvantages of the one-child family both for child and parent, and this discussion may both record and encourage a fall in the popularity of families of this kind. The motives at work may include something more fundamental than adherence to a new fashion; a change in convention may be taking place under the stress of experience. Among the generations of people who became parents before, say, 1925 the great majority had been, as children, members of large families and knew their disadvantages. They were anxious to avert those disadvantages from themselves and their children. In recent years, for the first time, the married couples who were engaged in building up their families included a large proportion of people who knew from personal experience that very small families also have their disadvantages. In this sense, it is possible, and even probable, that there has been a reaction against the very small family.

138. Changes have also been taking place in the material factors affecting parenthood. We have described ... how the

changes since the 19th century in industrial techniques, standards of living and educational institutions served on the whole to increase the relative advantages of the very small family. It is probable that this was the general trend up to the outbreak of war in 1939. Since then, a number of new influences have come into play. On the one hand, the increasing shortage of houses has placed an additional disadvantage in the way of potential parents; and lack of domestic help may also have exerted an influence over a section of the population in the same direction.

139. On the other hand, in a number of ways there have been since 1939 developments favourable to parenthood. First there is the trend of employment; the proportion of workers unemployed, after having varied since 1921 between about 10 per cent. in good years and about 20 per cent. in bad, fell to a very low level in the early years of the war and has remained very low ever since. The heavy unemployment of the interwar period must have affected the attitude to parenthood not only of the workers who at any one moment were out of work but also of the far larger number for whom it was an ever-present threat. The larger the number of children in the family for an unemployed man, the greater was the privation suffered by the whole family; thus for people threatened with unemployment there was a powerful incentive to restrict the size of the family. During the years since 1940 in which unemployment has been extremely low, the fear that it will one day recur has not, of course, disappeared; but it has become a less pressing preoccupation, and its force as a factor discouraging parenthood has probably been much less than it was in the inter-war years.

140. Secondly, the economic developments of wartime, combined with certain features of government policy, have changed the distribution of the real income of the community to the benefit of parents. The prices of food and clothing have been held down by price control and subsidies, allied to the control of production (the Utility scheme); while money incomes in general have risen considerably. The consumption of many articles is now limited by the extent of the ration rather than, as formerly, by ability to afford the purchase. Since the right to rationed goods is on a basis of so much per head, the parent of several children is to this extent in a

position superior to the childless; and his family also benefit from the special priorities in relation to milk and other foods for mother and children. Further, in the lower and lower-medium income groups the distribution of money incomes (after payment of tax) between parents and non-parents has become less unfavourable to parents, as a result of the extension of income tax, with its system of children's allowances. to cover a far larger proportion of the population. The distribution of money income has been further changed to the advantage of parents by the introduction (August 1946) of a national system of family allowances giving 5s. a week to each child after the first in each family. Taken altogether, these measures brought about in the lower income groups a considerable change in the relative standard of real income of parents and others; and the fact that the change may not have been consciously recognised to any great extent does not mean that it may not have influenced the number of births. In the last year or so, it is true, the trend has been the other way; the abolition of the subsidies to clothing, for example, was adverse to the relative position of parents, and food prices have also been rising; but these changes have not so far undone more than a small part of what was gained before. An increase in the costs of parenthood is also implicit in the raising of the school leaving age, but this has not yet had time to exert a serious counterbalancing influence.

141. There are many other possible influences, material and psychological. The austerities of wartime, for instance, restrictions on travel and expensive pleasures, the stronger sense of community arising out of wartime perils, the diffused hopes of building a better civilisation, these and many others may have contributed. It would be rash to dismiss such imponderable considerations as of trifling importance. Moreover, the effect of the conjuncture of several different influences working in the same direction is likely to be greater than could be computed as the sum of their separate effects.

142. There still remains, however, another possible influence of a different type, which deserves closer analysis. This is the possibility that the lowering of the age at marriage may itself be a cause of larger average family size. It is an undoubted fact that the people who marry younger have larger families, on the average, than the people who marry later. This however

is not in itself conclusive. The people who marry specially young differ considerably in their distribution by occupational group, and probably also in temperament, from those who marry later. The mere fact of having married early will not, of course, have conferred on the large numbers who have done so in recent years the different social background and temperament of those who in former times married specially early. Results obtained by assuming that the relation between age at marriage and family size remains the same when the average age at marriage changes are therefore liable to be seriously misleading.

64. Retail Shopping

Co-operative Independent Commission Report (Manchester, Co-operative Union Ltd., 1958), pp. 44–7.

A retail revolution began in the 1930s and accelerated after 1945. This revolution involved the rapid growth of multiple stores (like Marks and Spencer) which competed successfully for working-class patronage against the co-operatives. The pattern of retailing, however, is still very complex, with an abundance and variety that is only now being threatened by rationalization. The supermarket, in the American sense, though on the increase, by no means dominates against the smaller independent grocery and the co-operative.

Ten years after the report from which this extract is taken, the co-operative modernized their shops, under a new symbol, and on a nation-wide scale. Their 25,000 outlets made them bigger than rivals like J. Sainsbury, Fine Fare, and Tesco. Many shops were converted to supermarkets, and the size of the movement won for them attractive price concessions from manufacturers.

A successful retail organisation needs not only enough shops, and those skilfully located; the shops themselves must be of a high quality. Rising standards have made the consumer, working-class as well as middle-class, a great deal more exacting than her predecessor of even two decades ago. She will no longer endure blowzy buildings and dowdy display. She demands clean, modern premises, an attractive window-display, bright lighting, an effective and convenient lay-out, and generally a smart and contemporary air. The more progressive multiples give the first priority to such matters; and they set a high standard for their competitors.

Some Co-operative societies have attained, if indeed they have not exceeded, these high standards. The evidence can be seen in the growing number of modern supermarkets; the new food-halls on many post-war housing estates; the new or re-built department stores in the centre of the New Towns and blitzed cities; the smaller department stores which some societies are building in the growing suburban shopping centres; and the many imaginative renovations of older premises. All these deserve high praise, and achieve a standard exceeded by no other retail organisation.

But we have found the picture to be a terribly patchy one, and the fine new or renovated shops to be only a small proportion of the whole. If we ask what is the 'image' of a Co-operative shop in the public mind, the answer will not be a supermarket or new department store. It is more likely to be a ponderous, unrestored, and unimaginative grocery-cum-butchery-cum-drapery cluster, built in the early 1900s, still operating counter-service, the window-display old-fashioned, the exterior clumsy and badly in need of paint, the interior frowsy and unattractive. This is no doubt unfair. Yet there are still far too many such Co-operative premises, especially in parts of Scotland and of the North of England; and they give the whole Movement a name for backwardness and drabness.

There is, of course, an explanation. The Co-operative Movement is now paying the penalty for having been the first of the large-scale retailers. Having covered the country with shops before ever the multiples came on the scene, it now has a larger legacy of obsolete premises than its younger rivals. This is the price of having started first....

Moreover, excessive financial conservatism may prove a suicidal policy in the face of the younger generation of consumers, who, perhaps more discriminating and fastidious than their parents, demand a much higher standard of shop and service. If they do not find it in the Co-operative Movement, neither loyalty nor dividend will in the end prevent them from deserting to competitors....

A recent inquiry into the pattern of sales of women's and girls' clothing in urban areas shows that 74% of the clothing sold by Co-operative shops is for women aged over 30, and only 11% for women aged between 15 and 29. (The remainder is accounted for by clothing for girls aged 5 to 14.) In the case of the multiple shop firms, 52% of the clothing sold is for the over-30s, and 35% for the 15–29 age group; for department stores the respective figures are 64% and 27%. The same survey estimates that the expenditure per head on clothing by women aged 15–29 is 66% higher than the expenditure per head by women over 30. The Co-operative bias towards the older woman shopper, and the relative failure to attract the heavy-spending teenage and younger woman, stand out all too clearly.

65. Poverty in the Sixties

K. Coates and R. Silburn, *Poverty: the Forgotten Englishmen* (London, Penguin, 1970), pp. 47–55.

In the 1950s there was widespread belief that poverty had been banished from Britain by the Labour government's welfare achievements, redistributive taxation, and rising affluence. In the long-term history of working-class standards of living, there was much truth in this picture. The definition of poverty was changing, from the vague 'subsistence' (Rowntree and Beveridge) to a wage rate in the range of up to 40 per cent above the basic rates of payment made by the Supplementary Benefits Commission (the old National Assistance Board). On this definition, in 1960 still 14 per cent of the total population of Britain was living in poverty. Coates and Silburn investigated the St. Ann's area of a prosperous city, Nottingham, to find that the poor tended to live in ghettoes: of 413 households investigated, 156 (37 per cent) were in poverty; and of 1,395 people, 547 (40 per cent). The figures for the whole country had been estimated at 17·9 per cent of households and 14·2 per cent of population living in poverty.

Firstly, how many people in this one community are poor by modern standards, and what proportion of the community do they comprise? Secondly, do these people fall into any particular categories which might make one set of remedies more appropriate than another? The largest group of poor households we discovered in our Nottingham inquiries was that of the retirement pensioners... Indeed, because most of the aged were living alone, the actual number of people in these households was relatively small, and only 15 per cent of our total poor population fall into this category. Although many of these elderly people were among the very poorest we encountered, particularly if they were dependent upon the basic old age pension and did not apply for the supplementary pension to which most would have been entitled, any further generalization about their circumstances is extraordinarily difficult. . . .

The second largest category of poor households we interviewed included by far the largest number of persons (accounting for 50 per cent of the poorer population). This consisted of those families with breadwinners at work whose incomes did not entitle them to receive public relief . . . there are two elements here, which alone or in combination cause poverty. The first is found where the breadwinner's wage is very small, too small perhaps to support even the smallest family. The second is seen when the numbers of dependants in the family stretch the income further than it can go. Of course, both elements combine in the case of a large family trying to live on a very small income. We interviewed people who fell into each of these groups.

It is a commonplace that dependent children impose some strain on almost every family budget; at the same time, much of the discussion about family poverty has seemed to assume that only the very largest families are normally seriously affected. Quite obviously, the more dependent children there are in a family, the more likely it is that the family will be in poverty. However, the fact is that this risk becomes significant at a very much earlier stage in a family's growth than many people seem to realize. The argument is not only about the families of eight, nine, or ten children, but in a depressingly large number of cases, it is about the families with only two or three children. Thus, in St Ann's, nearly one in five of the one-child families, over a third of the two-child families, and approaching half the three-child families, were in poverty. This poverty became more aggravated as the family grew larger, so that all the families with seven or eight children were in poverty, as were four of the five families with six children. On the other hand it can be stressed that of the 214 families with children who were interviewed, only 24 had five or more children, and of these 24, 8 were not in poverty. In brief, the important discussion about family poverty is *not* primarily about 'problem families' of feckless breeders; consequently remedies in terms of birth-control, sterilization or other even zanier and less humane eugenic recommendations are not of great relevance. . . .

These facts about family poverty should encourage the rest of us to examine more closely the question of working-class wage levels. What is quite evident is that, even in a city as relatively prosperous as Nottingham, there are large groups of workers whose basic wages are extremely low: many of these men depend upon crucifyingly long hours of regular overtime to secure themselves a decent income. The size and character of the problems of the low-paid worker will be examined more fully later; at this point it suffices to say quite baldly that the most important single cause of poverty is not indolence, nor fecundity, nor sickness, nor even unemployment, nor villainy of any kind but is, quite simply, low wages.

Obviously, it is the category of low-paid workers, or workers with large families and inadequate wages, which is most prone to feel the effects of what Rowntree described as the 'cycle of poverty', in which families find themselves, at different points in their lives, above or below the poverty line. But we were compelled to notice another important contributive element to this cycle, when we began to scrutinize the data we collected about wage levels in Nottingham.

Almost two-fifths of the people employed in Nottingham are women and girls, as compared with a proportion of slightly over one third, nationally. Nottingham is a centre of light engineering, in which far larger proportions of women workers are normally employed than in heavy industry. It has a thriving pharmaceutical industry, where thousands of girls find jobs. Although mechanization has changed the constitution of the labour force in the tobacco industry, a high proportion of the workers at Players are women, and indeed a few years ago, half the people working there were women and girls. Traditionally, of course, Nottingham was a centre of the lace trade, which employed many thousands of women. As the fashion for lace faded out, so the hosiery industry, and new factories for the making-up of garments, expanded, absorbing the girls who were displaced from lace-making. Ericsson's telephones, now part of the Plessey group, employs more than 5,000 workers, a quarter of whom are women. The transport, distributive, banking and educational services, together with public utilities, employ two fifths of the workers in the Nottingham area, and half these are women. There is thus a constant and unremitting shortage of female labour.[4]

The effect of the numerous job opportunities for women and girls is, of course, to augment family earnings wherever women are able to hold jobs at the same time as their menfolk. This fact accounts, in great measure, for much of the apparent 'prosperity' of Nottingham. But it also implies something else. Wherever women workers are employed in large numbers, their presence has a tendency to depress the levels of earnings of the men working alongside them. Where a family has two breadwinners instead of one, it will, of course, normally be better off, even if both earn individually somewhat less than a single breadwinner might command elsewhere. But if it becomes the norm to rely on two sets of earnings, there is an additional dimension to add to Rowntree's cycle. Quite simply, during all the time that women are rearing small children, their inability to work will, whenever their husbands are low paid, thrust the young family either below, or perilously close to, the poverty line. That they will rise above it again after the children are old enough to 'look after themselves' (at ages which will vary greatly according to the degree of stress under which the family finds itself, as well as the cultural influences under which it lives) does not mitigate the fact that the degree of insecurity in such conditions is greater than that in areas where men's wages are cushioned from the pressure of female competition. Of course, the answer to this problem is, quite simply, equal pay for equal work. This will overcome the adverse pressures on men's earning power at the same time that it brings a greater element of justice into the working conditions of women. But until this simple gain has been registered, 'prosperity' will continue to be replaced by periods of great stringency, whenever and wherever young married women from such areas become mothers.

The third general category of poor people consists of families without a male breadwinner. In our sample we encountered twenty-one fatherless families. The greatest unhappiness

[4] See F. A. Wells, 'Industrial Structure' in *Nottingham and its Region*, British Association, 1966.

and hopelessness can be found in this group. Premature widowhood, or, more frequently, desertion, leave a mother alone to support dependent children. For these mothers, there is not only the emotional anguish over losing their husbands (and let it also be said that in some cases such a loss may bring a certain emotional relief), but all the brute practical difficulties of earning enough money to keep the family. This is more complicated than merely finding a job; it is often far more difficult to find someone to look after the children while their mother is out at work. If she has relatives near by then maybe the family can rally round and suitable arrangements can be made. But for the mother who is quite alone, the situation is much more overwhelming, and usually she is obliged to rely upon help from the Ministry of Social Security. Here again, we must beware of generalizations, for while it is true that many of these families are considerably distressed and even severely demoralized, we did encounter some very much more positive cases. One in particular sticks in mind, of a woman who, although apparently quite alone and friendless, rejoiced in her anonymity; she had, she told us, moved into St Ann's precisely because she knew no one there and, despite her difficulties, she obviously relished her freedom from what had been a particularly unhappy and complicated family situation. The most disturbing fact about this category of poor families is the large number of children in them; of the 94 individuals we found in this situation, 70 were children. It can only be supposed that they are suffering not only the pain of a broken home, but also all the other deprivations that poverty imposes.

We found very few households in which poverty was caused by sickness, the fourth general category with which we were concerned. The numbers of people involved amounted to only 5 per cent of the poor population. . . .

The fifth broad grouping of poor people consists of those on the dole. In St Ann's, 13 per cent of the poor population lived in households in which the breadwinner was unemployed.

66. Educational Needs

(a) *The Need for Scientists*
Scientific Man Power. Report of a Committee appointed by The Lord President of the Council (Cmd. 6824, 1945), pp. 6–9.

This was the first of a long series of reports looking to particular needs for trained people. The Committee included Sir Edward Appleton, Professor Blackett, and C. P. Snow. The war had proved to be a scientists' war; the problems of economic reconstruction and growth were also to be solved only with the help of the scientists.

(b) *Education in the Sixties*
The Newsom Report. Half our Future (London, H.M.S.O., 1963), pp. 4–5.

A series of reports set out the achievements and aims of education in England and Wales: the Crowther Report (1959) investigated the education of the age group sixteen to eighteen; the Newsom Report, that of the average pupil in secondary schools. Both reports recommended a school leaving age of sixteen, and already the proportion of children staying at school after fifteen was increasing. especially in the south and east. The Robbins Report (1963) recommended a large increase in higher education and the establishment of new universities.

(a)

12. We know that there were 45,000 scientists registered on the Ministry of Labour's Central (Technical and Scientific) Register at the end of 1945, and our existing capital must therefore be somewhere between 45,000 and 60,000. The Ministry of Labour inform us that, to the best of their belief, the Register covers between 80 per cent. and 85 per cent. of working scientists, and, in all the circumstances, we have come to the conclusion that it is unlikely that the nation has at its disposal to-day a force of more than 55,000 qualified scientists.

13. The assessment of future demand is no less problematical. We have studied the available results of a number of recent enquiries, among them one carried out by the Industrial Research Committee of the Federation of British Industries on

the probable demand for research and development workers in industry in the post-war years. As a result we have arrived at a figure of approximately 70,000 as the estimated minimum demand for scientific workers in this country and in the colonial service in 1950. Of this total roughly 30,000 represents teachers in the Universities and secondary schools.

15. It is only to the Universities that we can look for any substantial recruitment to the ranks of qualified scientists. The proportion that has come from other sources in the past is very small indeed and we do not favour any attempt to add a responsibility for producing a substantial number of pure scientists to the existing and prospective burdens of the Technical Colleges. Generally speaking, the university is an essential stage in a scientist's education and in any event the Technical Colleges will be hard put to it to produce the number of technologists that are required to support and apply the work of the scientists.

16. Before the war the British Universities were turning out on the average some 2,500 scientists each year. The rate of output fell in the late 30's but recovered during the war and the science faculties are now practically full. On the assumption that the faculties continued to be full but did not expand their output, they would turn out perhaps 12,500 scientists during the next five years.

24. An expansion in the output of qualified scientists involves problems that are to a great extent common to all faculties faced with a demand for a substantially increased output of graduates. It is hardly open to doubt that many faculties will find themselves in this position and, as we have no desire that science should receive exceptionally favourable treatment, we have, in the following paragraphs, spoken where it seemed to us appropriate, in terms of the development of the Universities as a whole.

(i) *The Talent Available*

25. We need to form an estimate of the proportion of the population that is inherently fitted to benefit from a university education. We attach very great importance to this question, as whatever happens, the quality of our university graduates must not be sacrificed to quantity. In few other fields are numbers

of so little value compared to quality properly developed. Character, temperament and wider qualities of mind are, of course, as important as intellectual acuity and the test of fitness for the Universities is not intelligence alone. Moreover, before it enters the University, intelligence must be trained and the associated personal qualities matured to a standard that we would not wish to see lowered. (In para. 39 below we deal with the effect of this factor on the supply of qualified entrants to the Universities during the next few years.)

26. We have surveyed the results obtained in recent years on the distribution of intelligence, as measured by 'intelligence tests', among the whole population and among samples of the members of certain Universities. We are encouraged to consider these results to be fairly reliable, especially in view of the wide experience gained during the war in the use of intelligence tests by the Services. The following results are especially relevant to our enquiry:

At present rather less than 2 per cent. of the population reach the Universities. About 5 per cent. of the whole population show, on test, an intelligence as great as the upper half of the students, who amount to 1 per cent. of the population. We conclude, therefore, that only about one in five of the boys and girls, who have intelligence equal to that of the best half of the University students, actually reach the Universities. It cannot be assumed that all of these have the other innate capacities necessary to a university career. It must be allowed, indeed, that many boys and girls of high intelligence would not desire a university career; yet there is clearly an ample reserve of intelligence in the country to allow both a doubling of the University numbers and at the same time a raising of standards.

27. There is also evidence that the great majority of the intelligent persons, who do not reach the Universities, are ex-pupils of the elementary schools. If university education were open to all on the basis of measured intelligence alone, about 80 per cent. would be expected to come from those children who started their education in the public elementary school and only 20 per cent. from those whose education had been in independent schools. In fact, at the present time only about 40 per cent. of university entrants are ex-pupils of elementary schools, whereas 60 per cent. are from independent schools.

Thus, among university entrants, elementary school pupils are only half those to be expected, and those from independent schools about three times as many as expected.

These figures, rough and incomplete as they are, do make clear that a high proportion of the reserve of potentially able students comes from families that are unable to afford the cost of higher education. If, therefore, it is not to be lost to the Universities, greatly increased financial assistance both at the secondary school and at the university level is essential.

28. We conclude, therefore, that there is available in our population a large reserve of innate intelligence and that, even allowing for the other factors, there are more potential graduates than we could hope to take into our universities by any degree of expansion practicable within the next ten years. This would still be true if the standard of intelligence at which the individual is regarded as suitable for a university education were materially higher than it is to-day.

(b)

7. Another fact, perhaps not often enough emphasized, is that the standard indicated by 'average' is rising all the time, and perhaps never more rapidly than in the last twenty-five years. As the life of our society becomes more complex, new demands are continually made on all of us; and this is as true in relation to our personal lives as it is in relation to the changing economic life of the country as a whole. The amount which men and women need to understand, and the range of experiences with which they are required to deal in all the daily business of living, are continually extending. The mysteries of one generation become the commonplaces of life to their grandchildren. In this sense standards do rise.

8. This is not often apparent, because we are seldom in a position to compare, directly, the achievements of pupils of one generation with those of another. In a later part of this report, in which the results are described of a series of tests designed to show the pupils' capacity to read with understanding, there is a clear record of improvement. A test score which even fourteen years ago would have been good enough to put boys or girls well into the above-average category would today put them firmly into the below-average group. Over the intervening years the general level of performance has risen. One

of the reasons why there is a quite proper anxiety over the general standards of literacy today is not that fewer and fewer people can read and write, but that more and more people need to do so with greater competence.

9. The point is, could many people, with the right educational help, achieve still more? If they could, then in human justice and in economic self-interest we ought, as a country, to provide that help. Any substantial recommendations affecting provision for half the population are bound to cost money. Are we prepared to foot the bill? We are conscious that, although there is a strong body of public opinion urging expenditure on education as a vital investment, the emphasis at present is almost invariably on the higher education of the most gifted. And with the prospect of a steady, long-term increase in the child population, the cost even of maintaining the existing services is mounting so rapidly that the competition for educational priorities is acute. We therefore think it essential to state at the outset the economic argument for investment in our pupils.

10. Briefly, it is that the future pattern of employment in this country will require a much larger pool of talent than is at present available; and that at least a substantial proportion of the 'average' and 'below average' pupils are sufficiently educable to supply that additional talent. The need is not only for more skilled workers to fill existing jobs, but also for a generally better educated and intelligently adaptable labour force to meet new demands.

11. In spite of popular belief to the contrary, technological advance—especially the introduction of automatic processes—is not leading to widespread unemployment among skilled workers or to the destruction of the level of skill. Skills may be changing and some individual skills become less important while new ones emerge, but the forecast made in 1956 in a government report on 'Automation', that on the whole the level of skill will tend to rise rather than fall, is being fulfilled. If anything, the progress of automation and the application of other technological developments are likely to be delayed by lack of trained personnel.

67. Racial Discrimination

W. W. Daniel, *Racial Discrimination in England* (London, Penguin, 1967), pp. 46–9, 86–9.

Racial discrimination is a recent problem. Before 1950 the non-European population was confined mainly to a few ports and there largely to the dock areas. Between 1955 and 1957, however, yearly coloured immigration was over 130,000, mainly from the West Indies; by 1960, significant numbers of Asians were also arriving. In 1960 the estimated total coloured population had reached one million (2 per cent of Britain's population): 525,000 West Indians, 200,000 Indians, 125,000 Pakistanis, and 150,000 others (mostly African); a third of the total were aged under 16, and more than half were British born. Until 1962 Commonwealth immigrants generally had unrestricted entry to Britain for themselves and their dependants, with the right to British citizenship after five years. In 1962 the government restricted the number of entrants, but not of the dependants of those already here. Though the numbers of voucher holders admitted under the Act (tightened up in 1965) was reduced to 8,500, considerable numbers of dependants continued to come. Enoch Powell, Conservative M.P., sparked off a fierce national debate by suggesting the virtual ending of further immigration, and government aid for the voluntary repatriation of immigrants already in Britain.

(*a*) IMMIGRANTS' VIEWS

In the one area where specific claims were checked by tests they were validated in 90 per cent of the cases. In all other cases where there is independent evidence of discrimination, particularly in the findings of the situation tests, the extent of discrimination claimed by coloured immigrants is shown to be substantially less than is indicated by the independent evidence. In fact the claims of the coloured people understate the extent of the problem. The experience of a Hungarian compared to the coloured person in all our situation tests demonstrates conclusively that the white immigrant experiences very much less discrimination than the coloured person.

Moreover, a striking element that does not emerge from the summary of categorized responses, but does emerge from a reading of individual responses, is the dramatic and often humiliating quality of the experiences on which they are based: from people moving from their seat when they sat next to immigrants on the bus, or putting change on the counter rather than into their hand or pulling white children away from their children in church; to verbal or physical violence in the streets or racialist slogans on the wall; to painful cases of discrimination in the basic requirements of life, that is to say work and homes, the evidence of which is developed in subsequent chapters.

Some of the quality of the experiences that had given rise to the beliefs described in this chapter are illustrated by the following extracts from people's replies.

Instances of personal abuse in everyday life made the greatest subjective impact:

> You see it every day. People say things 'You should be sent back home on a banana boat'. People's attitudes. Rude, unfriendly names like 'black bastards', 'niggers', and swearing.

> Because I am black people say all sorts of things like 'go back to the zoo'. They don't believe we were born by human beings. They believe we spring from monkeys.

> Motorists on the road swears at you with venom, 'black bastards; why don't you go back to your jungle'.

> On some parts of the street you can see written 'Go home Blacks'. On the buses you can hear them say the black niggers come and fill up all the bus.

> Even in the buses they won't sit against you. They stand up or move away. They don't want to live near us and when we move in they move away. They don't speak to us.

> I was hurt in a smash. The police came but refused to take me to hospital in his car, saying 'I won't take this coloured in my car'. People look at you on the roadside saying 'blackie bastard'.

Even in church or at sport they said they were not free from such humiliations:

> I thought the people would be more Christian-like. They way they treat us is not Christian; even in church they pull their children away from ours.

You can't associate with whites even in church. This is not Christian.

When playing cricket whites don't want to use the bath after you've used it.

Particularly hurtful was the way some white people were friendly in certain circumstances but hostile in others:

As far as the people are concerned they are friendly with the immigrants only when they work together but as soon as work is finished so is the friendship. They never notice you when they're with their kind.

Neighbours help you but can't speak to you in the street, afraid of what other whites will think.

Some of the Asians' more special problems can be illustrated thus:

Here the Indians are considered coloured. I thought coloured people were only Negroes. When I was in U.S.A. I was not considered as a coloured man. I think there is more colour prejudice in England than the U.S.A.

When I came to this country I used to wear a turban which is a religious custom among us, the Sikhs. I had great difficulty in getting a job and somewhere to live. My friends then advise me to take off the turban. This means I also have to cut my hair short. This is considered one of the greatest sins among our people and then one is not recognized as a Sikh. I did take my friends' advice and things did improve to some extent. But I am unhappy and cannot face my parents and my family in India.

English people still consider me as a foreigner and a coloured.

I am astonished by the immorality that surrounds me in this country. There is not a place for a good moslem to live in.

(b) British Views

Every informant mentioned language as the largest single difficulty connected with the employment of immigrant workers.

Local employers stated that although fluent English was not always essential in manual and semi-skilled work, the poor English of most immigrants impeded training or promotion and often made it necessary to engage an interpreter or use another member of their staff as one. Seventy-eight local employers stated specifically that they had to reject the majority of coloured applicants as unsuitable for employment because of their inadequate English and 103 employers claimed that the great proportion of those immigrants they did engage were suited only for the most straightforward and uncomplicated forms of labouring and manual work.

The representatives of employment exchanges and private employment bureaux also made this point at some length, and all of them said that because of lack of fluency, most immigrants seeking employment were, inevitably, compelled to seek positions at the unskilled or semi-skilled level. This applied, according to the private bureaux, to a high percentage of the coloured people who applied through them for general office posts, and who were unsuited for this type of work for no other reason, as one put it, than that 'they just cannot grasp what is going on because their English is just not up to standard'. Similarly trade union representatives stated that most of the difficulties arising between immigrants and employers, or immigrants and other workers, were due to the 'language problem', or the more general problem of a breakdown in communication between two sets of people who did not understand each other's customs, culture or standards.

The second difficulty, again widely referred to by all groups of informants, was that a high proportion of those coloured immigrants seeking employment either over-claimed their qualifications and experience or, alternatively, had learned their trade abroad and, although trained by their own standards, still fell short of the requirements demanded for employment in Britain. This lack of qualification and disparity of training standards as between Britain and Commonwealth countries often made it necessary to employ the immigrant at a level below that at which he felt he was qualified, or else impossible to employ him at all. This point was made with regard to both skilled technicians and professionally qualified people, and both employers and private bureaux gave examples of Indians with degrees from Indian universities being 'suited only for cost account jobs' because their qualifications were considerably below 'the standards set and expected in this country'.

The occupational characteristics more closely related to personal qualities, and the personal qualities themselves, are best illustrated by the following quotations:

> They never stick at anything. They move around—they're a sort of floating population. They don't seem to be very interested in permanency. I think a lot of their problems stem from this—employers have learned not to trust them.

> It's a well-known fact that they rarely stick at a job. They aren't made that way.

> They're all very casual—even off-hand. They seem unable to grasp the important things in life.

> They never stick at anything. They're always on the move.

> It takes a long time to get through to them, you know. They're not very quick at picking things up. You have to explain everything four or five times. Some are plain thick.

> They have very little ambition. Oh, they'll work hard on piece work but they don't want to better themselves jobwise.

> Some employers won't take them because they say they're slow to learn.

> They say they don't adapt.

> They're lazy. It's the tropical sun, it's in their blood. That makes them that way.

> They are lazy, arrogant, shifty, difficult and dirty. You can only get a day's work out of them if you stand over them.

> They are bone idle. You have to chase them all the time, keep them at it. They're insolent too.

> They are lazy, bone lazy. One Jamaican told me yesterday that he wouldn't move the barrels because it made his watch vibrate. They always have an excuse.

> A common reason is that they smell.

Cleanliness is another thing. Some can stink—I mean really stink. Workers object to being alongside them—it leads to all sorts of friction.

I have a lot of complaints about hygiene and things like that. It's a very tricky one to sort out and I don't like being involved in it. But it happens so often that you can't ignore it. I usually have a talk to the immigrants and tell them all about hot water and Lifebuoy soap.

I told one lot that cleanliness was next to Godliness. This upset them more than ever. They were Hindus. They accused me of trying to convert them.

They spit all over the place.

They spit a lot. This goes for all coloured people. It must be their tropical upbringing.

Their eating habits can be disgusting. This doesn't apply to all of them but a lot really can be quite foul at the table. A lot of other workers object to it, there is no doubt about it. I'm forever getting involved in that sort of thing. Employers mention it too. I have a talk with the immigrants and tell them how they are expected to behave. It doesn't always prove successful.

So many of them are rubbish.

They're always up to something, particularly the Jamaicans. If it isn't pilfering it's peddling dirty pictures. They're on to everything.

And then there's this sex thing ... it's psychological but it's no use pretending it's not there. Most employers are men and they have the idea that coloured men are more virile—so they're hostile to them.

The only time they're sharp is when they're thinking up the latest dodge. Apart from that they're either asleep or looking for sex.

They're sex mad, you know. We're always having trouble with that. You find them at it all over the place. You get frightened to open a door.

Lots of them try and get the girls to go on the streets. That's how they make their money. They only use the job as a front.

Two of them ran a brothel in one of our sheds.

We get bother because so many of them are thin-skinned —they have chips as big as Town Halls on their shoulders. Every time they are told off, they're convinced it's because they're coloured. It never occurs to them that white chaps get told off too.

68. The Urban Environment

Problems Created by Modern Technology. Town and Country Planning in Britain, Central Office of Information Reference Pamphlet No. 9. (London, H.M.S.O., 1968), pp. 25-7.

One of the major problems created by modern technology is that of coming to terms with the motor car. More people live in towns; more people own cars; more goods are carried by motor vehicles. The result is a major problem of relating traffic flows to the existing and planned road networks, and, at the same time, of preserving the town as a reasonable place of living.

The rapid increase in car ownership in recent years has been one of the most important social changes affecting town and country planning. In 1966 there were 9·5 million cars in Britain, compared with a total of 3·9 million in 1956. Nearly 70 per cent of all passenger-miles travelled in 1965 were in private motor vehicles—an increase of some 28 per cent over ten years. By 1980 car ownership is expected to have more than doubled its mid-1960 total. The 1966 sample census indicated that 45 per cent of households in Great Britain had a car: each household will own an estimated 1·3 cars by 2010. This situation has brought increased mobility and enjoyment to many people but has also engendered many problems, particularly in a congested country like Britain where the extent of unspoilt countryside is limited and where towns were not designed to accommodate so many vehicles. The growth in motor traffic has been described as the most disruptive force that has ever assailed British towns in peacetime.

About 90 per cent of the British people live in towns: half of them live and work in the seven great conurbations around London, Manchester, Birmingham, Glasgow, Leeds, Liverpool and Newcastle. In recent years people have moved from the inner areas of cities such as London and Glasgow to live in suburban areas on the outskirts: they commute to work in the central areas each day either by public or private transport.

The Buchanan report, *Traffic in Towns*, published in 1963 was a comprehensive study of the impact of a large and rapid increase in the growth of traffic and its probable effect on British towns by the end of the century. It set out some basic principles by which the efficient movement of traffic might be reconciled with the continuance of civilised urban living conditions. On the one hand, a primary network of main roads should be developed to carry vehicles between towns and other areas of development and on which traffic needs should take precedence. On the other hand Professor Buchanan maintains that in certain commercial, industrial or residential areas traffic should be subordinate to the needs of people living and working in them. These 'environmental' areas should be kept free of heavy traffic and general congestion: only vehicles having business in the areas should be allowed to enter them. The report shows that if the accommodation of a great deal of traffic and good environmental conditions are desired in urban areas at the same time, the cost of redevelopment is likely to be extremely expensive. Each town has its individual problems and decisions to make regarding (a) the capacity of its existing road system (which in most towns is inadequate for the likely traffic demand); (b) the scale of investment which can be undertaken for its extension and improvement to ensure the efficient movement of traffic on a large scale; and (c) the value a town's inhabitants place on such amenities as relative freedom from noise, fumes and visual intrusion in historic streets and squares or the element of danger and anxiety in busy shopping centres resulting from increased traffic.

Over the next few years, ... it is estimated that vehicular growth is likely to outstrip the rate at which urban road systems can be improved and extended. To make the best use of roads in congested areas local authorities are encouraged to draw up comprehensive schemes for traffic management. An estimated 68,000 cars enter central London each week-day carrying people to work, a high proportion of which are parked at the roadside, and a similar, although rather less acute, situation exists in other large cities. One-way streets, clearways and other devices have been introduced in many areas to improve the flow of traffic. Several experiments have been made, for example, in delivering goods to shops outside working hours to avoid the dislocation of traffic in town centres. An experiment is being undertaken in Glasgow to control traffic flow by automatic signals operated from a central point by computers. Street parking is often controlled by meters or other means which regulate short-term parking for shopping or business trips and which prohibit parking on busy roads. Off-street car parks are also being built. Methods of road pricing—charging directly for the use of roads according to the time spent and distance travelled on them—are being investigated. A planning bulletin has been issued giving advice on parking provision in town centres.... A number of towns, including Leeds, Norwich and Portsmouth, have schemes which aim to free streets from domination by traffic. In Coventry, the Barbican development in the City of London and the Bull Ring at Birmingham, traffic-free precincts have been established.

The Government is encouraging the improvement and expansion of public transport services by road and rail to relieve the congestion in cities, particularly that caused by commuters. In a White Paper, *Public Transport and Traffic*, it proposes that local passenger transport authorities be formed to draw up integrated plans for public bus and rail services and to manage them in their areas. Exchequer grants would be available, for the modernisation of public transport. A joint study by the Ministry of Transport and Manchester Corporation has been made into the comparative cost and environmental considerations involved in building and operating several rapid transit systems, including a monorail system. In the cities of Newcastle and Leeds traffic lanes have been reserved experimentally for buses in an attempt to increase the speed and reliability of the public service.

69.   The Aberfan Disaster

Report of the Tribunal Appointed to Inquire into the Disaster at Aberfan on October 21st, 1966 (London, H.M.S.O., 1967), pp. 26–8.

Not only does an industrial society create continuing environmental problems, it inherits massive problems from an earlier age when little or no thought was given to the problems of pollution. The Aberfan tragedy was of such magnitude that it marked a turning-point in public consciousness of the problems of the environment.

WHAT HAPPENED AT ABERFAN?

49. At about 9.15 a.m. on Friday, October 21st 1966, many thousands of tons of colliery rubbish swept swiftly and with a jet-like roar down the side of the Merthyr Mountain which forms the western flank of the coal-mining village of Aberfan. This massive breakaway from a vast tip overwhelmed in its course the two Hafod-Tanglwys-Uchaf farm cottages on the mountainside and killed their occupants. It crossed the disused canal and surmounted the railway embankment. It engulfed and destroyed a school and eighteen houses and damaged another school and other dwellings in the village before its onward flow substantially ceased. Then, in the words of the Attorney-General:

> 'With commendable speed, the work of attacking this seemingly ever-moving slimy, wet mass began as people strove to release the afflicted. Essential services were brought to the village and there began the unprecedented and Herculean task of recovery. People came in their hundreds from far and wide to lend their hands, whilst from the local collieries there hurried the officials and the sturdy experienced colliers to use their strength and skill as never before.'

But despite the desperate and heroically sustained efforts of so many of all ages and occupations who rushed to Aberfan from far and wide, after 11 a.m. on that fateful day nobody buried by the slide was rescued alive. In the disaster no less than 144 men, women and children lost their lives. 116 of the victims were children, most of them between the ages of 7 and 10, 109 of them perishing inside the Pantglas Junior School. Of the 28 adults who died, 5 were teachers in that school. In addition, 29 children and 6 adults were injured, some of them seriously. 16 houses were damaged by sludge, 60 houses had to be evacuated, others were unavoidably damaged in the course of the rescue operations, and a number of motor cars were crushed by the initial fall. According to Professor Bishop, in the final slip some 140,000 cubic yards of rubbish were deposited on the lower slopes of the mountainside and in the village of Aberfan, whilst the amount actually crossing the embankment is estimated, very approximately, to have been about 50,000 cubic yards.

50. The disaster tip—No. 7 in the series forming the tip complex on the Merthyr Mountain—contained a vast quantity of loosely-tipped, uncompacted material which had become over the years saturated with water in its lower parts. We must later consider in detail the march of events and the views advanced by the various expert witnesses, but it may be helpful at this early stage to quote the clear description of what happened given by one of the civil engineers—Mr. G. M. J. Williams—who so greatly assisted the Tribunal:

> 'On the morning of the 21st October, 1966, there were several movements within the tip ... and shortly after 9 o'clock there was apparently another movement, but in this case it took part of the saturated material past the point where liquefaction occurred. This initially liquefied material began to move rapidly, releasing energy which liquefied the rest of the saturated portion of the tip, and almost instantaneously the nature of the saturated lower parts of Tip No. 7 was changed from that of a solid to that of a heavy liquid of a density of approximately twice that of water. This was "the dark glistening wave" which several witnesses saw burst from the bottom of the tip. The upper part of the tip, not being saturated, did not liquefy but some of it would be carried forward floating on the liquefied material, whilst the rest dropped into the hole that had been left in the face of the tip. Being of the nature of a liquid the

whole mass then moved very rapidly down the hillside, spreading out sideways into a layer of substantially uniform thickness. As this happened, water was escaping from the mass so that the particles of soil regained their contact and the soil mass returned to its solid nature. Unfortunately, a part of the liquefied mass flowed over the Railway embankment and destroyed the schools and the houses in Moy Road.'

51. The day had begun windless and sunny, except for a belt of mist which filled the lower parts of the valley and prevented people in the village from seeing the tips on the mountainside. The men working at the top of the tip were above the mist and could see, projecting from it, the top of the colliery stack. Lessons at the junior school began at 9 o'clock. But at the senior school lessons did not begin until half-past nine, so that while the younger children were already engaged at their studies the older ones were making their way towards school. At about 9.15 a sound was heard, variously described as being like thunder, or of a low-flying jet plane or as though loose trams were running down an incline. Howard Rees, a pupil at the senior school, was making his way up the Moy Road towards the school when he saw what he described as a big wave of muck higher than a house coming over the old railway embankment and heading straight towards him. In this muck he could see boulders, trees, trams, bricks, slurry and water. To him it appeared to be 'moving fast, as fast as a car goes in a town' and to be accompanied 'by a rumbling noise like an old train going along'. This moving mass hurled itself against the two schools and the houses which stood between them. Outside one of those houses three of Howard's schoolfellows were sitting on a wall and he saw them buried, crushed and killed. He and some other boys who were on their way to school were fortunate to escape with their lives, though some of them were struck by flying pieces of material.

52. At this same time Mr. George Williams, a hairdresser, was making his way to his shop in Moy Road when he heard a noise 'like a jet plane' which he thought came from the tip. He could see nothing at that moment because of the fog, but soon saw the windows and doors of Moy Road houses crashing in '... like a pile of dominoes coming down'. He too was struck

by flying material and would have been buried had he not been protected by a piece of corrugated sheeting which acted as a shield to him until he was rescued by some Council workers. In a memorable phrase he told the Tribunal that after the tip fell everything went suddenly quiet, just like turning off the wireless, and he added, 'In that silence you couldn't hear a bird or a child.'

53. At the moment when disaster struck, Mr. Kenneth Davies, acting-headmaster of the senior school, the Pantglas County Secondary School, was getting ready to receive his staff and pupils for the lessons which were to begin at 9.30 a.m. His description of the terrifying events must be quoted. He was in the assembly-hall when:

> '... I heard a sound which to me appeared to be a jet plane screaming low over the school in the fog. Immediately following that there was a bang and the part of the school I was in shook, and some girl-pupils ... came running and screaming into the hall ... When passing the Needlework Room I noticed that the furthest corner had collapsed and the roof had started to collapse into the room as well... The Girls' Entrance was approximately two-thirds to three-quarters full of rubble and waste material ... I climbed onto the rubble in the doorway ... I was still looking for this plane, and when I looked directly in front of me ... I saw that the houses in Moy Road had vanished in a mass of tip-waste material and that the Junior School gable-ends, or part of the roof, were sticking up out of this morass. I looked down to my right and I saw that the Moy Road houses had gone ... Around the outside edge near the school where I was standing it would have been firm enough to stand on. I was standing on the outside edge.'

The witness in fact climbed up the side of the heap, which was some 25–30 feet high. With the remarkable discipline and heroic absence of panic which teachers and pupils alike exhibited at this dreadful time, the headmaster and his staff then evacuated the school and gathered their pupils at the front for a roll-call but its completion was prevented by a rush of filthy water which brought more waste materials with it. Through this swirling water the teachers guided and sometimes carried their pupils out of danger. 'The children were wonderful,' said Miss G. E. Evans, the deputy headmistress. So also, in our view, were the staff, and we here desire to record and re-iterate the tribute to their devotion to duty that we expressed during the hearing of the Inquiry.

WORK

70. The Manufacture of Biscuits

Woman, Wife and Worker (Social Science Department, London School of Economics and Political Science, London, 1960), pp. 20–3.

This study gave an interesting account of work in a light industry, the Peek Freans biscuit factory in Bermondsey, London, where a general labour shortage gave women the chance of employment. 4,000 workers were employed, 3,000 of them women and four-fifths of them married; only a quarter of the women were employed full-time. They worked to give their families a higher standard of living, but the job was less important to them than husband and children; absenteeism was 40 per cent higher than among male workers, amounting to an average of one and a half days out of eighteen worked.

A detailed study of the work of the two supervisors in charge of the packing department referred to earlier reflected the extent to which these expectations [that the supervisors would plan the job efficiently] were being met, and gave greater insight into the responsibilities of the supervisors at the time of study. Their day ran as follows:

7.30 a.m. saw them at their desk in the middle of a long workroom where, in the course of the day, some 208 women were at work packing biscuits into tins, and performing other ancillary jobs. As the day started a number of routine but necessary queries had to be dealt with. Some protective material for special handling was required by one woman; an absentee was reported and had to be replaced by someone from the relief team; two girls asked for leave of absence and a note requesting leave had to be written and despatched to the personnel department; it was discovered that a checker, a key-worker without whom a whole team would be held up, was not coming in and an urgent note had to go to the personnel department for a replacement. These preliminaries over, the supervisors walked round the department to see that nothing was amiss and to be available to anyone, trainees, checkers, or operatives, who wanted a word with them. These 'words' varied from an exchange of views about work left over by the evening shift, to a warning that tins were in short supply, and a query about the quality of an unusually palefaced batch of shortcake which finally went to swell the bags of broken biscuits sold at the factory shop.

Of a less routine nature was a short interview with four workers who were to be permanently transferred to another department—information not normally received well. The women concerned were then sent to the personnel department to get further details of this new work. Unfortunately, on this particular occasion, the department asking for help was not yet ready for them and within twenty minutes the four were back in the department in no mood to appreciate the problems of management. One supervisor had to keep them properly employed until the new department was ready for them. The completion of absentee figures and a brief consultation with a representative from the production planning department were two of several further jobs which had to be tackled before the 9.30 a.m. shift arrived. With the coming of this shift a number of matters had to be settled, and as the morning continued further queries came in from other departments: from the wages office, a pay claim; from the personnel department, requests for the temporary loan of women to be sent to another production department. Considerable trouble arose from a fault in the conveyor belt carrying away tins—a matter which could not be left as it might cause a serious blockage. During the morning the manager of the department looked in to sign the special wages sheet which had to be prepared in readiness for him.

It is easy to see how fully the supervisor's day was occupied and how failure to cope would soon lead to chaos. But hardly a day went by without special problems arising. On one occasion, about an hour before dinner break, a message arrived from the planning department to say that an emergency order of 1500 special tins had to be packed, wrapped and parcelled by noon on the following day. This meant a complete change of programme in certain sections of the room. One supervisor abandoned her other work to plan the new operation. She decided which teams were to be switched to the new work; phoned the electricians for some necessary adjustments; phoned the tinshop for extra tins; phoned the stationery department for additional supplies; transferred a woman immediately to begin lining tins so so that the afternoon shift could get off to a flying start when they came in; and transferred another girl to prepare the perforated tickets that were specially required for this particular order. When the afternoon women arrived, all was prepared and the job went through without a hitch.

One evening, just before 5.30, and as the day shift was leaving, an angry worker and an indignant checker, spoiling for a fight, surged up to one of the supervisors. In this particular storm in a teacup there were two possible danger points, either of which could have led to far more serious trouble. The worker might, with some justice, have felt she was unfairly treated: the checker, also with some justice, might have felt that the supervisor, her immediate boss, had failed to give her reasonable support. In retrospect it was easy to see that both these points must be satisfactorily met. At 5.25 p.m. at the end of a working day which had started ten hours earlier, it would not have been surprising if the supervisor, with many other matters, on her mind, had not fully grasped the situation. But she listened to what both had to say, and then by a skilful but perfectly proper deflecting of responsibility from the checker to herself, she was able to meet the woman's justified complaint without allowing the checker to lose face. This may seem a trivial incident, but the history of industrial disputes is strewn with situations intrinsically no more serious, which mishandling has converted from an incident to a crisis.

71. The Anglican Clergyman

R. Lewis and Angus Maude, *Professional People* (London, Phoenix House, 1952), pp. 144–6.

Once highly paid and with high social status, the Anglican clergymen after 1945 were threatened with poverty. The Church of England was slow to react to the financial needs of changing times, and only belatedly began to shift its considerable assets more into equities. The problem of finance was coupled with an increasing difficulty in recruiting, the two being interrelated.

The second question is whether the Church can hope to attract enough ordinands of good quality to do its work, and how they are to be financed. It costs about £1,200 to train an ordinand, and even scholarships and grants-in-aid from the Church will not prevent the young man or his family from having to find a substantial part of this sum for himself. After ordination, he may well have to start work on a salary of less than £300 a year. What is more important is that the average stipend of incumbents is about £465. Many priests exist on far less. In his *Challenge to the Laity* in 1947, the Archbishop of Canterbury requested that efforts should be made to bring all stipends up to a minimum gross figure of £500 a year. Many dioceses achieved this, only to find that the rise in the cost of living had raised the real requirement to something like £500 a year *net*. In November, 1951, the Archbishop recognized this change in a statement to the Church Assembly, in which he estimated that 'the former minimum figure of £500 gross to every incumbent represents in fact not more than about £430 a year, plus an official house to live in, but less unavoidable "official expenses". This figure leaves the incumbent very low in the list of skilled wage earners'. It might be added that out of this stipend many country priests have to run a car, in order to do their job at all; that many Church Councils do not even pay the rates and dilapidations on the vast old rectories which some parsons and their wives still have to live in (and to heat); that if the incumbent tries to let some of the house off to tenants he may find himself in a tangle of legal and administrative difficulties which can cause him grave embarrassment; that a married priest is expected to dispense much informal hospitality, which is not allowable against income-tax, and also to spend charitably at his own fêtes and bazaars; and finally that, if his parishioners seek to augment his stipend with a bountiful Easter offertory, the whole of it becomes subjected to income-tax.

Of course, men do not—and should not—go into the Ministry for gain, or for an easy competence. Poverty may even be good for enthusiastic young curates—although those who are most anxious to prescribe this medicine are seldom those who have sampled it themselves. But there comes a point at which not only the priest's family life, but his whole work, may become soured and hopeless through the constant struggle to make ends meet with a minimum of dignity. The stipends are not enough.

Nor should the cynical layman console his conscience with the belief that the Church has vast hidden stores of wealth which would solve the problem if they were properly applied. It has not. The finances of the Church have already been thoroughly and ably overhauled, and the process is still continuing. Neither is it true that a minority of clerics enjoy fat stipends which they will not share with their poorer brethren. In 1939, out of some 14,000 benefices only 264 were worth more than £1,000 a year; there are certainly fewer now, nor do many exceed this sum by much. Where a Bishop's income is officially stated to be £4,000 a year or over, this is invariably a gross figure from which very heavy expenses have to be deducted; where the diocese has assumed responsibility for the Bishop's palace and certain other expenses, the Diocesan's income is fixed at £2,000, which is not nowadays by any means excessive for a man of great responsibilities, who must travel much and offer hospitality to many. Some Deans, Archdeacons, and Canons scarcely seem to justify their competences (which vary widely, from about £600 to £2,000 a year); but many do admirable work, and the Church would certainly suffer if it could not provide reasonable conditions for its scholars and its ablest administrators. People may feel that there is no place for an Archdeacon Grantley in the twentieth-century Church; but is it good that all clergy should be reduced to the straits of Crawley and Quiverful?

* * *

The annual wastage in the Ministry through death and retirement is about 550 priests, and an annual recruitment of at least 600 deacons is needed. The ordinations in 1950 totalled only 425, but the annual number is still rising after its wartime slump, and the situation is not without its hopeful features. The Bishop of Southwell has said: 'One desperate expedient the Bishops have most resolutely ruled out. We will not be parties to lowering the standards of qualifications or training required. The Church does not want the leavings or failures of other professions—it needs the best; and the more it asks for the more it is likely to get.'

72. The Cotton Industry

R. Robson, *The Cotton Industry in Britain* (London, Macmillan, 1957), pp. 320–1.

A recurrent theme in industry is the decline in craftsmanship, both because it is less needed where machines have taken over what was once done by hand, and also because workmen are less skilled and less keen to be skilled. Mr. Robson, in his study of the cotton industry, states the case for the continued need of skill and 'know-how' in a traditional industry.

Machinery is more automatic than it used to be, *e.g.* the automatic loom dispenses with a great deal of skill and also conscientiousness on the part of the operative, while over the whole textile field more and more instruments are being devised to control operations, *e.g.* sizing, which were previously matters of individual judgment. On the other hand there remain a large number of operations at all stages of textile manufacture where the skill element has not been eliminated. In spinning, the adjustment of any of the various machines to obtain the best results from a given type of cotton is a skilled operation, particularly so in the case of the combing machine, where even the trained engineer needs long experience. In weaving, skill is required in such operations as maintaining correct yarn tension or in assembling a large number of threads. In finishing, even among the simpler operations, the correct speed in singeing, the correct pressure in bleaching to obtain a particular variety of white, the correct setting of the line in schreinering, the correct dye viscosity or the adjustment of roller pressure in printing can only be learned by experience. Bleaching is itself perhaps the simplest of processes, but since the finished product relies on handle and appearance rather than colour or design it becomes a skilled operation. Moreover, in those directions in which the textile industry is making technological advances, *e.g.* in the use of newer dyes, in blends of fibres or in new machines, a new element of skill is introduced which is only slowly rationalised as knowledge increases.

In general, skill in textile operations does not lie in whether the job can be done or not, but in the relative standard of performance which manifests itself in a higher output per head, in a more consistent standard of quality or in an ability to develop new styles and combinations of yarns, weaves and finishes. This difference becomes more marked as the need for skill increases, as it does in three main directions. First, in the direction of increasing technical difficulty. For example, the dyeing process can be put in grades of increasing difficulty and of increasing machine complexity, starting with sulphur colours and direct dyes and ending with vat dyes and resins, or in the case of weaving starting with plain, coarse cloths then fine cloths and ending with fancy weaves and mixed warps. Secondly, skill is needed in the maintenance of quality, *i.e.* of a consistent standard which can only be achieved by careful supervision and by careful machine adjustment and upkeep at every stage, as is required in the spinning of fine counts or in the maintenance of evenness of pick and control of tension in weaving, for example, typewriter ribbon cloths.

Thirdly, a higher degree of skill is required for a number of reasons in the production of fine goods. It then becomes necessary to avoid or remove faults and imperfections, *e.g.* in knotting yarns, since these imperfections are more readily discernible while, moreover, in some cases they may be harmful, *e.g.* to knitting-machines. Moreover, the production of fine goods, in general, requires more care on the part of the operative, since the yarn will not stand rough handling. Similarly, it requires more precision and a closer adjustment of machines, *e.g.* in spinning, a more exact adjustment to staple length throughout, or in weaving, much greater care to avoid breakages, or in preparation, in sizing and assembling the warps.

More generally, a higher standard of maintenance, inspection and servicing is required for the machines and a higher standard of performance from the operatives.

73.   The Factory Worker

F. Zweig, *The Worker in an Affluent Society* (London, Routledge, 1961), pp. 205–8, 210.

The research for this book was carried out at the works of the English Steel Corporation Ltd., in Sheffield, in 1958, and at four other plants (the Workington Iron and Steel Co.; Vauxhall Motors at Luton; the Dunlop Rubber Co. in Birmingham; and the Mullard Radio Valve Co. in Mitcham) in 1958–9. The core of the work consisted of interviews with some 672 men and women. The plants were large, well-run organizations in prosperous industries and the conclusions may not hold for other workers.

First we can witness a considerable rise in security-mindedness. The factory worker has a relatively high security of employment. The bitter memories of the past are fading away. If he is under forty he hardly knows what it is like to be unemployed. The worker has an established routine, both in working and living. His continuity of employment does not fall very much behind that of an office worker. He has a recognized niche and social position, he is attached to his workplace by many institutional arrangements such as pension rights or seniority rights and other fringe benefits which are constantly increasing. He stays put in his job if he can manage it. His greater measure of security makes him also conscious of security, as his stake is so much bigger and he has so much more to lose. Every threat to his standard of living, which has been raised so considerably, he views with great concern. The traditional type was reputed to have no concern for the morrow. 'I may be dead to-morrow', or 'To-morrow never comes', or 'To-morrow will take care of itself'—these were phrases commonly heard among the working classes. Now to-morrow seems to matter much more. This is also expressed in the worker's propensity to save, which is stronger than ever.

Next comes what is often called 'The revolution of rising

expectations'. His appetite is whetted, he wants more. He has a good life but he wants more of it. He wants to better himself, not so much by promotion but by higher wages. He is prepared to work shifts and overtime to make more. He wants his own car, more gadgets, often his own house. When asked about his possessions, he would often use the term: 'Next on my list is . . .' The traditional standard of living is outmoded. There is no more of 'What was good for my father is good for me'; it is rather 'I have many things which would be unthinkable to my father', or 'I have achieved something which I thought would be impossible for me'.

Linked with this is the steep rise in acquisitive instincts. The traditionally-minded workman was known for his contempt for money, which went very well with his contempt for the moneyed class. This contempt is largely disappearing. The worker to-day doesn't want to waste his money. His pattern of consumption turns to more durable goods; he spends his money more wisely and more economically. He wants to show something for his labour, something tangible which can be seen by everybody and which speaks clearly the language of success. He wants to show that he has not wasted his life, but achieved something which does not fall behind the standard of others. In this way a large section of the working-class population becomes a property-owning class. For the time being it is mostly property for the worker's own consumption—a house, a car, T.V., a washing machine, refrigerator, post-office savings, savings bonds, insurance rights—but the overspill to capital assets is already on its way. I came across workers with considerable nest-eggs up to £3,000, well invested. In this way a new type of a bourgeois worker appears on the stage.

Closely linked with these traits is his family-mindedness and home-centredness, as security, acquisitiveness and family-mindedness go well together. He seeks his pleasures and comforts at home more than ever. 'I am a fairly domesticated animal', was a typical remark which I heard. Family life assumes a romanticized image of happiness and joy. Family life stands, in his mind, for happiness, enjoyment and relaxation. As he sits by his fireside and watches T.V. he feels free and happy. The wife doesn't snap at him as she used to; the children are no longer seen crawling and messing about on the floor, shouting and screaming. The foul air, the vermin, the outside, smelling lavatory, the broken chairs have been removed as if by magic. Instead there is a nicely furnished house of his own, or a council house which is a near equivalent to his own property. His main hobby is decorating his home and he is busy with his brush all the year round: 'I never saw my father handling a brush, now it seems I have a use for my brush the whole year round.' These contrasts may be slightly over-drawn in relation both to the present and the past, but I believe they have validity in relation to the general trend.

Part of a worker's home and family-centredness is his intense interest in his offspring. If he has no ambition for himself, he has plenty for his children. 'In my days a man pushing a pram would have been a laughing-stock; now you see a great many men pushing their pram proudly,' said an older man. He not only pushes the pram, he often washes the children and gives them baths, he reads them stories, follows their school records, calls at the school on parents' day, tries to fix them up with a good job or apprenticeship. 'It is the finest thing there is to give the children every advantage', or 'My boy has everything he wants', or 'I scrubbed and scraped to give my children every chance', he may say.

This has an enormous effect on the father-image among the working classes. The bullying father or the father whose authority was used as a bogey has largely disappeared, and instead an older brother relation comes to the fore. The father is there to assist and help, to give guidance; but he is no more the master with a big stick. In my previous enquiries I often found the father-image distorted among the working classes and strongly imbued with the shadow of the Oedipus complex. The working-class child often had only the care and affection of the mother, while the father was an aloof figure. Now the powerful figure of the working-class 'Mum' is receding, as in many workers' families the father steps into her place or occupies an equal place beside her, the more so as she often goes out to work. Anyway the balance of affection is nowadays more equitably distributed between the two parents. The changing father-image is an important factor in the changing ethos of the working classes. Whether a child suffers under arbitrary authority at home or enjoys a kindly and reasonable guidance has a great bearing on his character and outlook.

74. The Effects of Automation on Labour

Department of Scientific and Industrial Research, *Automation in Perspective* (London, H.M.S.O., 1956), pp. 18–21.

If the first industrial revolution was based on mechanization—the use of power-driven machines—the second industrial revolution will be based on automation—the self-operating machine, programmed and self-adjusting. Automation to a large extent depends on the computer. The Flowers Report (Cmd. 2883, 1966) shows that 1,000 computers of various kinds were in use in Britain compared with 15,000 in the U.S.A. The large-scale use of automation will have important effects on the working man, his employment, skills, job satisfaction, and pay.

Employment

Automation very often saves operative labour and there is a fear that it will enable the demand for goods and services to be met with some workers unemployed. In practice, this danger need never become real, though beyond doubt there will be, as in the past, a steady movement of workers from one industry, one occupation or possibly one region to another.

If full employment continues to be general it will be fairly easy to introduce automation because displaced workers can soon obtain other work, though not always with their old firms or in their old occupations. Indeed, automation, by increasing productivity, can lower prices and so stimulate the demand for goods and for labour to make them.

So far automation alone has rarely created a problem of redundancy, either in Great Britain or overseas. A firm introducing automatic techniques can often avoid discharging surplus labour either because it is large and can transfer the surplus from one department to another, or because it does not replace labour lost through retirement or resignation, or because labour is scarce and only by adopting labour-saving methods can the firm expand production as fast as it wishes. . . .

Skills

At present more attention is given to the immediate effects of automation on employment than to the effects that will persist in the long run—notably its effect on the skills and on the satisfaction gained from work.

It seems clear that, taking the nation as a whole, automation will enlarge the demand for skill. In the first place it will increase the ratio of managers, supervisors and technicians to operatives and it will demand more skill of maintenance men. To that extent it will certainly raise the total amount of skill required by industry. But it will also change the skills demanded of the remaining operatives.

Broadly speaking, these operatives fall into three groups: machine-minders, monitors of automatic control systems and computer-operators in offices. A good example of the machine-minder is the worker who tends to complex and sometimes extensive transfer-machine. He loads and starts the machine, watches the operations for which he is responsible, inspects the machine and keeps it as fully employed as possible. He must understand what goes on in the machine without always seeing it. He has more to think about than the operator of a single machine-tool. His activities are more varied and continuous; he may cover more ground, has more points to watch, and must react promptly. He decides when to stop, and when not to stop the machine; a stoppage can soon affect the whole line and an oversight can quickly multiply scrap.

To make another example, an operator of automatic looms does not have to stop each machine to change spools. He (or very often she) loads the machine with a battery of spools, which are automatically fed in turn while the machine is running; so he can take on many more looms than formerly. He inspects each machine for faults, but he contributes little to the quality of the cloth because automatic looms work more smoothly than the older machines and there are fewer stoppages than before.

Monitors of control systems are of three types: some manipulate a process by remote control; others change the speed and quality of production using a panel of instruments;

and a third group, in the most modern plants, watch automatic instruments and give warning when faults are not put right with sufficient accuracy. In a modern continuous stripmill there are operators of all three types, but most are of the first or second; they push levers and buttons according to information given by instruments; they do not handle the sheet as in older mills, so they are much less exposed to danger and need much less physical skill and strength; but they must be conscientious and reliable because a mistake can halt the entire mill.

Monitoring of the third type is well illustrated in a modern oil-refinery. Operating conditions like temperature and pressure are centrally controlled through instruments that give a signal automatically when the error is too great so that the monitor, usually called the "panel-man", can make adjustments. He need only read instruments from time to time and does not suffer strain through close and continuous inspection.

A monitor has often to interpret readings from a group of instruments and relate his own readings to others. He must be able to work intelligently with technical specialists and to take prompt and sensible action when things go wrong.

* * *

The nature of the operative's work varies from one automatic process to another and according to how far a process is automatic. But, in general, automation makes the manual or craft skills less important and lays emphasis on new skills based on a knowledge of plant and equipment. It also stresses the ability to take in information, interpret it and act quickly and intelligently in emergencies. On partly automatic processes, operators may have to watch instruments continuously and this can cause fatigue; but once control is highly automatic, the danger is much less. There is another problem; control panels can become so big and complex that monitors find it hard to take in all the recorded changes. They work more efficiently and with less strain if the system of control is shown on the panel of instruments as a simple diagram which they can easily interpret.

Satisfaction from Work

Mass-production has created a vast number of unskilled, repetitive and monotonous jobs, which are remotely connected with the broad aims of production and are sometimes bereft of interest. Very often the pace of work is set by machines and this causes strain and dissatisfaction among operatives. The more a process becomes automatic without actually completing the change, the more stress is pace-setting likely to cause. But once workers cease to take an active part in the process, they are free to move at their own pace, and so will probably find their work less irritating. Also, when control is remote, they work away from the noise, dirt, smell or danger of the machines.

Obviously workers benefit from the disappearance of heavy physical tasks and from the low risk of accidents on automatic processes. They also gain satisfaction from their new responsibilities and skills, from their concern with what goes on in whole processes (or large parts of processes), and from frequent co-operation with maintenance men and technical specialists. Against these benefits must be set the loss of close social contacts, which have often made the monotony of mass-production bearable. They also lose, even more than workers on repetitive jobs, the pleasure of working with tools or materials. But these drawbacks may be less important than the chance of working with technically trained colleagues in what is, in effect, a new kind of team, and of sharing with them responsibility for keeping a whole machine-line or process running. What may prove difficult for them is the alternation between quick and intelligent response in emergencies and inactivity for most of the remaining time.

75. The Civil Service

The Civil Service. The Fulton Report,
vol. I (Cmd. 3638, London, H.M.S.O.,
1968), pp. 9–11.

Much dissatisfaction with the efficiency of government led to this Report which proposed fundamental changes in recruitment, training and organization. The six defects identified by the Report were: the insistence on the 'all-round administrator', who was moved from key post to key post every two or three years; the number of classes which impeded efficiency, 47 general classes and 1,400 departmental classes; specialist groups like scientists who did not get the authority and responsibility they deserved. The service was somewhat cut off from the world outside; it did not use as well as it might the abilities of the people who composed it.

THE CIVIL SERVICE TODAY

1. The Home Civil Service today is still fundamentally the product of the nineteenth-century philosophy of the Northcote–Trevelyan Report. The tasks it faces are those of the second half of the twentieth century. This is what we have found; it is what we seek to remedy.

2. The foundations were laid by Northcote and Trevelyan in their report of 1854. Northcote and Trevelyan were much influenced by Macaulay whose Committee reported in the same year on the reform of the India Service. The two reports, so remarkable for their bluntness and brevity (together they run to about twenty pages in the original printing), have had such a far-reaching influence that we reproduce them in full in Appendix B. [not printed here].

3. These reports condemned the nepotism, the incompetence and other defects of the system inherited from the eighteenth century. Both proposed the introduction of competitive entry examinations. The Macaulay Report extolled the merits of the young men from Oxford and Cambridge who had read nothing but subjects unrelated to their future careers. The Northcote–Trevelyan Report pointed to the possible advantages of reading newer, more relevant subjects, such as geography or political economy, rather than the classics. But as

the two services grew, this difference between the two reports seems to have been lost. There emerged the tradition of the 'all-rounder' as he has been called by his champions, or 'amateur' as he has been called by his critics.

4. Both reports concentrated on the graduates who thereafter came to form the top of each service. They took much less notice of the rest. In India, the supporting echelons were native, and the technical services, such as railways and engineering, were the business of specialists who stood lower than the ruling administrators. At home, the all-round administrators were to be supported by non-graduates to do executive and clerical work and by specialists (e.g. Inspectors of Schools) in those departments where they were needed. A man had to enter the Service on completing his education; once in, he was in for life. The outcome was a career service, immune from nepotism and political jobbery and, by the same token, attractive for its total security as well as for the intellectual achievement and social status that success in the entry examination implied.

5. Carrying out the Northcote–Trevelyan Report took time; there was long debate. Over the years other committees and commissions have considered various aspects of the Civil Service. Many new specialist classes have been added to the system, notably the scientists, engineers and their supporting classes. There is now an impressive amount of detailed training. Many other modifications have been made. The reports of the main committees and commissions are summarised and discussed in a note published in Volume 3 [not printed here].

6. Nevertheless, the basic principles and philosophy of the Northcote–Trevelyan Report have prevailed: the essential features of their structure have remained.

7. Meanwhile, the role of government has greatly changed. Its traditional regulatory functions have multiplied in size and greatly broadened in scope. It has taken on vast new responsibilities. It is expected to achieve such general economic aims as full employment, a satisfactory rate of growth, stable prices and a healthy balance of payments. Through these and other policies (e.g. public purchasing, investment grants, financial regulators) it profoundly influences the output, costs and profitability of industry generally in both the home and overseas markets. Through nationalisation it more

directly controls a number of basic industries. It has responsibilities for the location of industry and for town and country planning. It engages in research and development both for civil and military purposes. It provides comprehensive social services and is now expected to promote the fullest possible development of individual human potential. All these changes have made for a massive growth in public expenditure. Public spending means public control. A century ago the tasks of government were mainly passive and regulatory. Now they amount to a much more active and positive engagement in our affairs.

8. Technological progress and the vast amount of new knowledge have made a major impact on these tasks and on the process of taking decisions; the change goes on. Siting a new airport, buying military supplies, striking the right balance between coal, gas, oil and nuclear-powered electricity in a new energy policy—all these problems compel civil servants to use new techniques of analysis, management and co-ordination which are beyond those not specially trained in them.

9. The increase in the positive activities of government has not been solely an extension of the powers and functions of the State in an era of technological change. There has also been a complex intermingling of the public and private sectors. This has led to a proliferation of para-state organisations: public corporations, nationalised industries, negotiating bodies with varying degrees of public and private participation, public participation in private enterprises, voluntary bodies financed from public funds. Between the operations of the public and the private sectors there is often no clear boundary. Central and local government stand in a similarly intricate relationship; central government is generally held responsible for services that it partly or mainly finances but local authorities actually provide. As the tasks of government have grown and become more complex, so the need to consult and co-ordinate has grown as well.

10. The time it takes to reach a decision and carry it out has often lengthened. This is partly because of technological advance and the resulting complexity e.g. of defence equipment. Another reason is that the public and Parliament demand greater foresight and order in, for example, the devel-

opment of land, the transport system and other resources, than they did in the past.

11. Governments also work more and more in an international setting. The improvement in communications and the greater interdependence of nations enlarges the difficulties as well as the opportunities of government.

12. To meet these new tasks of government the modern Civil Service must be able to handle the social, economic, scientific and technical problems of our time, in an international setting. Because the solutions to complex problems need long preparation, the Service must be far-sighted; from its accumulated knowledge and experience, it must show initiative in working out what are the needs of the future and how they might be met. A special responsibility now rests upon the Civil Service because one Parliament or even one Government often cannot see the process through.

13. At the same time, the Civil Service works under political direction and under the obligation of political accountability. This is the setting in which the daily work of many civil servants is carried out; thus they need to have a lively awareness of the political implications of what they are doing or advising. The Civil Service has also to be flexible enough to serve governments of any political complexion—whether they are committed to extend or in certain respects to reduce the role of the State. Throughout, it has to remember that it exists to serve the whole community, and that imaginative humanity sometimes matters more than tidy efficiency and administrative uniformity.

14. In our view the structure and practices of the Service have not kept up with the changing tasks. The defects we have found can nearly all be attributed to this. We have found no instance where reform has run ahead too rapidly. So, today, the Service is in need of fundamental change. It is inadequate in six main respects for the most efficient discharge of the present and prospective responsibilities of government.

76. Dockers

The Devlin Report, 'Into Certain Matters Concerning the Port Transport Industry' (Cmd. 2734, 1965), pp. 16–19.

Conditions of work in the docks had always been hazardous and casual. The dockers first came prominently before the public with their successful strike of 1889, but they remained badly organized until Ernest Bevin amalgamated many small unions into the powerful *Transport and General Workers' Union*. The most serious problem of dock work was its casual nature, and in 1947 an attempt was made to solve this by a register of workers and a weekly minimum wage to be paid for attendance, even if there was no work to be done. But only 25 per cent of dockers were in permanent employment, and even with a minimum pay, the free call system, by which gang foremen picked out the men for daily employment, led to much bitterness and strife. More work was lost in strikes on the docks than in any other major industry. These problems were those which the Devlin Committee tried to remedy.

Bad Time-keeping

39. When does bad time-keeping become a 'practice'? We think the answer must be when a particular sort of absence during working hours has become so widespread that it is no longer dealt with by the ordinary processes of industrial discipline against the individual. If in such circumstances a man were 'prosecuted', he would defend himself on the ground that he was only doing what everybody else did and what had come to be tacitly accepted by employers. And if in such circumstances he was punished, it would arouse resentment that might lead to serious unrest. The two chief examples of bad time-keeping that amount to a practice in this sense are 'welting' in Liverpool and 'spelling' in Glasgow.

40. Welting in Liverpool is the practice whereby only half a gang is working at any given time; for each half it is one hour on and one hour off. During his hour off the man does what he likes; he is resting or smoking or having a cup of tea. To this extent welting encompasses the recognised tea-breaks, morning and afternoon.

It is said that the practice originated at the time when men were required to work long hours in refrigerated holds. The effect of it has to be considered in the light of the fact that Liverpool is not primarily a pieceworking port. The men have always relied for their earnings on working a great deal of overtime and the employers have encouraged this by offering a large overtime premium. On weekdays after five o'clock it is double the ordinary time-rate as compared with time-and-a-half, in many other ports. The result is that by working from 5.0 to 7.0 p.m., the ordinary overtime hours, a man can earn the equivalent of half-a-day's pay and add 50 per cent to his daily earnings. There are correspondingly higher rates at weekends. The welt can be explained, if not excused, by long hours. Many men are regularly working a 60 hour week and it can be said that this would be far too long for the ordinary man if he were not relieved by the welt.

41. We cannot get any firm figures of loss of time. Employers —or at any rate their head offices—probably do not know exactly what is going on; and if they do they are not always candid about it, for there is no doubt that the individual employer must bear a great deal of responsibility for having allowed the practice to grow as it has. A leading employer gave us as his estimate that 20 per cent of the time nominally being worked is unproductive. Others who are in a position to judge agree that this figure is probably close to the mark.

42. It is not easy to establish the facts either about the motives for welting or its consequences. The trade union side disputes the employers' assertion that the object of welting is to prolong the day's work into the overtime period. Whether or not this is the object, it is certainly the result, and welting can go on in the overtime period as well, thus prolonging it further. Over the last five years the Liverpool docker has averaged four more hours' overtime a week than the London docker, although he has put in fewer hours' work during the normal working week than has the Londoner. It is also argued by the trade union side that in many cases the foreman permits the welt only on the understanding that the half-time that is worked is at such a pace as to be as productive as whole time working at a normal pace. There may be some truth in this; it may be the case that those who are actually at work under the welt work at 'piecework stroke' for time-rates. To suggest, however, that half a gang by working harder can shift as much

as a full gang working at normal pace is to imply that normal pace is very slow indeed.

43. Even if it could be accepted that Liverpool dockers who practice the welt maintain an output equivalent to the full gang working at 'time-work stroke', there remains the loss in output compared with what might be achieved by the full gang working at 'piecework stroke'. Over the last five years weekly earnings in Liverpool have averaged only 70 to 75 per cent of those in predominantly pieceworking London despite the much higher proportion of overtime working at Liverpool. Since there is no reason to suppose that costs are much higher in London than in Liverpool, it must be supposed that this difference in earnings is balanced by a roughly equivalent difference in weekly output. If the Liverpool docker worked at the pace of the Londoner throughout the working day the turn-round of ships in Liverpool would be considerably more rapid.

44. It seems, therefore, that two consequences of the welt can be established with some assurance. First, it breeds indiscipline. The fact that men choose their own time for slipping in and out of work with the connivance of their supervisor must undermine the latter's control. Secondly, although much of the direct cost of the practice may be borne by the men in lower earnings, the time taken on each job is extended, the turn-round of ships is slowed down, and the capital equipment of the port is used at well below capacity. These costs are not borne by the docker, but by the users of the port of Liverpool and, through them, by the nation.

45. The employers feel that the most hopeful cure for the welt is the extension of piecework. Piecework was introduced into the port in 1942 but it has not yet really caught on. It may be that the rates are not good enough to compete with the attraction of overtime at double time; certainly the Liverpool docker has yet to be convinced that they are. But the evidence is that in those parts of the port where piecework is effective the welt disappears. This is what one would expect; for with the time-rate as a minimum guarantee as it is in London, an hour of welting would in effect become a dead hour unpaid for. But the position in Glasgow, which is a piecework port, suggests that piecework would not be a complete cure.

46. A complete cure is unlikely without much firmer discipline. Although the trade union side has put forward considera-

tions, which we have recorded above, to show that the welt and its effects are not as black as they are sometimes painted, their leaders are entirely against it, as they always have been, and are prepared to co-operate in stamping it out. They point out that the initiative lies with the employers and that it is the duty of the foremen to report cases of welting. They blame the employers with some justification for having allowed the situation to become as bad as it is.

Whoever is to blame for the present situation, we agree with the employers that it cannot be remedied overnight simply by issuing and enforcing an ultimatum that welting is to cease and that every case of it will be reported to the Board for disciplinary action. The practice is too ingrained for this and we think that an ultimatum would probably result in mass indiscipline, against which the authority of the unions (which, as we shall explain later in this Report, has recently been considerably weakened) would not prevail. We think that, as we have said in the case of irresponsibility generally, the better course is to hope that under conditions of regular employment and with perhaps the extension of pieceworking, it will become possible for firm action to be taken with good prospects of success. Except in this respect, we do not think that a further measure of decasualisation has anything to contribute to the solution of the problem, for the welt is not a malpractice created by the casual system.

47. What is welting in Liverpool is spelling in Glasgow; the practice is the same. It began in Glasgow during the last war as a result of excessively long working hours which were continued into peacetime. A man could be booked for as much as 70 hours in the week and even now in the 40 hour week for as much as 64. The difference between Liverpool and Glasgow is that, Glasgow being a piecework port, the practice may not cost the employers anything directly in wages. The whole gang shares in the piecework rates. But the other evil consequences which we have referred to in Liverpool apply equally in Glasgow.

The union leaders are strongly against the practice; it was they and not the employers who raised the point with us. The union leaders blame the employers for failing to co-operate effectively in stamping out the practice. They say indeed that employers encourage it since they do not have to pay men while they are spelling. They say that with half the gang away safety regulations are ignored and that it is by dangerous means that output is maintained and even increased.

48. The employers deny all this strongly. They say that between 1948 and 1956 they initiated four serious attempts to abolish spelling and that all of them failed. But the reasons the employers give for the failure certainly do not entirely absolve them from blame.

The first is the system of daily engagement. In Glasgow there is a free call and no continuity rule. At least there is no formal written continuity rule; we believe that in practice there is some continuity but at any rate there is nothing to prevent a man from refusing at the free call to continue with the job he was doing the day before. This means, the employers say, that an employer who disciplines men for unauthorised absences will find them avoiding working for him. We can understand the reluctance of an employer not to risk losing the services of a good man because he has taken unpaid leave of absence, but if that was acceptable as a general principle, there would soon be no industrial discipline at all.

49. Welting and spelling are highly organised forms of bad time-keeping. In less organised forms bad time-keeping can become a practice in the sense that late starts, early finishes and prolonged tea-breaks can become so regular and so frequently condoned that attempts at enforcing punctuality would be resented, perhaps to the point of unofficial action. We have not attempted to survey bad time-keeping in all ports; we have had evidence of it from the ports of London and Bristol. In both ports the employers calculate that an hour a day is lost in this way—that is a time loss of at least 10 per cent. In London it is said that the most marked loss of time is in early finishing. It is said that the quays are deserted after 6.15 p.m. although overtime lasts till 7.0 p.m. Where overtime is not being worked there are also early finishes though they are not so marked.

50. The unions in reply to this, while contending that the picture of quays being deserted at 6.15 p.m. is exaggerated, admit that most men stop work well before 7.0 p.m. In the Royal Docks, they agree, most ships are finished by 6.30 p.m. But they point out that the docks are very large, that there is very little transport within them and that a man may have to

walk a long way to the dock gates; from the Royal, they say, it take half-an-hour.

Such cases as there are of men leaving before 6.30 p.m. the unions say, are due either to the nature of the work or to the defective organisation of it. The nature of the work obviously does not permit everybody to finish on the stroke of the clock. When goods are being off-loaded from the ship on to the quay for storage in sheds or delivery to waiting lorries, it must always be difficult to organise the finish; and when there is a deadline such as 7.0 p.m., the finish must be early rather than late. There comes a point when the supervisor or foreman will say that no more cargo is to be discharged because there would not be time to handle any more on the quay or to load it on to waiting lorries. When the ship's gang has fastened down the hatches, are they then to hang about? Or it may be that through poor organisation the employer has under-estimated the number of lorries that could be loaded that evening and then there is no more work or at least no more piecework.

OPINION

77. A Social Revolution?

  (a) C. R. Attlee, *As It Happened* (London, Heinemann, 1954), pp. 162–3.
  (b) R. H. S. Crossman, *New Fabian Essays* (London, Turnstile Press, 1952), pp. 5–6.

Many people at the time thought that the post-war Labour government had carried out a social revolution; to an historian they appear rather to have confirmed 'the welfare state' whose origins can be traced back to the nineteenth century. The extension of affluence to the mass of the population had perhaps more revolutionary effects than legislation.

Clement Attlee (1883–1969) had a conventional middle-class background, practised as a lawyer, worked in East End settlements, lectured at the London School of Economics, survived war service, and entered Parliament for Limehouse in the East End. After various junior offices he was elected leader of the Labour Party after George Lansbury in 1935. During the Second World War he was Churchill's deputy, and then Prime Minister 1945–51. R. H. S. Crossman enjoyed a distinguished scholastic career, led the Labour group on Oxford City Council 1934–40, was Lord President of the Council in 1966, and Minister for the Social Services 1969–70.

(a)

The Labour Party came to power with a well-defined policy worked out over many years. It had been set out very clearly in our Election Manifesto and we were determined to carry it out. Its ultimate objective was the creation of a society based on social justice, and, in our view, this could only be attained by bringing under public ownership and control the main factors in the economic system.

Nationalisation was not an end in itself but an essential element in achieving the ends which we sought. Controls were desirable not for their own sake but because they were necessary in order to gain freedom from the economic power of the owners of capital. A juster distribution of wealth was not a policy designed to soak the rich or to take revenge, but because a society with gross inequalities of wealth and opportunity is fundamentally unhealthy.

It had always been our practice, in accord with the natural genius of the British people, to work empirically. We were not afraid of compromise and partial solutions. We knew that mistakes would be made and that advance would be often by trial and error. We realised that the application of socialist principles in a country such as Britain with a peculiar economic structure based on international trade required great flexibility.

We were also well aware of the especially difficult situation of the country resulting from the great life-and-death struggle from which we had emerged victorious. But, in our view, this did not make change in the socialist direction less necessary. On the contrary, it was clear that there could be no return to past conditions. The old pattern was worn out and it was for us to weave the new. Thus, the kind of reproach levelled at us by Churchill, that, instead of uniting the country by a programme of social reform on the lines of the Beveridge Report, we were following a course dictated by social prejudice or theory, left us completely unmoved. We had not been elected to try to patch up an old system but to make something new. Our policy was not a reformed capitalism but progress toward a democratic socialism. Furthermore, our experience in the war had shown how much could be accomplished when public advantage was put before private vested interest. If this was right in war-time it was also right in peace. I therefore determined that we would go ahead as fast as possible with our programme. When I formed the Government I told Dalton that it would be his task to nationalise the Bank of England, as well as to deal with the financial position of the country.

(b)

My contention is that this absence of a theoretical basis for practical programmes of action, is the main reason why the post-war Labour Government marked the end of a century of social reform and not, as its socialist supporters had hoped, the beginning of a new epoch. To say this is not to condemn it. The development from the liberalism of 1906 to the modern Welfare State had to be completed; so did the transformation of the Empire into the Commonwealth. But neither of these processes was explicitly socialist. The principle of national self-determination, which Labour has fulfilled in its Indian policy and on which it is working in the Colonial Empire, is essentially liberal; and the planned Welfare State is really the adaptation of capitalism to the demands of modern trade unionism. The mixed economy evolved in the decade between 1940 and 1950, has not abolished competitive free enterprise, but adapted it to meet the social and economic demands of organised labour. What was achieved by the first Labour Government was, in fact, the climax of a long process, in the course of which capitalism has been civilised and, to a large extent, reconciled with the principles of democracy. By refusing to accept the Marxist philosophy, we have almost succeeded in disproving its prophecies of inevitable conflict. This is a historic achievement, but the fact remains that, in achieving it, the Labour Party is in danger of becoming not the party of change, but the defender of the post-war *status quo*.

Philosophy begins where pragmatism fails. When the common-sense socialist has come to the end of his programme and there are no longer a number of obvious reforms which men of goodwill broadly agree should be carried out, it is time to sit back and reflect.

78. The End of Imperialism

Keesings Contemporary Archives (London, 1960), pp. 17269-70.

The period of this book sees the beginning of modern imperialism in the grab for Africa in the 1880s, and its end in 'the wind of change' of the 1960s. The British Empire was actually enlarged after the First World War by various ex-German and ex-Turkish colonies. The white dominions acquired formally their independence, under a symbolic crown, in 1931. India also agitated for this status. Egypt had already become independent in 1922. The Empire went to war again in 1939, but after 1945 it was rapidly dispersed. India was granted independence in 1947. The first British African state to win independence was Ghana, the old Gold Coast, in 1957, and the rest of Britain's African colonies became independent in the sixties. In South Africa (and Rhodesia) a small white minority still hold power over black and 'coloured' peoples. In 1961, later than this speech, but when Macmillan was still Prime Minister, South Africa withdrew from the Commonwealth. It was widely believed in Britain that the decline of Empire and of British power in the world generally had led to severe stresses and a malaise in British society. By the later sixties there was a widespread disillusionment with what was left even of the Commonwealth, and gradually the objective of much opinion was 'entry into Europe' (membership of the Common Market).

Harold Macmillan was Conservative Prime Minister from 1957 to 1963. Between the wars he had been an outstanding critic of governments' failure to tackle the problem of unemployment, and after the Second World War he helped to modernize Conservative policies. He coined at least one other famous phrase, 'You've never had it so good', to describe the affluence of the British electorate.

'As I have travelled through the Union I have found everywhere ... a deep preoccupation with what is happening in the rest of the African continent. I understand and sympathize with your interest in these events, and your anxiety about them.

'Ever since the break-up of the Roman Empire one of the constant facts of political life in Europe has been the emergence of independent nations. They have come into existence over the centuries in different shapes with different forms of government. But all have been inspired with a keen feeling of nationalism, which has grown as nations have grown. In the

20th century, and especially since the end of the war, the processes which gave birth to the nation-states of Europe have been repeated all over the world. We have seen the awakening of national consciousness in peoples who have for centuries lived in dependence on some other Power.

'Fifteen years ago this movement spread through Asia. Many countries there, of different races and civilization, pressed their claim to an independent national life. To-day the same thing is happening in Africa. The most striking of all the impressions I have formed since I left London a month ago is of the strength of this African national consciousness. In different places it may take different forms, but it is happening everywhere. The wind of change is blowing through the continent.

'Whether we like it or not, this growth of national consciousness is a political fact. We must all accept it as a fact. Our national policies must take account of it. Of course, you understand this as well as anyone. You are sprung from Europe, the home of nationalism. And here in Africa you have yourselves created a new nation. Indeed, in the history of our times yours will be recorded as the first of the African nationalisms.

'And this tide of national consciousness which is now rising in Africa is a fact for which you and we and the other nations of the Western world are ultimately responsible; for its causes are to be found in the achievements of Western civilization in pushing forward the frontiers of knowledge, applying science in the service of human needs, expanding food production, speeding and multiplying means of communication, and, above all, spreading education.

'As I have said, the growth of national consciousness in Africa is a political fact and we must accept it as such. I sincerely believe that if we cannot do so, we may imperil the precarious balance of East and West on which the peace of the world depends.

'The world to-day is divided into three great groups. First, there are what we call the Western Powers. You in South Africa and we in Britain belong to this group, together with our friends and allies in other parts of the Commonwealth, in the United States, and in Europe. Secondly, there are the Communists—Russia and her satellites in Europe, and China, whose population will rise by 1970 to the staggering total of

800 million. Thirdly, there are those parts of the world whose people are at present uncommitted either to Communism or to our Western ideas. In this context we think first of Asia and of Africa.

'As I see it, the great issue in this second half of the 20th century is whether the uncommitted peoples of Asia and Africa will swing to the East or to the West. Will they be drawn into the Communist camp? Or will the great experiments in self-government that are now being made in Asia and Africa, especially within the Commonwealth, prove so successful, and by their example so compelling, that the balance will come down in favour of freedom and order and justice?

'The struggle is joined, and it is a struggle for the minds of men. What is now on trial is much more than our military strength or our diplomatic and administrative skill. It is our way of life. The uncommitted nations want to see before they choose. What can we show them to help them choose aright? Each of the independent members of the Commonwealth must answer that question for itself.

'It is the basic principle of our modern Commonwealth that we respect each other's sovereignty in matters of internal policy. At the same time we must recognize that, in this shrinking world in which we live to-day, the internal policies of one nation may have effects outside it. We may sometimes be tempted to say to each other, "Mind your own business." But in these days I would myself expand the old saying so that it runs, "Mind your own business, but mind how it affects my business, too."

'Let me be very frank with you, my friends. What Governments and Parliaments in the United Kingdom have done since the war in according independence to India, Pakistan, Ceylon, Malaya, and Ghana, and what they will do for Nigeria and the other countries now nearing independence—all this, though we take full and sole responsibility for it, we do in the belief that it is the only way to establish the future of the Commonwealth and of the free world on sound foundations.'

79. The Role of the Middle Classes after the War

R. Lewis and A. Maude, *The English Middle Classes* (London, Phoenix House, 1949), pp. 282-3.

The middle classes since the war have widely felt that they have been victimized by the Labour party, and not sufficiently protected by the Conservatives. Their wealth and status have been eroded; their values challenged; their ranks inflated from below. Research by social scientists has suggested, however, that the middle classes have maintained their income and social status, and have been able, even, to exploit the welfare services of the State. The issue is still an open one.

If Labour forfeits middle-class support at the next election, it will be by disappointing middle-class hopes; hopes that it would, while doing something to redress past injustices, act wholeheartedly in the 'national interest' without regard to class advantage. If it is sent by the middle classes into the wilderness, it will be sent specifically to *learn* that lesson, not to return five years later in a bitter and revengeful mood. For it is clear that the middle classes already accept Labour as one of the alternatives at Westminster, to be called back from time to time, perhaps to have as preponderating an influence on the new post-war period as Conservatism had on the period after the First World War. But to achieve this it must govern for the whole nation, for minorities as well as majorities, for all classes, not for one. The prosperity of the middle classes is entirely bound up with that of the whole nation; and they are the only section of the community which is as yet deeply conscious of its relationship with the whole. The working class can refuse to think ahead, perhaps on the theory that there is always Communism if it does not get what it wants; (whether or not Communism will give it what it wants is neither here nor there). But the middle classes cannot refuse to face the future.

No doubt the predominant middle-class idea of a 'national' policy between the wars was simply one which enabled the country's economy to survive without a political revolution—

that is to say, one which kept the working class just sufficiently contented to prevent the miseries of the poor from violently intruding themselves upon middle-class complacency. And it is no doubt equally true that such a policy could not again be sustained in Britain. In so far as it demanded grossly disproportionate sacrifices from the poorest section of the community, the 'national' policy may be held to have been somewhat hypocritically misnamed. But it *did* enable the country's economy to survive the slump, and to equip a vast war machine on top of that. And at the core of middle-class distrust of the Labour Government in 1947–8 was the fear that its policy would continue to favour one class *without* saving the country—that, indeed, it might accelerate disaster.

The middle classes have been oppressed by other fears which are not wholly selfish. The great and real problem of ensuring the liberty of the individual against the encroaching power of the State has been widely discussed—not always with wisdom and moderation, but on the whole soberly enough to reveal that issues thought settled since the nineteenth century are still wide open; the warnings of men like Lord Hewart are being remembered. As middle-class difficulties have increased, realization that security is not the only goal of human society has begun to take effect. At the very moment when the State is promising social security to all, the goal of many generations proves a mirage; for never since the fall of the Roman Empire has the sense of insecurity been greater. Thinking on these subjects is becoming more realistic: if Conservatives can crystallize it into a social purpose above the materialism (albeit scented with the perfume of sentimentality) which has been, so far, the answer of the Labour Party to so many human problems, then Conservatism will win back—and retain for as long as the sense of purpose remains fresh—the 'floating votes' lost in 1945.

So far, we may seem to have been pre-occupied—like the *Economist*—with the future of the upper-middle class. But the future of the lower-middle class is no less important; also it is, if anything, even less easily predictable. It is possible, at least to the eye of faith, to descry niches in the Socialist temple into which each of the upper middle-class groups of professional people and 'managers' could be squeezed, without the certainty of losing all the characteristics of a social class. But the position of members of the lower-middle class is less happy. Working-class men and women (especially women) do not like them. They are composed of families which have in many cases but recently risen from the ranks of the working class; but they have not yet risen so far that they feel secure against the risk of falling back into those ranks. Consequently they tend to raise barriers, and to adopt defensive tactics, which cannot but give offence. They are, as we have seen, the people who seek to protect their children against the 'contamination' of the State-aided primary school; they are also the people who seek most earnestly to equip their daughters for a non-manual employment.

80. A Changing Morality

G. M. Carstairs, *This Island Now, the Reith Lectures for 1962* (London, The Hogarth Press, 1963), pp. 54–5.

Carstairs spoke as Professor of Psychological Medicine at Edinburgh and his views provoked a hornet's nest of criticism, and a continuing debate. Critics have denounced 'the permissive society'. Roy Jenkins, Labour Chancellor of the Exchequer, in a counterblast, described 'the civilized society' in which the individual would be able to lead a responsible life, but one which also offered a wide choice of acceptable behaviour. Professor R. Hoggart has suggested that in two respects British society was changing attitudes which stretched back to the Protestant Reformation: towards sexual experience, and towards work where competitive, individualistic effort was losing its appeal. These changed attitudes were reflected particularly in the actions of students, many of whom protested vigorously against the society in which they lived.

There is more social mobility in Britain today than ever before, but it is still limited, and still stressful for those who do break through. At the same time the popular press, radio, television and films perpetually flatter their audiences by assuming that they are familiar with a level of material comfort and sophistication which seldom corresponds to their actual experience. The real advances in standards of living have been outstripped by this constant stimulation of material aspirations.

Much of the delinquency shown by working-class youth can be viewed as protest behaviour, a protest not so much against their present hardships as expressing their feeling that there are a lot more good things in life which they would like to have but which are still beyond their reach. It is a gesture of defiance against a society which appears to ignore their predicament, a gesture which is more likely to be made when relationships with their parents are faulty. In fact, all too often such youths lack support where they need it most. Edward Blishen, a singularly compassionate schoolteacher, has described a subculture of our society, that of the urban secondary modern school, where one child in three seemed never to have received the assurance of parental affection.

This poverty of human relationships can be found elsewhere, but it seems most frequent in city slums; it is perhaps the legacy of a squalid century of substandard living. In the more distant past, even if parental affection was lacking, working-class children grew up in a community which had strong views of right and wrong; this morality perhaps owed more to the solidarity of a group which had shared rough times than to the formal Christian ethics in which better-class children were instructed. Today, however, both the popular and the church-going types of morality have slipped into disuse. Popular morality is now a wasteland, littered with the debris of broken convictions. Concepts such as honour, or even honesty, have an old-fashioned sound; but nothing has taken their place. The confusion is perhaps greatest over sexual morality; here the former theological canons of behaviour are seldom taken seriously. In their place a new concept is emerging, of sexual relationships as a source of pleasure, but also as a mutual encountering of personalities in which each explores the other and at the same time discovers new depths in himself or herself. This concept of sex as a rewarding relationship is after all not so remote from the experience of our maligned teenagers as it is from that of many of their parents. It bears no resemblance at all to the unromantic compromise between sensuality and drudgery which has been the lot of so many British husbands and wives in the past sixty years. Its full realization could only be possible in a society where women enjoyed social and economic equality with men. We have not yet known such a society, but during this century we have moved a long way towards it.

81. The Cult of Youth

C. Booker, *The Neophiliacs* (London, Collins, 1969), pp. 35-8.

Social analysts have been at a loss about the changing character of English society since 1945. What did change? What caused the change? Was it good or bad? The following is a typical attempt at analysis which certainly exaggerates the speed of change and its magnitude but which is very perceptive about the direction of change. The implied criticism of America is typical of opinion which saw generally in America a threat to both the economic and cultural integrity of European civilization.

Anyone who looks at the evolution of English social history over the twenty years following the Second World War must be struck by the profound change that took place in and around the year 1956.

During the ten years between 1945 and 1955, the outward character of English life, like that of almost every country in Europe, was shaped above all by the memory and aftermath of the war. . . .

Out of this comparative placidity, however, at the end of 1955 Britain suddenly entered on a period of upheaval. Within twelve months it was outwardly marked by a trail of signs, storms and sensations: the coming of commercial television, the rise of the Angry Young Men, the Suez crisis, the coming of the rock 'n roll craze, and even, after a period of comparative quiescence, the beginnings of a crime wave.

This particular period of upheaval lasted for about two and a half years. When it was over, Britain was a changed country. In common with other countries of the West, she had entered on the whole of that transformation which has since become such a dominant factor in the lives of all those who have lived through it. Its major ingredient has been a material prosperity unlike anything known before. With it have come that host of social phenomena which initially, in Britain and Western Europe, were loosely lumped together under the heading of 'Americanisation'—a brash, standardisation mass-culture, centred on the enormously increased influence of television and advertising, a popular music more marked than ever by the hypnotic beat of jazz and the new prominence, as a distinct social force, given to teenagers and the young.

Above all, with the coming of this new age, a new spirit was unleashed—a new wind of essentially youthful hostility to every kind of established convention and traditional authority, a wind of moral freedom and rebellion. Reflected in innumerable ways, from the language of advertising and the teenagers to the new kinds of buildings that from the middle-Fifties on began to rise in London and many other cities, there took place a pronounced shift of focus from the past to a sense that society was being carried rapidly forward into some nebulously 'modernistic' future.

... And indeed the first centres of change, the first scattered signs of the transformation to come, arose in English life long before 1956.

One was the appearance shortly after the war, in some of the poorest slum districts of South London, of the first 'Teddy Boys' who, with their more sinister cousins the 'cosh boys', soon became one of the minor curiosities of post-war England. In view of the later widespread assumption that the 'teenage phenomenon' was nothing more than a by-product of affluence, it is worth emphasising that these earliest fashion-conscious teenagers were so far from being affluent that in many cases, in order to pay for their elaborate dress, they had to resort to petty larceny.

Just why in the late Forties this vogue should have arisen; why, in 1948, the slum boys of South London should have adopted as their uniform the long 'Edwardian' coats and tight trousers which had originally been adopted for a short while by young men-about-Mayfair at the time when the New Look was introduced, will for ever be a mystery. But it was also clear from their studied poses, their deadpan expressions and curious hairstyles, that another major influence in arousing their self-consciousness was the 'tough' dream world portrayed by Hollywood gangster films and Westerns.

A second omen of what was to come was another minority

cult bound together not so much by dress as by music. The craze for reviving the traditional jazz styles of the Twenties, which had spread from America during the war, and which by the late Forties was established in a number of West End clubs, suburban pubs and provincial dance halls, flourished on a very different social level to that which had given birth to the Teddy Boys. Most of the traditional jazz fans, with their reverence for the legendary Chicago of the Twenties and a long-lost New Orleans waterfront, were middle-class. One minor postwar phenomenon was the sharp rise in the number of entrants, mainly from all levels of the middle-class, to art schools; and there was a strong link between these art students and the traditional jazz cult, which also attracted a significant number of somewhat self-consciously unconventional former public schoolboys, such as the Etonian Humphrey Lyttelton (himself also a former art student) and George Melly. Only three things did the revivalist jazz fans have in common with the Teddy Boys of South London—their youth (even though most of the musicians themselves were in their twenties or thirties), their sense of apartness from conventional society and their reverence for a particular Romantic image of America.

A third, rather more general indication of what was to come was provided in 1951 by the Festival of Britain Exhibition. Against the drab background of a city of bomb-sites, peeling paint and still boarded-up windows, the Exhibition was deliberately conceived as a look into the Britain of 'tomorrow'. Today it is hard to recall the first impact of those startling technological constructions in metal and glass, that 'modernistic' world of concrete piazzas, abstract sculptures and brightly-coloured plastics—for many of them were designed by the same young architects who were to be the creators of the office-blocks, the high-rise flats and pedestrian precincts that were the commonplaces of the New England of the future.

A fourth, less public omen was the gathering at the Institute of Contemporary Arts in 1952 of a small group of young artists, architects, designers and critics to hold regular discussions on a number of things which were just beginning to exercise a curious fascination for them—things which hitherto would hardly have been considered as falling within the artistic purview, such as the mass-media, advertising, pop music, science fiction, violence in the cinema and car styling (it was after one of the meetings of this 'Independent Group' in 1954, spent in looking at blown-up projections of advertisements, that one member of the group, Lawrence Alloway, coined for such things the term 'pop art').

82. A New Kind of Society?

T. R. Fyvel, *Intellectuals Today* (London, Chatto and Windus, 1968), pp. 21–3.

This is another attempt to analyse contemporary society, impressionistic and opinionated, to be compared with the contribution of Titmuss with its precise statistical base. Fyvel is representative of a great mass of contemporary literature on changes in society. Each of his generalizations can be challenged, but the author is trying to make sense, like Masterman and Orwell for earlier periods, of changes that worry and engage him. It would be necessary to go back to the period of the industrial revolution, before 1850, to find another period when there was so much questioning of society.

The bourgeois society of Western Europe which arose out of the French and industrial revolutions has come to an end—not only the structure of bourgeois society but many of its values have gone. In its place we have the shifting moral and cultural values of the new affluent society. I think many people of middle age and middle-class background find it hard to take in this change. We have bred into our bones the values of that solid Western upper middle class which set its stamp on the European class system, on the idea of our school and university élite education, even on the appearance of our cities where every street and suburb had until recently an easily recognizable class character. However, if we are honest we must admit that in the retrospect which all of us now have, this bourgeois society of our parents and grandparents had two great weaknesses. Even in a relatively rich country like Britain, only about 15–20 per cent of the nation participated in the dominant culture; the urban working classes in particular were left out. And, this culture depended on the middle classes having plenty of resident domestic servants. (As late as 1931 there were a million of them in England and Wales.)

The end of bourgeois society came when its own technological advance clashed irretrievably with its class exclusiveness. The affluent society came into being when capitalist mass production had reached such a pitch that the wheels of the economy could only be kept going if all citizens of the society, men, women and children of all classes, every one, were turned into consumers on the largest possible scale. Most Western capitalist societies have advanced at least some distance towards this goal. In Britain the decisive point was probably reached in the mid-fifties, when all signs suddenly pointed to boisterous new forms of mass consumption and mass entertainment; this was also the case on the Continent.

* * *

THE AFFLUENT BACKGROUND

Among the distinct and recognizable features of the new affluent society is the creed that 'everyone must consume'. In Britain one sees it illustrated visually in that huge mass march of the British working classes, out of their drab nineteenth-century streets and slums towards the ideal of new suburban life with a small house, a small garden, a small car and a large television set (but perhaps only small bliss). In the face of this irresistible drive to create consumers in the mass, the old bourgeois exclusiveness of the upper and middle classes, based on large houses and plenty of domestic servants, has simply crumbled away as the servants departed into factories and shops, to become themselves earners and spenders; and as the terrace houses of the old middle-class families were divided into flats for the new. Indeed, behind the glittering barrage of advertising and publicity campaigns, the historic institutions and other concerns of middle-class England have often seemed strikingly pushed into the background, somehow reduced in authority, have become dated.

Out with the old, in with the new. To cope with the demands of regulating consumption, the affluent society requires a new, enlarged middle class of administrators, scientists, technologists, communicators and the rest, which in numbers is far larger than was the old-style *bourgeoisie* of independent entrepreneurs and professional men. And if the old sense of sturdy English upper middle-class independence is gone, there is a new professionalism in its place.

To train up this new class, the education system of the affluent society has constantly to be re-vamped and enlarged—in fact, it comes to be regarded explicitly as an instrument for affecting class and technical change. This expansion must also make for more egalitarian education governed by merit; never mind if old forms are retained. I have found it interesting to watch the transformation of Oxford and Cambridge from institutions for the confirmation of special status (as they were in my time) to institutions for its acquisition.

THE ECONOMY

83. **The Post-war Balance of Payments Problem (1950)**

H. D. Henderson, *The Inter-war Years and other Papers* (Oxford University Press, 1955), pp. 413–14.

If the nineteenth century had been one of continuous economic stress for the British economy, the twentieth century has been one of continuous strain. Why? Here is an analysis, made in 1950, by a distinguished economist of the main problem of the post-1945 period, problems centring on the balance of payments in an economy so dependent on international trade.

During the past generation the economy of Great Britain has been subjected to strains the severity of which is not I think sufficiently appreciated. There have been two major world wars. The first of these accelerated and intensified certain inevitable changes which would in any case have given rise sooner or later to problems of adjustment of some difficulty. With the advantage of after-knowledge we can see that it was inevitable that industrialization would spread gradually throughout the world, and that this would prejudice the position of some of our older-established exporting industries, of which the Lancashire cotton industry is the most obvious example, which had done an immense and rapidly growing trade in the days when Great Britain had a virtual monopoly of industrial exports. The disturbance of normal trade relations during the First World War speeded up this change, which might otherwise have come about more gradually. The industries which suffered from inability to regain their pre-1914 export markets happened to be highly localized industries, concentrated in areas in the economic life of which they played so prominent a part that most trades in their neighbourhood suffered from their depression. Thus there arose the problems of depressed areas and depressed industries, which were quite foreign to our previous experience, and which called for measures of readjustment to which the ideas that then prevailed were not attuned. Moreover, our attempts to solve those problems were embarrassed by the onset during 1929 of a world depression of unparalleled severity and obstinacy which dislocated the established mechanism of international exchange, and left behind it increased difficulties for most types of international trade.

The Second World War has left us with a problem of readjustment of another sort, the problem of restoring equilibrium in our balance of external payments. Many other countries were faced with a similar problem after the First World War; and the difficulties of those countries played a major part in the collapse of the free international economic mechanism to which I have referred. But in Great Britain, though we emerged from the First World War with a greatly weakened balance of payments position, and with a much smaller margin available for fresh overseas investments, we had still a margin on the right side of the international account, at any rate until the very end of the inter-war period. We could thus accept the decline of British export trade, and seek a remedy for the unemployment it caused in increased production for the home market.

Such policies are not open to us today. We have now to establish a balance between our export earnings and our expenditure on imports radically different from that with which we made shift before the war. We have not only to establish this radically different balance; we have to maintain the years that lie ahead, in times of bad trade as well as in times of good trade, or at least as an average over bad times and good; and we have to do this under world conditions which remain far less friendly to the expansion of international trade than were those of the Victorian age.

This has been our central problem in economic policy since the end of the war, and it is likely to remain our central problem for a long period to come.

84. The Problem of Inflation

> *Growth in the British Economy. A Study of Economic Problems and Policies in Contemporary Britain* (Political and Economic Planning, London, Allen and Unwin, 1960), pp. 21–3.

If the balance of payments problems was the major external constraint on the British economy after 1945, inflation has been the major internal constraint, especially in the 1960s. Inflation particularly affects the allocation of resources, and whether or not inflation has been mainly demand-induced (from rising real wages) or supply-induced (from the rising cost of inputs), it has been recognized by both the main political parties as a major obstacle to economic growth. Neither party, however, has been able successfully to restrain inflation, and the Labour government's wage-freeze was almost certainly harmful in its effects.

INFLATION AND THE ALLOCATION OF RESOURCES

Increased production would be the most satisfactory way of solving Britain's economic problems, but in circumstances where production could not be immediately increased, the allocation of resources has been a vital matter. With defence spending that has been no more than has been necessary on political and military grounds, and social service expenditure that has not been greater than has been needed to maintain a healthy and well-educated population, then only one other item in the national accounts can logically be blamed for the failure to devote sufficient resources to investment and the strengthening of the foreign exchange reserves: personal consumption expenditure. If other cuts, in overseas investment, or in defence spending, or in the social services, were not acceptable, then the only possible conclusion is that personal consumption ought to have been restrained in order to put the economy into a more healthy position. But a vague appeal to 'take in belts' is not good enough: it is necessary to be quite clear about what has happened to consumers' expenditure since the war, and what is involved in any proposal for restraint in the interests of more investment.

Comparisons with other countries show that consumption

has been taking a higher share of the national income in Britain than in any other comparable country. At the same time a considerable measure of restraint of consumption has already taken place since 1938. Consumption in the United Kingdom has not advanced at as rapid a rate as production because of the extra calls that have been made on national resources in comparison with 1938. But it still takes a high proportion of the gross national product, not much lower than in 1938, which was by no means a model year. It has not been fully adjusted to the radically different circumstances of post-war years, and runs at a higher level than in most Western industrial nations. It is concluded that if all the other calls on resources are accepted, or at least recognised to offer only a limited scope for reduction, then consumption expenditure has not been restrained sufficiently to keep the economy in a healthy condition and to permit a satisfactory steady rate of growth.

What has prevented the Government from controlling consumption at a level that would have left room for the re-establishment of a surplus on the balance of payments, for defence programmes, overseas investment, and an adequate rate of investment at home? The answer is to be found in the failure to control sufficiently the pressures of demand in the economy, and to solve the problem of costs and prices continuing to rise even in years when the pressure of demand has not been high. The Government has available a whole armoury of monetary and fiscal measures for the regulation of demand, and some physical controls could also be used. The varying combinations of these measures that have been used in the last fourteen years have not succeeded in completely preventing strain on resources and consequent price increases and balance of payments crises. The policy of deflation, which can only succeed by reducing the level of employment and weakening the bargaining power of the unions, is bound to be regarded with disfavour in many sections of the community, and particularly among workers. Many people have therefore hoped for a policy that could cure inflation directly, by co-operation with the unions in ensuring that the average percentage wage increase did not exceed the average percentage increase in productivity.

Such a policy of co-operation need not be confined to preventing inflationary wage increases, but could be extended to positive efforts to achieve a faster rate of economic growth. Indeed, the best way to ensure that the percentage increase in wages does not in any year exceed the percentage rise in productivity is to increase more rapidly each year the level of output per head. In this way price stability can be achieved with a high rate of growth instead of by putting the economy into a straitjacket. Although this co-operative solution may be accepted as superior to the policies used in the post-war period, it would be idle to assume that having stated the ideal solution it can immediately be put into effect. All that can be proposed is that all efforts should be made, by the Government, by the employers, and by the unions, to increase co-operation in economic life. This process is bound, however, to be a slow one, having to overcome deep-seated prejudices and to break down institutional barriers, and it will be many years before it can make a significant contribution to the better functioning of the economy.

85. A National Plan for Economic Growth

The National Plan (Cmd. 2764, 1965), pp. 1–5.

The Labour government had two policies for the future of the economy: a 'prices and incomes policy' and a National Plan. The National Plan aimed not at controlling the economy so much as at encouraging and harmonizing; it was a matter of aspiration, not a matter of practical politics. The Department of Economic Affairs would collect estimates of production from businessmen, match their consistency and encourage increased growth. Thus businessmen could operate in a surer context of government support. It is fair to say that the National Plan did not have much operational significance.

1. This is a plan to provide the basis for greater economic growth. An essential part of the Plan is a solution to Britain's balance of payments problem; for growth cannot be maintained unless we pay our way in the world.

2. For too long the United Kingdom has suffered from a

weak balance of payments. Periodic crises have led to sharp checks to economic expansion and productive investment; these in turn have left us vulnerable to further balance of payments difficulties when expansion was resumed. It is the Government's aim to break out of this vicious circle and to introduce and maintain policies which will enable us to enjoy more rapid and more sustained economic growth.

3. This will only be possible when the underlying weaknesses of the economy have been removed. The policies set out in this Plan are designed to do this; but they will take time. The Government have therefore found it necessary to deal with the balance of payments deficit by a series of short-term measures; some operate directly on the balance of payments, some make room for increases in exports and import-replacement, some ease the pressure of demand at home in overloaded sectors. These measures should provide time in which the policies for sustained growth can become fully effective.

4. In 1964 the overall balance of payments deficit rose to an exceptionally high level. It is hoped that at least half the deficit will have been removed this year. But there is still a substantial imbalance to be removed, and the debt incurred in 1964 and 1965 will to be repaid during the rest of the decade.

5. The task of correcting the balance of payments and achieving the surplus necessary to repay our debts, while at the same time fostering the rapid growth of the economy, is the central challenge. We must succeed if we are to achieve all our objectives of social justice and welfare, of rising standards of living, of better social capital, and of a full life for all in a pleasant environment. We must succeed if we are to take our proper place as a great trading nation, with ties with the Commonwealth, with Europe (particularly the countries of the European Free Trade Association), and with America. Commonwealth Prime Ministers have agreed to hold meetings to discuss the expansion of Commonwealth trade and to consider the extent to which each country's production and plans could meet requirements in other member countries. We must pay our way in an increasingly competitive world in which we have for too long been losing ground steadily to other industrial countries. This is a trend that we must not allow to continue.

6. The Plan is designed to achieve a 25 per cent increase in

national output between 1964 and 1970. ... Industries were asked what 25 per cent national growth from 1964 to 1970 would mean for them. The co-operation received has been excellent.

17. Perhaps the most encouraging result of the inquiry was in the field of *exports*. The replies from industry suggested that these could grow by about $5\frac{1}{2}$ per cent a year in volume. This is substantially faster than the average of about 3 per cent a year over the past decade. It is also substantially faster than the forecasts given by industry to the National Economic Development Council in its 1962 inquiry. The changing geographical and commodity composition of our exports makes it reasonable to expect a faster expansion in future, as a growing proportion of them is now going to the more rapidly expanding markets and is in the more rapidly expanding lines; there has already been some acceleration in the growth of exports in recent years. Some of the reasons given by industries for expecting a further acceleration are outlined in Chapter 7 (paragraphs 61 to 65). While the growth of world trade in manufactures may slow down in the immediate future, in the longer run the rate of expansion should not be significantly lower than in the past decade when it has averaged rather more than 7 per cent a year. There should thus be scope for a considerable expansion in our exports. Nevertheless, the estimates received from industry must be treated with caution; projections are particularly difficult in this field. The forecast rate of expansion will certainly not come about automatically. To obtain it important changes in attitudes and policies are required, a point which was made clear by the industries concerned. For the purpose of the Plan a rate of increase of about $5\frac{1}{4}$ per cent a year has been taken in the volume of exports. This is approximately the rate required to achieve our balance of payments objective....

18. The inquiry suggested that national *productivity* (output per head) could grow by 3.2 per cent a year between 1964 and 1970. ... This is substantially faster than the average growth over the past 10–15 years, but less than the rate of 3.4 per cent required to achieve the growth programme. In the light of past trends (*see* paragraph 8 above), it should be possible to improve on industry's forecast, and to do substantially better given new policies to raise industrial efficiency and

economise manpower; but again fundamental changes in attitudes are required.

19. Since the productivity increase forecast by the Industrial Inquiry is not yet quite enough to achieve 25 per cent national growth, there is an apparent *'manpower gap'* with the demand for extra labour of 800,000 exceeding the 400,000 likely to become available without changes in policies. The resulting gap of 400,000 could be reduced to about 200,000 by successful regional policies. No great significance can be attached to this precise figure, given the difficulties of forecasting supply and demand for labour five years ahead; and it is not large in relation to a total labour force of over 25 million. But it is substantial in relation to the *growth* of the labour force.

20. The inquiry revealed the need for large *movements of labour*, with three major sectors—agriculture, mining and inland transport—requiring some 400,000 less workers; other industries, including aircraft, textiles, clothing and footwear, 200,000 less; while other sectors were estimated to require an extra 1,400,000 workers, the major claimants being mechanical and electrical engineering, construction, public administration, health, education and other services. There have been large movements of labour in the past. But with total manpower going up very slowly in the next five years it is particularly necessary to get labour redeployed from where it can be spared to where it is needed. It is important that this redeployment should be planned so far as possible in advance.

21. If the growth programme is to be achieved, *investment* by manufacturing industry as a whole, and by the construction industries, must be increased more rapidly than in the past. Otherwise there is a danger of insufficient increases in productivity and manpower shortages, as well as of insufficient capacity to export and to compete with imports. Import studies carried out by the Economic Development Committees and by the National Economic Development Office have brought out the way in which insufficient capacity, or insufficiently labour-saving capacity, has led to large increases in imports of manufactures when demand has increased substantially in the past; examples are chemicals, paper and board, steel, cement, machine tools.

22. The Industrial Inquiry and the work of the Economic Development Committees have helped to pin-point *specific*

areas of strain and the problems that have to be resolved for the growth programme to be achieved.

86. The Issue of Planning

J. Jewkes, *Ordeal by Planning* (London, Macmillan, 1948), pp. 205–8.

There has been a long tradition of *laissez-faire* in Britain, and even when the problems of the mid-twentieth century economy seemed to indicate the need for planning, there were voices like those of J. Jewkes, a liberal academic economist, and, later, Enoch Powell, a Conservative M.P., which argued for 'freedom' in the market place as well as in the ballot box.

THE DESTRUCTION OF INDEPENDENT THOUGHT AND CRITICISM

In the long run the planned economy destroys the independent habits and attitudes through which alone freedom can be preserved. As private property diminishes in importance through penal taxation, the lowering of the rate of interest and the growing relative importance of State property, fewer and fewer people are in the independent position in which they can fearlessly criticise Government policy without risking their livelihood and the security of their family. The number of people grows whose incomes wholly or partly depend upon keeping their mouths shut and their thoughts private. The planned economy always involves a great increase in the number of Government officials who can hardly criticise their employer without risking their chances of promotion. Business men operating in what is left of the free economy know only too well that there are innumerable ways in which outspoken critics of official muddles can be penalised. They may tell in private their stories of planning inefficiency but, in self-defence, they dare go no further. And some professional classes, such as accountants and lawyers, often stand to gain, at least for a time, out of the conditions which exist under extensive Government intervention.

Independence is further undermined by the deliberate destruction or the progressive atrophy of voluntary organisations and associations. These forms of co-operation are not 'plannable instruments' and must, therefore, be frowned upon in the planned system. Voluntary associations are the lifeblood of free society; they have in the past led to much of our progress in education, social insurance and health services because they have left the way open for groups of like-minded people to experiment with new ideas and to criticise existing methods by showing the way to do better. They are hardly likely to survive in an environment in which it is assumed that the State has taken upon itself the responsibility, often to the deliberate exclusion of private effort, for all social services.

The planned economy must finally destroy the very instruments of free speech. The burden thrown upon the legislature by the enormous mass of work involved in a planned economy inevitably drives the executive to restrict the freedom of debate in the Houses of Representatives. When resources have to be allocated between rival uses, the claims of the instruments of free speech will be relegated to second or third place. Harassed by the interminable complexities of their own system, the planners must finally be driven, in order to keep economic life in operation at all, to cut through their knots by making arbitrary decisions and stifling unwelcome criticism.

Perhaps, however, for the mass of the people the whole atmosphere of independence and freedom is most insidiously destroyed by the proliferation of minor officials, essential for the working of the plan, each of whom is charged with certain powers over our everyday actions. These officials are no better or worse than any of us. Most of them may be conscientiously anxious to carry out their duties and to use their powers with discretion and understanding. But the system which brings them into existence is dangerous. They are conscious of their power, they (and those who are subject to them) recognise the inconvenience of recourse to appeal against the exercise of that power. These are the conditions which may multiply petty tyranny of the most obnoxious kind. The network of power many extend quietly without it being remarked. The Prime Minister revealed in February 1947 that seventeen Ministries have power to authorise inspections involving the entry into private houses and premises without a search warrant. It later was admitted that 10,916 Government officials were authorised

THE ECONOMY

to carry out inspections and investigations without a search warrant. The 'snooping' called for in enforcing regulations leads to the creation of a new body of plain-clothed police whose work may differ little from that of the *agent provocateur*. This is the sordid atmosphere which breeds the anonymous informer and everywhere sets one man against another.

87. The Development of Nuclear Power

Report from the Select Committee on Nationalised Industries, vol. I, Report and Proceedings (House of Commons, 28 May 1963), pp. 114-15.

The new industrial revolution is to be based on new sources of power (nuclear energy) and new forms of economic organization (automation). Britain pioneered nuclear power stations, and the extract summarized the programme up to 1963.

360. The constant theme in the development of the nuclear power programme in this country has been the need for a third fuel to supplement existing conventional sources of energy. According to the Command Paper published in February, 1955, which set out the original programme, nuclear power to generate electricity was coming at a time when the country's growing demand for energy, and especially electricity, was 'placing an increasing strain on our supplies of coal and makes the search for supplementary sources of energy a matter of urgency. This programme was for a capacity of 1500–2000 megawatts of nuclear power by 1965; the objective in fact adopted the industry was 1800 megawatts. Apart from the need for a third fuel, the cost of nuclear power from the first station was expected to equal that of conventional power and to fall rapidly as experience of the new processes broadened.

361. In March, 1957, the fuel situation and technical progress that had been made in nuclear generation prompted the Government to treble the size of the original programme: the plan was now for 5,000–6,000 megawatts of nuclear capacity by the end of 1965. Later in 1957 restrictions on capital expen-

diture caused the programme to be stretched out to the end of 1966.

362. In June, 1960, the enlarged nuclear programme was again considerably stretched out; only 3,000 megawatts of nuclear capacity would now be reached by 1965, and 5,000 megawatts not until 1968. As a new Command Paper explained, the fact that coal had become more plentiful and the supply of oil had also improved removed 'the need on fuel supply grounds for an immediate and sharp acceleration in the rate of ordering nuclear capacity'. Moreover, the high hopes entertained in 1955 of the comparative costs of nuclear and conventional power had not been realised. In these circumstances the Government decided to place orders for nuclear stations at the rate of roughly one every year. Although the programme had been cut down, the Command Paper repeated that the country's growing energy demands would call for 'more and more supplies of nuclear power'. As the Chairman of the Generating Board put it to Your Committee, the revised programme took account of the fact that 'unquestionable nuclear power will be needed in the 1970's to supplement conventional fuel resources....'

364. Two nuclear power stations at Berkeley and Bradwell began sending out power to the Grid in 1962. Six more stations at Hinkley Point, Trawsfynydd, Dungeness, Sizewell, Oldbury and Wylfa in Anglesey are being built. They are planned to come into operation, roughly at the rate of one a year, up till 1968. This programme together with another station planned to begin in 1964 and with the programmes of the Atomic Energy Authority and the South of Scotland Electricity Board will produce a capacity of 4,900 megawatts of nuclear power by 1968 to conform to the objective set out in the Command Paper of about 5,000 megawatts of capacity in 1968'. All these stations use the type of reactor which the Atomic Energy Authority pioneered at Calder Hall, known as the magnox reactor because its fuel element consists of metallic uranium contained in cases made of magnesium alloy.

365. 'The industry has no doubt about the long-term need for a programme of nuclear power; what differences there have been have concerned the size of the programme.' These words in a memorandum submitted by the Electricity Council reflect some of the chequered history of the nuclear power pro-

gramme. Your Committee have already shown how the programme from its modest beginnings in 1955 was much inflated in 1957 and has since then been scaled down on two occasions. The Central Electricity Authority took no part in the detailed preparation of the original White Paper and were given only a month or so to comment on it in draft. Nevertheless, on the information available at the time, they accepted the proposals as a reasonable approach to the development of nuclear energy. In 1956–57 a Committee which was appointed to review the nuclear programme, prepared an appreciation of three possible levels of expansion of the programme. The largest programme was adopted although the industry preferred the middle level of expansion.

88. The Special Case of Agriculture

Report from the Select Committee on Agriculture, 1968–9 (London, H.M.S.O., 1969), pp. vi–viii.

Agriculture, having been allowed to decline before 1914, was gradually backed by the government, tentatively between the wars, and massively after 1945. This report explains the functions of the Ministry and its costs.

THE FUNCTIONS OF THE MINISTRY OF AGRICULTURE, FISHERIES AND FOOD

2. The Agriculture Act 1947 gave the Ministry the objective of 'promoting and maintaining ... a stable and efficient agricultural industry capable of producing such part of the nation's food and other agricultural produce as in the national interest it is desirable to produce in the United Kingdom, and of producing it at minimum prices consistently with proper remuneration and living conditions for farmers and workers in agriculture and an adequate return on capital invested in the industry.' In a speech on 13th November 1968 to the Farmers' Club the Minister of Agriculture, Fisheries and Food defined the department's agricultural policy as having four principal objectives:

(a) to ensure an economically efficient allocation of resources between agriculture and the rest of the economy and within agriculture;
(b) to improve the stability of prices and incomes of farmers;
(c) to increase farming efficiency;
(d) to meet certain social objectives.

Of the four aims of the Ministry we have concentrated our attention on the first two. We have not specifically considered the department's duty to protect the interests of consumers, its responsibility to maintain the well-being of rural communities nor its functions relating to forestry.

The Place of Agriculture in the National Economy

3. Agriculture produces between 3 per cent. and $3\frac{1}{2}$ per cent. of the gross national product and employs a similar percentage of the working population. In common with all other industries, it has to operate within the framework of the economy as a whole. However, its overall profitability, and the relative profitability of its various sectors, are far more directly affected by decisions of government than that in most other industries. This is the result of greater dependence on direct Government support and of the methods by which that support is given. It is difficult to measure the value of Government support to an industry given by tariffs or quota protection compared with that given by subsidies. The Minister in his speech to the Farmers' Club also stated that 'agriculture gets about the same or possibly a little higher protection than manufacturing industry as a whole.... Agriculture is not in an exceptionally privileged position.' Its position is, however, distinctive in that its principal support is direct, in cash from the Exchequer. Theoretically when the allocation of resources to different industries is assessed, it should not matter whether supports rests on tariffs or on subsidies. In practice, the fact that the Treasury has to provide the money for subsidies does colour decisions and does result in agriculture being controlled more directly and in more detail than most other industries....

* * *

THE ECONOMY 339

6. The cost of agricultural support is borne in part directly by the Exchequer as payments to farmers for price guarantees and production and improvement grants and in part by the consumer. It has, for example, been estimated to us by a witness from the Treasury that 75 per cent. of the cost is borne by the Exchequer and 25 per cent. by the consumer, the consumer's contribution being almost entirely for the maintenance of milk prices. The estimated total cost of Exchequer support for 1968–69 is £286 million. Of this, £145·7 million is for the implementation of price guarantees and £97·3 million for relevant production grants whose cost is included in the determination of guarantees; a further £30 million (which is not taken into account in the determination of guarantees) is also estimated for other grants and subsidies and £11·1 million for administrative expenses. Finally, there is an estimated £2·2 million for the special benefit of agricultural producers in Northern Ireland. The total cost of support declined from the 1961–62 peak of £343·2 million or 9·32 per cent. of total Civil Expenditure in that year to £229·1 million or 3·82 per cent. of total Civil Expenditure in 1966–67. In terms of both real cost to the taxpayer and as a proportion of public spending the cost of agricultural support thus fell substantially from the level of the early 1960's. The increase since 1966–67 has been due to both the higher price guarantees and the higher production grants needed to enable the industry to meet the targets of the Selective Expansion Programme. The increase has also reflected the rise in agricultural costs, partly as a result of devaluation. Your Committee has not examined in detail the estimates made in the Annual Review and Determination of Guarantees, but the total cost of supporting the Review products represents 3·5 per cent. of the estimated cost of all supply services other than defence for 1968–69. Although this is a relatively small proportion of total Treasury expenditure, it is nevertheless a large sum of public money.

89. Monopoly and Competition

Beer. A Report on the Supply of Beer
(Monopolies Commission. H.M.S.O., 1969),
pp. 95–7.

Monopolies had not threatened the British economy in the past, in spite of the allegations of a small 'anti-trust' literature. Indeed excessive competition and smallness of scale had been seen as a weakness of the economy. With the growth of large-scale business, however, there were fears of monopoly and business action 'against the public interest'. The Monopolies Commission was set up in 1948 to decide when the public interest was threatened and to recommend against proposed mergers and existing monopolies. The following case on the supply of beer, is typical of the analyses of the Commission.

334. In 1967 the total quantity of home-brewed and imported beer delivered for consumption in the United Kingdom amounted to approximately 32m. barrels. ... There are a few small brewers, responsible in aggregate for a negligible proportion of this total who own no licensed retail premises; subject to this exception virtually all of the total was delivered by brewers who, either directly or through subsidiary companies, own both managed and tenanted licensed premises to which they deliver beer. We are primarily concerned in this reference with the supply of beer within the United Kingdom for retail on licensed premises. Where a brewer delivers beer to his own managed premises it is perhaps arguable that he cannot be said to supply himself and that in such a case, the beer is not at any stage supplied for retail on licensed premises. We do not think it necessary to resolve this technicality since it is quite clear that, whether or not beer supplied to the public through brewers' managed houses should be deducted in calculating the total amount of beer supplied within the terms of the reference, virtually all of the total amount is supplied by brewers who own tenanted licensed premises and who supply beer to the tenants concerned on terms which restrict the tenants as to the sources from which they may buy beer and, effectively, as to the brands of beer they may buy, stock and sell.

THE ECONOMY 341

335. We therefore conclude that conditions to which the 1948 Act, as amended, applies prevail as respects the supply of beer within the United Kingdom for retail sale on licensed premises because at least one-third of all the beer so supplied is supplied by persons who so conduct their respective affairs as to restrict competition in connection with the supply of beer inasmuch as, being the owners of licensed premises, they prescribe the brands of beer which shall and shall not be sold on such premises.

338. In the United Kingdom as a whole there are at present nearly 140,000 licensed outlets,* of which nearly 110,000 have on-licences* and the rest off-licences. Of the on-licensed outlets about 75,000 are public houses (including some 2,000 hotels with full on-licences), about 26,000 are clubs (mainly registered), the remainder being restaurants, residential hotels etc., with limited licences.† Brewers own about 58,500 (78 per cent) of the public houses and about 9,000 (30 per cent) of the off-licensed premises,‡ but no clubs and very few of the outlets with limited on-licences; and altogether the outlets owned by brewers amount to nearly 50 per cent of all outlets. Brewer ownership is largely concentrated in England and Wales where they own 86 per cent of the public houses and 34 per cent of the off-licensed premises, or 53 per cent of all outlets;§ in Scotland they own only 15 per cent of the outlets and in Northern Ireland one brewer owns one outlet. The most important elements in the 'free' retail trade consist of about 26,000 clubs (which are effectively on-licensed) and about 16,000 off-licensed outlets;¶ the rest of the free trade is made up of about 14,400 public houses (of which half are in England and Wales), about 2,100 other outlets with full on-licences (hotels, British Railways bars etc.), and about 8,000 outlets with limited on-

* i.e. including registered clubs, which we regard as effectively on-licensed.

† These figures are necessarily approximate, since licensing conditions differ as between England and Wales, Scotland and Northern Ireland.

‡ Nearly 5,000 of the approximately, 30,000 off-licences are believed to be held for technical reasons by wholesalers who do not, in practice, conduct a retail trade. Thus the brewers own nearly 36 per cent of the off-licensed retail premises.

§ As to the existence of local monopolies (in the sense that particular brewers own a high proportion of houses in particular areas).

¶ i.e. excluding the off-licensed wholesalers.

licences. Purchases by 'free' outlets account for about 34 per cent of brewers' total home sales of beer; purchases by clubs account for nearly two-thirds of this (i.e. 20 per cent). We have no comparable figures for other products sold in licensed premises but inasmuch as a higher proportion of wines and spirits than of beer is supplied through off-licensed outlets it is likely that the 'free' retail outlets have a larger share of the wines and spirits trade than of the beer trade.

339. Rather more than one-quarter of the licensed premises owned by brewers (i.e. about 14,000 public houses, and about 4,000 off-licensed premises) are 'managed' houses; or in other words in nearly 13 per cent of all outlets in the United Kingdom the brewers themselves act as retailers of beer and of the other products sold there. The rest of their houses are let to tenants who, in effect, conduct the retail trade on the terms and conditions laid down by the brewer-landlords. The free retail trade buys most of its beer direct from the brewers, who also sell beer direct to one another; thus, although there are some independent bottlers and wholesalers, the greater part of the wholesaling trade in beer is in the hands of the brewers. As to products other than beer we again have no precise statistics, but since the supply of virtually all the wines, spirits, cider and minerals required by their tenanted as well as their managed houses is channelled through the brewers, who also have some trade with free houses, it is clear that a very substantial part of the wholesaling trade is in their hands. As to production of these other goods, brewers now control the sources of about 40 per cent of the cider supplied in this country but only certain minor sources so far as spirits and minerals are concerned; they are also importers of wines. In all these fields the brewers meet strong competition from independent producers (or, in the case of wines and some spirits, importers) but their interests have increased substantially in recent years, partly by mergers with formerly independent suppliers, partly by setting up or adding to their own production or importing businesses.

90. Wages and Earnings in Shipbuilding and Engineering

H. A. Clegg and R. Adams, *The Employers' Challenge* (Oxford, Blackwell, 1957), pp. 41–5.

A dispute in the shipbuilding and engineering industry in 1957 led to the biggest strike in twenty years. Clegg and Adams explain just how complicated the wage structures were in both industries, and how difficult it was, therefore, to negotiate. The dispute, essentially, lay in the determination of the employers to resist 'inflationary' wage claims; the result, an inquiry which awarded pay-increases slightly less than the unions' demands.

It is notorious that both industries have complicated patterns of wages and earnings. So much so that some would deny that they can properly be called 'structures'. It does not follow, however, that it is impossible to describe them in a manner which is simple, but nevertheless reveals their main outlines.

In both industries national negotiations are primarily concerned with two rates, one for skilled and the other for unskilled workers. The rates for semi-skilled workers are then determined in relation to one of these. In engineering this part of the process is usually left to local arrangement. Some crafts enjoy a differential above the skilled rate either by national agreement, tradition or local arrangement. London and some other districts provide a margin above the national rates for all their employees. The national decision on the two rates, however, raise or lower the whole structure of basic rates.

Piece-work is common in both industries. In engineering the national agreement lays down that prices must allow a worker 'of average ability' to earn enough to exceed the time rate by a given percentage. The fixing of piece-prices is left to the individual works. In practice the prices are usually negotiated by shop stewards. In shipbuilding the prices for the members of the Boilermakers' Society are laid down in district lists. Most of the lists were drawn up many years ago and require revision. The employers allege that the union is unwilling to permit the reduction of 'loose' prices to compensate for increases in 'tight' prices, and allowances have therefore to be made to permit workers on the latter jobs to reach reasonable earnings.

Besides piece-work there are other systems of payment by results under which bonus payments above the time rate are related by a variety of formulae to the output of an individual or team. The returns of the Ministry of Labour for 1956 estimate the proportion of workers paid under all systems of payment by results in both industries to be 52 per cent. The employers' figures are rather higher.

Still other workers receive 'lieu' rates to encourage them to work at piece-work speed although the output of their job cannot be measured.

There are also what the engineering agreements refer to as 'merit rates commonly applied'. These are additional amounts paid to groups, departments or the whole of a works, possibly to overcome a shortage of workers, or to meet some other labour problem.

Like other industries, engineering and shipbuilding provide extra payment for overtime, week-end work, night-work, shifts and so on.

This does not exhaust the means whereby an engineer or a shipyard worker may raise his earnings above the national base rate, but the list includes the most important devices, and is sufficient to explain why the majority of workers in both industries earn more than that rate. The engineering employers estimate that during the week of the earnings survey of June 1956 only a small number of workers received no more than the national rate. Amongst fitters (the largest of the crafts) the proportion was 2·27 per cent, and amongst labourers (excluding foundry labourers) 14·82 per cent.

Average earnings in the two industries thus exceed the basic rates by a considerable margin. The Ministry of Labour's figures for earnings for the last pay week in October 1956, give average weekly earnings in the engineering group of industries as 251s. 10d. and in shipbuilding and ship-repairing as 261s. 11d. This compares with engineering base rates of 168s. 4d. for craftsmen and 142s. 4d. for labourers, and shipbuilding rates of 170s. 6d. for craftsmen and 142s. 6d. for labourers. Another comparison is with the Ministry's figure for all industries, which was 237s. 11d. The excess of engineering over the national average cannot be explained by higher overtime working, since the Ministry's figure for average hours in engineering was 48·1, and for all workers 48·5. Average hours in shipbuilding, however, at 50·1, exceeded the national average.

91. Industrial Relations at Fords

Report of a Court of Inquiry into the Causes and Circumstances of a Dispute between the Ford Motor Company, Ltd., Dagenham and Members of the Trade Unions Represented on the Trade Union Side of the Ford National Joint Negotiating Committee (London, H.M.S.O., April 1963), pp. 10–12.

The motor car industry has become in Britain, as in the U.S.A., the barometer of the economy's health. However, in spite of its high wage structure, it has a bad record for strikes. The Court of Inquiry of 1963 inquired into the reasons for the troubled industrial relations at Dagenham.

INDUSTRIAL RELATIONS IN THE DAGENHAM ASSEMBLY PLANT

27. This new building, however, is regarded by the Company as the main source of their troubles. It had been realised that when production commenced in this new building on 1st April, 1959 there would be problems to be faced concerning the integration of the labour forces drawn from the Fords and Briggs Plants. The Briggs or Body Group employees had built up over the years many customs and practices of their own and a similar situation existed with employees from Fords or Chassis Group. Being aware of these conditions the Assembly Plant Management, particularly bearing in mind the recommendations of the Cameron Report, felt that this was an opportunity to make a fresh start and to create a new spirit of trust and confidence. They set out to work closely with the Joint Works Committee and with the Convener and through this co-operation the Committee was established as a real Joint Body, with status and a sense of responsibility. The Convener was encouraged to have frequent contact with the Personnel

Manager and many informal exchanges of view took place to the benefit of both sides.

28. In the Company's view, the opening of the new Assembly Plant could have become a model illustration of the value of co-operation and consultation if the efforts of the Management and the attitude of the Convener who had come from the Chassis Group had been parallelled by those of all the Shop Stewards. The Company however stated that the efforts of well meaning Trade Unionists then in office to work in the general interest were thwarted by a group of militant employees, some of whom were to be included among those discharged by the Company on 31st January, 1963. At the end of the first year's operation of the Joint Works Committee the post of Convener had passed to one of these militant Shop Stewards, a Communist, who had come from the Body Group and he was supported on the Committee by other Stewards with a similar outlook. Immediately, the Company statement continued, the previous spirit of co-operation had disappeared, and it became obvious that the new régime was not so much interested in maintaining good relationships as to creating disruption and achieving concessions from the Company in complete disregard of the circumstances and justification of the cases being submitted. At this time there were many stoppages of work and other forms of disruptive action taking place in the Plant, some of which were actively supported by the Shop Stewards on the Joint Works Committee, who appeared to be reluctant to encourage observance of the Procedure unless they were able to achieve some concessions in the process. In the Company's view there was no excuse for the Procedure not being used. Indeed items could be, and were, progressed beyond the Foreman and Superintendent to the Personnel Department and then to the Joint Works Committee or District Official level, but at any of these stages, or even without the Procedure having been used at all, a stoppage of work or overtime ban would take place. Sometimes these breaches of Procedure happened without warning; at other times Stewards informed Supervision that they were going to take place.

29. The Company pointed out that the Assembly Plant is to all intents and purposes a continuous production process from the start of the Paint Shop to the end of the Prepare-for-Sale area where the cars are actually completed, and since the

preceding operations in the Body Group are also dependent upon an uninterrupted flow in the Assembly Plant, any action taken even by a comparatively small group of employees can speedily result in a complete stoppage of production in both Plants. The effect of these disruptions in the Assembly Plant can be seen from the record of the Plant during 1962 as compared with other Plants belonging to the Company. In the Company's Plants outside Dagenham the time lost through industrial disputes in the course of the year was half-an-hour per man, whereas the loss at Dagenham excluding the Assembly Plant was 15 hours per man, and in the Assembly Plant alone the time lost was 78 hours per man. The total man hours lost in the Assembly Plant had been 100,000 in 1960, 184,000 in 1961, 454,000 in 1962 and none in 1963 up to the time when the Company's statement was presented to the Court. Since the Plant opened the Company had had 69 strikes and also 114 overtime bans. This record did not of course show the number of threatened stoppages or overtime bans which did not take place and which would easily have doubled these figures. Many of these were removed on the intervention of Supervision or the Personnel Department and sometimes by the intervention of Shop Stewards or the Convener. The Company mentioned that the effect on Supervision of these constant threats had been quite demoralising. A Supervisor knew that if he remonstrated with an employee or merely moved him to another job, this could stop the Plant, and if he tried to achieve normal efficiency from employees who were not cooperating, the same thing could happen.

92. Collective Bargaining

In Place of Strife (Cmd. 3888, 1969), pp. 9–11.

Profound changes have occurred since the war in the assumptions which govern the relations of men and management. It has become clear that strikes endanger the livelihood of men who are not involved, and in various ways threaten 'the public interest' (for example, by their effect on exports). Moreover a high percentage of strikes (over 90 per cent) are unofficial, a reflection of the inadequate relations between trade union officials and the shop stewards, the voluntary shop floor leaders. The responsibility for good industrial relations, it is now recognized, lies not only with the trade unions, and with management, but also with government. And so both Socialist and Conservative governments have attempted to reform trade unions and the system of industrial relations.

In Place of Strife, a paper largely inspired by Mrs. Barbara Castle, Secretary of State for Employment and Productivity (1968–70), recognized that the existing arrangements were not satisfactory and proposed certain changes of machinery for settling industrial disputes. A Commission of Industrial Relations was to be set up to investigate ways of improving collective bargaining: collective agreements were to be registered and made binding; a compulsory conciliation pause was proposed for strikes; the Secretary of State was to be empowered to require the union contemplating a strike to hold a secret ballot of members; unofficial strikers were to be fined. So great was the opposition of the trade unions that the proposed legislation was withdrawn. A similar, but more stringent bill was introduced by the Conservatives in 1971.

16. Finally, our organised system of collective bargaining has not got to grips with a number of economic and social problems. As the Donovan Report indicated, it has often failed to provide for effective and acceptable collective bargaining arrangements covering matters of common concern to employees and employers. Little has been done to reform outdated and generally condemned procedural agreements—such as those now existing in the engineering industry. Too often employees have felt that major decisions directly concerning them were being taken at such a high level that the decision-makers were out of reach and unable to understand the human

consequences of their actions. Decisions have been taken to close down plants without consultation and with inadequate fore-warning to the employees. Outdated social distinctions between hourly-paid employees and those on staff conditions have perpetuated. At the same time, some employees have opposed and obstructed the spread of collective bargaining to new sections of the work force, especially those increasing numbers employed in 'white-collar' jobs. Unions too have often failed to involve their members closely enough in their work, or to tackle with sufficient urgency the problems of overlapping membership and unnecessary rivalry, which always diminish their effectiveness and sometimes their reputation. Many employer's relations with unions have been greatly complicated by the large number of unions that may have members in a single factory.

17. The combined effect of such defects is to increase the feeling of many employees that they have no real stake in the enterprise for which they work. There are of course other factors too. Britain is passing through a period of rapid technological change. New processes and methods of production are combined with changing patterns of company ownership and management structure. Established jobs and ways of work are disappearing to be replaced by unusual and unfamiliar tasks in surroundings often equally strange. This naturally reinforces feelings of insecurity among employees and even management itself, and results in lack of co-operation and resistance to change, especially if systems for dealing with legitimate grievances and problems of all kinds do not adapt themselves to the demands placed upon them. Efficiency suffers and the community pays.

18. Yet there can be no reversal of the forces of change. On the contrary, the Government has taken action to accelerate change. This is necessary if Britain is to survive and prosper. But it means that we must make sure that employees have the opportunity to participate in influencing the direction of change, that we must overhaul arrangements for dealing with the consequences of change as they affect all who work in industry, and that we must remedy the defects described in the preceding paragraphs. This requires policies to secure four objectives:

(i) the reform of collective bargaining;
  (ii) the extension of the role and rights of trade unions;
  (iii) new aids to those who are involved in collective bargaining; and
  (iv) new safeguards for the community and individuals.

The next four sections of this White Paper explain the major measures proposed by the Government to deal with each of these questions.

The Reform of Collective Bargaining

19. Collective bargaining is essentially a process by which employees take part in the decisions that affect their working lives. If it is carried on by efficient management and representatives of well-organised unions, negotiating over a wide range of subjects, it represents the best method so far devised of advancing industrial democracy in the interests of both employees and employers. It offers the community the best opportunity for securing well-ordered progress towards higher levels of performance and the introduction of new methods of work.

20. Yet as the Donovan Report has shown this is far from being the situation in the economy as a whole. Even where collective bargaining is well developed it has many defects. Very often there is a marked difference between the formal collective bargaining system and what actually happens. It is often supposed that formal industry-wide negotiations are the only important method of collective bargaining; but in practice an increasing amount of bargaining, and an increasing proportion of the wage packet, is settled outside the 'formal system' by informal understandings and arrangements between shop stewards and managers or foremen at work-place level. Yet this concentration on 'informality', and the network of shop floor arrangements and understandings that result from it, create serious problems. Few clear principles and standards are developed to settle shop floor grievances. Managements and unions tend to yield to immediate pressures, especially when applied by minority groups who can exploit their strategic position at the expense of their fellow trade unionists. Anomalies develop in wage payment systems. There

is no stable or equitable relationship between payment and performance. These who are dissatisfied, strike in breach of agreed procedures for dealing with their grievances, because they are ineffective and do not deal rapidly and equitably with the problems of the shop floor.

21. In some industries and firms, managements and unions have made a joint effort to keep their machinery for negotiations and for handling grievances under review, and have made changes to remedy defects and to meet today's needs. But in too many cases these arrangements are still seriously inadequate. Many were designed at a time when procedures for negotiation were needed only at industry level, while all that was necessary at company or plant level was provision for resolving disagreements over the interpretation of the industry-wide agreement. Many procedures are too slow, too informal and too uncertain to be fully effective in the face of rapid technical change. The people who have to work them, on both management and union sides, often have inadequate training. While procedures must operate flexibly to deal with the wide variety of demands upon them, it does not follow that in modern conditions flexibility can best be achieved by informality and complete reliance on good relations. When employees do not have a clear idea what procedures require in given circumstances, they are likely to ignore them as irrelevant. Flexibility can usually be best obtained by designing formal machinery which can work flexibly.

22. The lack of comprehensive, mutually agreed procedures encourages arbitrary behaviour by managements. Indeed, many of the 'wildcat' strikes which cause so much concern today are the result of management's mistaken belief that it has the right to impose changes on its workpeople without full and adequate consultation and then invite them to go through 'procedure' afterwards for the remedy of any grievances. This is to show a complete misunderstanding of what good procedures should be designed to do, namely, to secure the co-operation of employees through their representatives in the changes that affect their working lives. Until this approach is adopted there can be no fundamental solution to the problem of unofficial strikes.

23. Most important of all, perhaps, the disparity between the formal system and the realities of shop floor life is often not

fully appreciated or even understood by senior management in the enterprises where it occurs. As the Donovan Report said, the assumptions of the formal system still exert a powerful influence over men's minds and prevent the development of effective and orderly methods of collective bargaining. Too often senior management continues to regard industrial relations as a matter for employers' association officials or lower levels of management, rather than as one of its primary responsibilities. On the union side many national leaders continue to uphold the assumptions of increasingly ineffectual industry-wide negotiating structures.

* * *

27. The major responsibility for solving the problem lies with management. Given the right help and encouragement most unions will readily respond to proposals for improving the machinery of collective bargaining. The initiative must lie with employers, and notably the boards or chief executives of undertakings, for they are best placed to set in train the detailed study of existing systems and their defects and to make the right kind of positive approaches to trade unions. The best way forward will often be the negotiation of formal, comprehensive and authoritative company or factory agreements. Negotiation at these levels, through a single negotiating body instead of piecemeal negotiations, is likely to secure the introduction of adequate procedures for the settlement of disputes, and the extension of collective bargaining into other matters which directly concern employees, such as discipline and redundancy arrangements. In the context of such agreements, a good management will find itself able to manage its undertaking more effectively, and good trade union representatives will be better placed to promote the interests of their members without unnecessary conflict between different groups.

93. The Low-paid Worker

A National Minimum Wage. An Inquiry
(Dept. of Employment and Productivity.
H.M.S.O., 1969), pp. 17–19.

This inquiry of 1969 shows the industrial and regional distribution of low-paid workers. It is interesting to note that the concentration is in 'the depressed areas' of the 1930s. The inquiry takes up the old theme that a minimum wage would protect the lowest-paid workers and ensure them a reasonable standard of living.

54. It appears that to a considerable extent, low-paid workers are concentrated in cotton weaving and spinning, woollens, and in a number of smallish industries (all covered by Wages Councils) with low median earnings as well as low earnings for the lower paid. A second group of industries have relatively large numbers of low earners, although they do not necessarily form a large proportion of their total labour force. These industries comprise National and Local Government, Construction, Shipbuilding, Motor repairs, Other machinery and Marine Engineering. Increases for low-paid workers in industries such as these could have extensive repercussions upon the earnings of large numbers of workers in the same industry and in other industries.

55. Of the industries not covered by the 1960 survey the most important from the point of low incomes were agriculture, which was the subject of a separate survey, and retail distribution. Both industries are covered by Wages Boards or Councils. The report of the National Board for Prices and Incomes on the pay of workers in agriculture in England and Wales (Cmnd. 3199) found that in 1966 average weekly and hourly earnings were below those in any other industry for which information was available. About a third of male general farm workers earned under £12 a week including overtime, and this was associated with long hours. In retail distribution about 7 per cent of adult male full-time workers earned under £12 a week in October 1966 and nearly 30 per cent earned under £15. Allowing for changes since 1960 these figures put the industry on a par with those listed in Appendix V. Average

hours worked in retailing are however on the low side so that on an hourly earnings basis the industry ranks higher than it does on a weekly basis. ...

57. It is customary for younger workers to be paid less than the adult rate and it follows, therefore, that a national minimum which applied irrespective of age would particularly benefit such workers. This is illustrated in the following table. This shows the variation of average earnings with age among full-year workers in 1964/65, based on an analysis made by the Ministry of Social Security, covering everyone with P.A.Y.E. tax deduction cards who worked for at least 48 weeks in the year.

Age	Men	Women
	£	£
18–19	548	422
20–24	793	507
25–29	978	577
30–34	1,070	542
35–39	1,126	527
40–44	1,137	528
45–49	1,141	537
50–54	1,119	527
55–59	1,075	531
60–64	939	521
65–69	938	

58. It will be seen that for men, earnings increase rapidly with age in the twenties and into the early thirties, reach a peak in the late forties, and thereafter decline slowly. For women the pattern is quite different, the peak earning age being the late twenties. In both cases a national minimum wage would benefit particularly the youngest workers, if it were applied without distinction as to age, and also older workers provided they retained their jobs.

REGIONAL VARIATIONS IN EARNINGS

59. It is relevant to consider the distribution of low earnings between regions because of the impact the introduction of a national minimum might make upon the economies of regions with a particularly high concentration of such people. There are two ways of looking at the regional distribution of employees with low earnings. One is to see what proportion of employees in the country earning less than a stated amount are in different regions, and to compare the figures with the proportions of all employees who are employed in those regions. The other is to look at the distribution of earnings within regions. The Working Party examined both types of data, using figures drawn from the Ministry of Social Security sample of full-year employees with tax deduction cards. . . .

* * *

61. . . . taking £600 a year for illustrative purposes, . . . the areas with above-average proportions of both men and women earning less than this amount were Northern region, East Anglia, South-Western region, Scotland and Wales. Thus a national minimum would have a particular impact on the regions containing the main Development Areas. The areas with above-average proportions of women, but not of men, were Yorkshire and Humberside, North-Western, East Midlands regions. However, it is clear that the differences between regions are very small in comparison with the differences between the percentages for men and those for women in any region.

62. The Working Party were unable to examine the position in Northern Ireland because of the absence of comparable statistics. However, the level of average earnings in Northern Ireland is lower than in any part of Great Britain, which suggests that the introduction of a national minimum might present particular problems there.

94. Overseas Aid

(a) *Assistance from the United Kingdom for Overseas Development* (Cmd. 974, March 1960), pp. 5–6.
(b) *Aid to Developing Countries* (Cmd. 2147, 1963), pp. 8–9.
(c) *Overseas Development: The Work in Hand* (Cmd. 3180, 1967), p. 79.

The major theme of international relations since the war has been the gap between the 'haves' and the 'have-nots', the developed and un- or under-developed economies of the world. In the nineteenth century capital flowed to the undeveloped areas of the world in response to market impulses; today it is doubtful whether the market will provide by itself the capital necessary for the transformation of the backward economies of the world. And so various aid schemes have grown up whereby the developed countries aid the underdeveloped with capital and technical assistance.

ASSISTANCE FROM THE UNITED KINGDOM FOR OVERSEAS DEVELOPMENT

AIMS AND METHODS

(a)
The Aims

1. The Queen's speech on 27th October, 1959, contained the following passage:

> 'The improvement of conditions of life in the less-developed countries of the world will remain an urgent concern of my Government. They will promote economic co-operation between the nations and support plans for financial and technical assistance.'

The Government of the United Kingdom has thus reaffirmed its continuing interest in the welfare of the poorer countries overseas and its intention to go on helping them and to co-operate in international plans directed to that object.

2. Malnutrition is the common lot of millions in the poorer countries of the world. Housing is frequently primitive, trans-

port and communications are very limited, basic social services that we take for granted are almost non-existent. Two-thirds of the world's population live in countries which face problems like these. In India, for example, there are eight times as many people as in this country: their average income per head, about £25 a year, is one-sixteenth of ours. There is a growing determination in the poorer countries to rise out of poverty through economic development. They are making great efforts themselves, and the larger part of the money they are spending on capital development comes from their own resources. But additional resources, and in particular foreign exchange to buy plant and materials from abroad, are an essential supplement. There is a challenge to the advanced countries to help them to help themselves.

* * *

Private Investment

4. The United Kingdom is prominent among the small group of industrially advanced nations from which funds for investment abroad are provided by private individuals and companies. These funds are invested through a variety of channels, and one organisation, the Commonwealth Development Finance Company, has been specially established to facilitate the investment of private capital in the Commonwealth. Information about the extent of United Kingdom private investment overseas is at present not precise, and measures have recently been taken to improve our knowledge. On the information at present available it is estimated that the total has averaged over the last seven years, £300 million a year. This estimate includes sums averaging about £55 million a year that have been raised on the London market by the Governments of independent Commonwealth and Colonial countries and by public corporations and companies.

5. The investment of this money is frequently accompanied by the availability of United Kingdom knowledge and experience of industrial and commercial business. Intangible assistance of this kind cannot be evaluated, but it is an important additional contribution.

6. A large amount, in total, of commercial credit is extended by United Kingdom exporters to overseas importers of United Kingdom goods in accordance with normal trade

practice. Much of this is under guarantee by the Export Credits Guarantee Department; and much of it is in respect of goods supplied to the less developed countries. Such credit is, in general, however, of a short or medium-term nature and is not therefore included in the estimate of £300 million for the amount of private investment overseas.

7. A large part of the £300 million is invested in the more developed Commonwealth countries, where a basis for industrial expansion already exists and where their rate of growth—influenced for example by immigration—is more rapid than they can finance from their own resources and offers an attractive prospect for investment. It is estimated that something of the order of £100 million a year has been invested in less-developed areas. This includes some re-investment of locally earned profits but information about such re-investment is incomplete and the figure of £100 million may be too low.

(b)

17. In the past the British private investor has played a leading part in the economic development of the world; for example, the development of America, North and South, owed much to the funds invested from this country. British private investment still plays a vigorous part in the growth of the developing nations, and we take pride in the fact that the level of our investment is, in relation to our national income, among the highest in the world.

18. British private investment overseas has in recent years been running at something like £300 million a year, largely in the Commonwealth. The greater part of this is direct investment, but the estimate also includes sums raised on the London market by the Governments of Commonwealth countries, and by overseas corporations and companies. Over the past few years new issues by Commonwealth Governments have averaged about £30 million a year, and loans raised so far in 1963 have included issues by the Governments of Malaya, Jamaica and certain of the smaller West Indian islands, as well as the Nigerian Ports' Authority.

19. Much of our private investment is attracted to the more developed countries of the Commonwealth; but something like £150 million a year, or an amount which is not far below the expenditure on aid from public funds, is invested in the de-

veloping countries, again mainly in the Commonwealth. (The Commonwealth Development Finance Company has been specially established to facilitate the investment of private capital in the Commonwealth and continues to expand its operations. Its shares are held by many commercial, financial and industrial interests.) This a substantial item in our balance of payments; but it is one which provides valuable help for the developing countries, particularly as the investment of British capital is often accompanied by the knowledge and experience which enable a skilled judgment to be brought to bear on the importance to this feature of private investment from overseas.

20. There has been a tendency in recent years for the flow of private capital to the developing countries to level off or even decline, and a number of ways of correcting this have been suggested. The United Kingdom has supported the attempt which has been made in the O.E.C.D. to draw up a convention for the protection of foreign investors, in the hope that a clear statement of the pre-requisites for attracting and retaining private capital would encourage the introduction of measures to this end. We are also participating in a study, in the Development Assistance Committee of the O.E.C.D., of the feasibility and desirability of multilateral investment insurance. Further, we support the proposal of the International Bank to set up a centre for conciliation and arbitration on investment disputes. It is not clear what the outcome of these moves will be; but we hope that a wider understanding of the issues involved will result in an increase in the flow of private capital to the developing countries.

21. In addition to private capital investment, a large amount of commercial credit is extended by British exporters to overseas importers of our goods, in accordance with normal trade practice. Much of this is under guarantee by the Export Credits Guarantee Department; and much of it is in respect of goods supplied to the developing countries. This credit is of a short or medium-term nature, apart from the guaranteeing of longer credit where this is necessary to match foreign competition, or where appropriate in the light of the highly selective criteria governing the issue of guarantees for the purchase of capital goods for large projects, and ocean-going ships.

(c)

239. The Ministry directly controls or assists a number of home-based scientific units which help with the problems of the under-developed countries. Out of the provision for science and technology in 1966/67, nearly £650,000 is for the laboratories and institutes directly under the Ministry's control or financed by the Ministry which, in their several fields, combine a number of closely related functions. They undertake some fundamental research; they also carry out applied research and technological development up to the point of commercial application (arrangements have been made with the National Research Development Corporation for the development of some of their projects in this way). They answer enquiries from overseas, some of which require special investigations; they advise the Ministry; they disseminate information; they take in trainees from overseas; they provide staff for advisory visits; and they act as home bases from which staff can be seconded for overseas employment. It is the policy of the Ministry to expand the activities of these units as funds become available. They are: *The Tropical Products Institute; The Anti-Locust Research Centre; The Tropical Stored Products Centre; The Tropical Pesticides Research Unit; The Tropical Pesticides Research Headquarters and Information Unit; The Tropical Section of the Road Research Laboratory;* and *The Overseas Division of the Building Research Station.* (The last two organisations are financed by the Ministry but are not part of its establishment.)

95. Britain and Europe

A European Free Trade Area (U.K. Memorandum to the Organization for European Economic Corporation, Feb. 1957, H.M.S.O.), pp. 3-4.

In March 1957 France, West Germany, Italy, Holland, Belgium, and Luxembourg, 'the Six', signed the Treaty of Rome to form the Common Market, the European Economic Community. Britain chose not to associate herself with the Common Market at first, partly out of a desire not to be involved in Europe, but also because of her Commonwealth commitments. Britain's answer, at first, was a European Free Trade Area (EFTA), but the success of the Common Market finally convinced Harold Macmillan and his government to apply, unsuccessfully, for membership in 1960. The French were successful in keeping Britain out of the Common Market during the 1960s.

UNITED KINGDOM VIEWS ON A EUROPEAN FREE TRADE AREA

1. Her Majesty's Government earnestly desire a successful outcome of the continued efforts in the post-war world to strengthen the cohesion and promote the prosperity of Western Europe. They have therefore given careful thought in recent months to the widely held view that our principal economic need in Europe is to remove existing barriers to trade and develop a single market for manufactured goods. With a population of 250 million, there is clearly a great opportunity which Europe can seize provided that the free circulation of goods is not impeded by tariffs and quantitative restrictions throughout Europe. The need to take large and constructive measures to remove these barriers to trade is the more urgent in view of the pace of technical development which increasingly demands larger markets in order that its full benefits may be obtained. During the past decade a great deal has been done in the Organisation for European Economic Co-operation and elsewhere to lower trade barriers in Europe, but Her Majesty's Government believe that what is needed now is a much bolder approach.

2. With these considerations in mind Her Majesty's Govern-

ment are glad that the negotiations which were set in train in June 1955 for the establishment of a Customs and Economic Union consisting of France, Germany, Italy, Belgium, Holland and Luxembourg are now approaching a successful conclusion. There are, however, substantial reasons why the United Kingdom could not become a member of such a Union. These arise in particular from the United Kingdom's interests and responsibilies in the Commonwealth. If the United Kingdom were to join the Customs and Economic Union, the United Kingdom tariff would be replaced by a single common tariff with the other member countries against the rest of the world. This would mean that goods entering the United Kingdom from the Commonwealth would have to pay duty at the same rate as goods coming from any other third country not a member of the Customs and Economic Union, while goods from the Union would be admitted free of duty. Her Majesty's Government could not contemplate entering arrangements which would in principle make it impossible for the United Kingdom to treat imports from the Commonwealth at least as favourably as those from Europe.

3. At the same time it is of great importance in the view of Her Majesty's Government to establish free trade over as wide an area as possible within Western Europe. It was for this reason that Her Majesty's Government strongly supported the decision taken by the Council of O.E.E.C. in July 1956 that a study should be made urgently to discover whether other member countries of the Organisation could be associated with the Customs and Economic Union.

4. The possibility of such an association has now been examined by Working Party No. 17 of the Council. Her Majesty's Government believe, with the members of that Working Party, that it is fully practicable for the United Kingdom and many other O.E.E.C. countries, including the countries which are proposing to create a Customs and Economic Union, to enter a Free Trade Area. The members of this Free Trade Area would undertake to eliminate, in respect of each other's products, protective duties and other restrictive regulations of commerce, including quantitative restrictions. They would be free to keep their own separate and different tariffs on imports from outside the Area, except that the countries which were also members of the Customs and Economic Union would in

due course establish a common external tariff; they would also be able to vary these tariffs, subject to any international agreements by which they are bound from time to time.

5. Her Majesty's Government's own examination of this problem had led to the conclusion that, provided foodstuffs were excluded from its scope, a Free Trade Area in Europe could be established. On 26th November 1956 the Government informed the House of Commons of their intention to enter into negotiations in O.E.E.C. with this in view. Now that the practicability of a European Free Trade Area has been confirmed by the Working Party, Her Majesty's Government hope that other countries will also declare, either before or in the course of the forthcoming meeting of O.E.E.C., their willingness to negotiate on these lines.

6. Her Majesty's Government recognise the desirability of associating with the development of the Free Trade Area as many countries in Europe as possible. At the same time it is of the essence of the Free Trade Area that the obligations undertaken by its members progressively to remove tariffs over the whole field of trade which it includes should be reciprocal. Accordingly it is the view of Her Majesty's Government that the Free Trade Area should be open to all the member countries of O.E.E.C. who were prepared to accept its obligations. The position of any such countries which felt unable to accept the obligations of a Free Trade Area would need to be examined.

7. Her Majesty's Government believe that the establishment of a Free Trade Area would in no way prejudice the high level of employment which Europe generally has enjoyed, and that it should be an objective of member countries to maintain a high and stable level of employment. This would be an essential condition for the effective working of the Free Trade Area.

8. Her Majesty's Government advocate the establishment of a Free Trade Area in the belief that it will raise industrial efficiency by the encouragement it will afford to increased specialisation, large-scale production, and new technical and industrial developments. It should therefore strengthen the economy of Western Europe as a whole; and as a result of its formation the members of the Area should be in a stronger position progressively to remove restrictions on imports from countries outside the Area. The proposal aims at the total re-

moval of barriers to trade over a significant proportion of world trade. It does not involve the creation of any new barriers against trade with the rest of the world. On the contrary Her Majesty's Government consider it indispensable that the Free Trade Area should be formed in such a way as to be wholly consistent with the existing objective of a collective approach to the widest possible system of multilateral trade and payments. It should also be fully consistent with the obligations of member countries as contracting parties to the G.A.T.T., the provisions of which would continue to govern the relations of the G.A.T.T. members of the Area with other G.A.T.T. countries.

PART IV: STATISTICAL APPENDIX

PART FOUR: STATISTICAL APPENDIX

I

Economic and social statistics provide a necessary framework for the understanding of economic and social history. Economics as a science is concerned with explaining three broad problems: what is produced, how it is produced, how the goods produced are distributed. Since the resources that go into the production of goods and services are scarce in relation to the demand for them, economic activity is a rationing activity; economic decisions are choices between alternatives, with the understanding that the choice to produce one good means forgoing the production of some other good; this forgone alternative is defined by economists as 'the opportunity cost' of the good produced. The aim of economic activity is consumption; and that economy is the most efficient which produces most consumer goods per head of population and which produces them at a faster rate of growth. There are two ways in which production decisions can be made in any economy: by command, by the decision of some central authority, as in socialist economies; or by allowing each individual to exercise his choices for goods through demand in the market, as in a free market economy. The former type of economy, some people claim, is more equitable; the latter type, most economists claim, is more efficient. And one of the great social problems of the free market economies has been to balance efficiency with equity, and to make the working of a market economy more acceptable by seeing that those who obviously do not benefit from it—the poor, the old, the sick—get assistance from the State. Britain since 1870 has had a free market economy, modified both by the growth of the public sector (that part of the economy managed by government) and of the welfare state (the provision of social services). Until the industrial revolution man had always been very poor, struggling for survival in a world in which production grew very slowly, and, in consequence, also population. Since the middle of the eighteenth century, however, first in Britain, and then in other industrializing countries, output and population have grown at an unprecedented rate, and man has gradually escaped from the bondage of poverty. To an important extent the economic and social history of modern Britain is the history of economic growth and increasing welfare, and, with that, of a changing way of life. The growth and welfare can be documented by statistics (of population, production, employment, trade, etc.); the change in the way of life is more difficult to document and is probably impossible to quantify. This statistical appendix is to enable the student to supplement the literary sources with statistics which both describe and explain the changing conditions of life in Britain since 1870.

II

The following statistics are divided into functional sections with explanatory notes on terms and sources at the end of each section. The following terms, which are explained, should be noted by the student: national income; gross national product (G.N.P.); birth, death, and infantile mortality rates; primary, secondary, and tertiary sectors (agriculture, industry, and services); gross and net output; index numbers; balance of trade, balance of payments, terms of trade. The student should also know the meanings, as used by the economists, of the following terms: consumption, production, and distribution; supply and demand; cost and revenue; money and banking; the trade cycle; investment and employment. These can be conveniently found in any general elementary textbook of economics. Generally the student should understand how the market economy works; on the one hand through consumers exercising demand (i.e. the willingness to buy at a certain price) and manufacturers producing supply (i.e. the willingness to produce and sell at a certain price); on the other hand, through the various factors contributing to production (labour and capital, for example) being rewarded by incomes (i.e. the process of distribution) which enable them to participate as purchasers in the market, and thus begin 'a circular flow' of spending and producing.

List of Tables and Charts

A. *National Income and Government Expenditure*
   I. The Growth of U.K. National Income, 1870–1959.
      Fig IA. Chart.
   II. Rates of growth of Real National Income per Head per Annum, 1855–1964.
   III. Percentage Distribution of Government Expenditure by Function, 1890–1955.
   IV. Major Social Services, 1890–1951.
   V. Explanations of National Income Estimates.

B. *Population*
   I. Home Population, 1871–1961.
   II. U.K. Birth, Death, and Infantile Mortality Rates, 1870–1968.
      Fig. IIA. Chart.
   III. Expectation of Life at Birth and at Age 1, 1870–1966.
   IV. Age Distribution, 1901 and 1961.
   V. The Distribution of Population in England and Wales in Urban and Rural Areas, 1871–1968.
   VI. The Growth of Conurbations, 1871–1961.
   VII. Explanation of Population Statistics.

C. *Manpower*
   I. Distribution of Employment by Industries (G.B.), 1871–1961.
   II. Distribution of Employment by Sectors (G.B.), 1801–1965.
   III. Unemployment (U.K.), 1870–1960.
   IV. Unemployment by Regions, Inter-War Years.
   V. Numbers and Members of Trade Unions (U.K.), 1892–1965.
   VI. Industrial Disputes (U.K.), 1888–1965.
   VII. Explanation of Manpower Statistics.

D. *Production Statistics*
   I. Structure of the British National Product, 1801–1965.

II. Censuses of Production: Output of Individual Industries, 1907–1963.
III. Index of Total Industrial Production, 1900–1966.
   Fig. IIIA. Chart.
IV. Staple Industries
   (i) Cotton Production.
   (ii) Coal Production and Employment.
   (iii) Iron and Steel Production.
V. New Industries
   (i) Cars and Commercial Vehicles.
   (ii) Electricity Generated.
VI. Transport
   (i) Railways.
   (ii) Motor Vehicles in Use (G.B.), 1904–1965.
VII. Agriculture
   (i) Acreage of Crops (G.B.), 1870–1960.
   (ii) Livestock Numbers.
      (a) Numbers of Sheep, 1871–1954.
      (b) Numbers of Cows and Heifers in Milk and Total Cattle, 1871–1955.
VIII. Explanation of Production Statistics.

E. Foreign Trade
   I. Directions of U.K. Trade, 1900–1965.
   II. Composition of Domestic Exports, 1870–1966.
   III. Balance of Payments, U.K., 1870–1965.
   IV. Explanation of Foreign Trade Statistics.

A. National Income and Government Expenditure

I. The Growth of U.K. National Income, 1870–1959

Year	National income current prices £m.	National income per head current prices £	National income per head at 1913–14 prices £
1870	923	29·53	26·84
1875	1,085	33·04	29·77
1880	1,079	31·16	29·68
1885	1,124	31·21	34·29
1890	1,405	37·48	41·19
1895	1,449	36·94	44·51
1900	1,768	42·96	45·70
1905	1,832	42·62	46·84
1910	2,078	46·26	48·19
1911	2,294	49·77	50·27
1920	5,787	123·60	47·72
1925	4,091	90·79	47·79
1930	4,076	88·87	49·93
1935	4,238	90·42	55·82
1940	7,141	148·07	73·30
1945	8,285	168·46	66·06
1950	10,710	211·59	61·33
1955	15,416	302·46	71·67
1959	18,931	364·16	77·98

Source: Deane and Cole.

II. Rates of Growth of Real National Income per Head per Annum, 1855–1964

1855/64–1885/94 . . .	2·0
1865/74–1895/1904 . . .	1·9
1875/84–1905/14 . . .	1·5
1899–1913	0·1
1924–1937	1·1
1954–1964	2·2

Source: Deane and Cole.

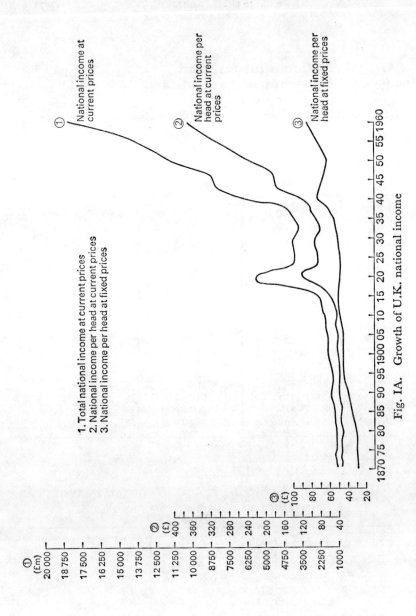

Fig. IA. Growth of U.K. national income

III. Percentage Distribution of Government Expenditure by Function, 1890–1955

(Percentage of G.N.P. and total government expenditure at current prices)

	1890	1900	1910	1920	1928	1938	1950	1955
Administration								
G.N.P.	1·1	0·8	1·0	1·2	1·1	1·1	1·5	1·1
Expenditure	12·1	5·9	8·1	4·5	4·5	3·3	3·9	3·0
National Debt								
G.N.P.	1·6	1·0	0·9	5·4	6·7	4·0	4·4	4·2
Expenditure	18·2	7·0	7·4	20·4	27·9	13·4	11·2	11·5
Law and Order								
G.N.P.	0·6	0·5	0·6	0·5	0·7	0·7	0·7	0·7
Expenditure	6·9	3·5	4·7	2·1	2·3	2·4	1·7	1·9
Overseas Services								
G.N.P.	—	0·1	0·1	—	—	0·1	1·5	0·5
Expenditure	0·3	0·4	0·4	0·2	0·1	0·2	3·9	1·3
Military and Defence								
G.N.P.	2·4	6·9	3·5	8·6	2·8	8·9	7·2	9·6
Expenditure	26·7	48·0	27·3	32·6	11·4	29·8	18·5	26·1
Social Services								
G.N.P.	1·9	2·6	4·2	6·8	9·6	11·3	18·0	16·3
Expenditure	20·9	18·0	32·8	25·9	39·7	37·6	46·1	44·6
Economic Services								
G.N.P.	1·0	1·9	1·8	3·3	2·6	2·9	4·9	3·2
Expenditure	11·0	13·0	13·9	12·8	10·7	9·5	12·6	8·6
Environmental Services								
G.N.P.	0·3	0·6	0·7	0·4	0·7	1·0	0·8	1·1
Expenditure	3·8	4·3	5·3	1·6	2·9	3·2	2·1	3·0
All Services								
G.N.P.	8·9	14·4	12·7	26·2	24·2	30·0	39·0	36·6
Expenditure	100	100	100	100	100	100	100	100
Total government expenditure								
(£m.)	130·6	280·8	272·0	1,592·1	1,094·7	1,587·0	4,539·0	6,143·0
(As a percentage of G.N.P.)	8·8	14·4	12·8	26·1	24·2	30·1	39·5	37·3

Source: Peacock and Wiseman.

IV. Major Social Services, 1890–1951

(Current expenditure and percentage of G.N.P.)
(£m.)

	1890	1900	1910	1923	1936	1949	1951
Education							
Expenditure (£m.)	11·5	19·3	33·5	87·4	115·1	267·1	344·5
% G.N.P.	0·8	1·1	1·1	2·2	2·5	2·5	3·0
Public Health and National Health Service							
Expenditure (£m.)	1·4	2·8	4·7	44·4	65·3	403·7	448·8
% G.N.P.	0·1	0·1	0·2	1·1	1·4	3·8	4·0
Housing							
Expenditure (£m.)	0·2	0·5	1·5	16·5	43·8	67·2	74·1
Assistance and Extended Benefit							
Expenditure (£m.)	9·1	12·3	16·1	34·3	94·0	68·0	92·8
Non-contributory Pensions							
Expenditure (£m.)			7·4	92·3	87·0	108·8	101·3
Social Security							
Expenditure (£m.)				41·2	129·0	398·2	428·0
% G.N.P.				1·0	2·9	3·8	3·8
Nutrition and School Meals							
Expenditure (£m.)						63·0	65·0
New Services							
Expenditure (£m.)					3·9	77·8	87·3
Total							
Expenditure (£m.)	22·2	34·9	63·2	316·1	538·1	1,454·2	1,641·8

Source: Hicks.

V. Explanation of National Income Estimates

The national income in any year is the total net earnings received by the various factors employed in the production of goods and services. These factors are regarded conventionally as labour, enterprise, capital, and land, and they receive wages, profit, interest, and rent. The gross national product, usually written as G.N.P., is the total value at current prices of all final goods and services produced in one year by a nation's economy, before deductions are made to allow for the depreciation or consumption of capital goods. Generally the national income is thought of as the total value over one year of all 'the goods and services becoming available to the nation for consumption or adding to wealth'. In estimating national income, the economic activity of a country is measured in three ways: as income (by adding together everyone's income); as consumption or expenditure (by adding together everybody's consumption or expenditure); as production (by adding together the value of all goods and services made by the various sectors of the economy). (For methods of compilation see E. Devons, *An Introduction to British Economic Statistics*, Cambridge University Press, 1961, ch. IX.) Allowance must be made, of course, for international transactions which add to, or abstract from, a country's income. Over time, also, allowance must be made for price changes so that increases in money incomes without corresponding increases in production do not show up as increases in real national income. The importance of national income estimates are to provide a convenient yardstick of an economy's performance. They are much used, therefore, by governments as a basis for making economic decisions about the economy, and by businessmen in making decisions about business activity. Official statistics exist only for the period since 1938, but estimates have been made for earlier periods, including continuous estimates for the period since 1870.

Sources for Section A:

    A. L. Bowley, *Wages and Income since 1860* (Cambridge University Press, 1937).
    P. Deane and W. A. Cole, *British Economic Growth, 1688–1959* (Cambridge University Press, 1967).
    P. Deane, 'Contemporary Estimates of National Income in

the Second Half of the Nineteenth Century', *Economic History Review* (1957).

C. H. Feinstein, 'Income and Investment in the United Kingdom, 1856–1914', *Economic Journal* (1961).

U. K. Hicks, *British Public Finances. Their Structure and Development, 1880–1952* (Oxford University Press, 1958).

J. Jeffreys and D. Walters, 'National Income and Expenditure in the United Kingdom 1870–1952'. *Income and Wealth*, Series V (1956).

A. T. Peacock and J. Wiseman, *The Growth of Public Expenditure in the United Kingdom* (Princeton University Press, 1961).

A. R. Prest, 'National Income of the United Kingdom, 1870–1946', *Economic Journal* (1948).

B. Population

I. Home Population, 1871–1961
('000)

	U.K.	England and Wales	Wales	Scotland	Northern Ireland
1871	27,431	22,712	1,413	3,360	1,359
1881	31,015	25,974	1,572	3,706	1,305
1891	34,364	29,003	1,771	4,026	1,236
1901	38,237	32,528	2,013	4,472	1,237
1911	42,082	36,070	2,421	4,761	1,251
1921	44,027	37,887	2,656	4,882	1,258
1931	46,038	39,952	2,593	4,843	1,243
1941	48,216	41,748	2,626	5,160	1,308
1951	50,225	43,758	2,599	5,096	1,371
1961	52,709	46,105	2,644	5,179	1,425

Source: Annual Abstract of Statistics.

II. U.K. Birth, Death, and Infantile Mortality Rates, 1870–1965
(per 1,000 of population)

	Births	Male deaths	Female deaths	Infantile mortality (children under 1 year of age)
1870/2	35·0	23·3	20·8	150
1880/2	33·6	20·8	18·6	137
1890/2	30·6	20·7	18·6	145
1900/2	28·6	18·4	16·3	142
1910/2	24·6	14·9	13·3	110
1920/2	23·1	13·5	11·9	82
1930/2	16·3	12·9	11·5	67
1940/2	15·0	15·5	11·9	59
1950/2	15·0	12·7	11·2	30
1960/2	17·9	12·4	11·2	22
1965	20·2	12·2	10·6	22

Source: Annual Abstract of Statistics.

STATISTICAL APPENDIX

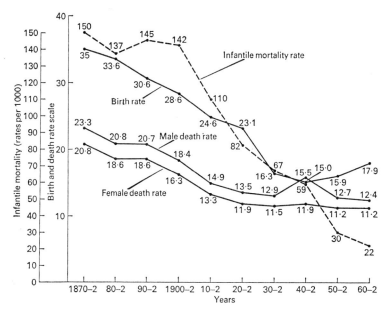

Fig. IIA. U.K. vital statistics. Birth and death rates per 1,000 of population

III. EXPECTATION OF LIFE AT BIRTH AND AT AGE 1
(England and Wales)

	Birth		Age 1	
	Male	Female	Male	Female
1871–80	41·4	44·6	48·1	50·1
1891–1900	44·1	47·8	52·2	54·5
1910–12	51·5	55·4	57·5	60·3
1930–32	58·7	62·9	62·3	65·5
1950–52	66·4	71·5	67·7	72·4
1966	68·4	74·7	68·9	74·9

Source: *Registrar General's Statistical Review of England and Wales*, 1966, Part II, Tables, Population (1968).

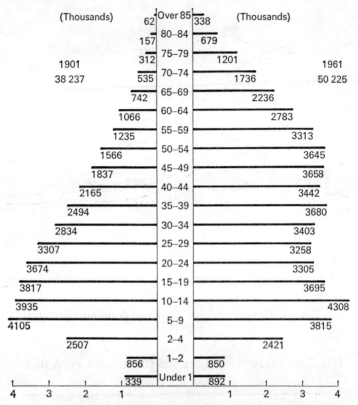

Fig. IV. Age distribution (U.K.) 1901 and 1961
Source: *Annual Abstract of Statistics.*

V. The Distribution of Population in England and Wales in Urban and Rural Areas, 1871–1968

('000)

	Urban	Rural
1871	14,041	8,671
1901	25,058	7,469
1931	31,952	8,000
1961	36,872	9,233
1968	38,325	10,268

Source: *Annual Abstract of Statistics.*

VI. The Growth of Conurbations
('000)

	1871	1901	1931	1961
Greater London	3,890	6,586	8,216	8,183
South-East Lancashire	1,386	2,117	2,427	2,428
West Yorkshire	1,064	1,524	1,655	1,704
West Midlands	969	1,483	1,933	2,347
Merseyside	690	1,030	1,347	1,384
Tyneside	346	678	827	855

Sources: Mitchell and Deane.
Annual Abstract of Statistics.

VII. Explanation of Population Statistics

Statistical information about population comes from two sources: the Census of Population, which has been taken at ten-yearly intervals since 1801 (except 1941); and the reports of the Registrars General, which give weekly, quarterly, and yearly statistics of birth, deaths, and marriages. The Census is an accurate and detailed counting of the population once every ten years; for intervening years, estimates are made from the reports of the Registrars General. Changes in total population are the result of deaths, births, and migration. The *crude death-rate* for any year is the number of deaths per 1,000 of the population; a more significant measure is the number of deaths at each age—the *specific mortality rate*; particularly significant is the *infant mortality rate* (the deaths of babies under one year to the number of births during the year, again expressed as deaths per 1,000). The *crude birth-rate* for any year is the number of births per 1,000 of the population; more significant meaures are the *general fertility rate* (the number of births per 1,000 females of child-bearing age) and *specific fertility rates* (birth-rates for each age or age group of females of child-bearing age). Relating birth- and death-rates gives a measure of the balance between fertility and mortality, helping to explain why population increases or decreases; the simplest indication is the difference between births and deaths, but a more significant measure is that which estimates the ability of a population to replace itself—*the net reproduction rate*, which is the number of female children which will be born to a newly-born female who is subject to the specific fertility and mortality rates of her generation; the *net reproduction rate* is a measure of mother-replacement, and indicates whether a population will grow from its own resources. Migration, into or out of a country, obviously adds to or subtracts from the population of that country, and must be taken into account in explaining total population movements. Two other measurements are important for the understanding of population change: *life expectancy*, the expectation of life, either at birth or after one year of life; and *age distribution*, the numbers of people in each age group between birth and 85 years of age or over.

Sources for Section B:
  Annual Abstract of Statistics (General Statistical Office, H.M.S.O.).
  N. M. Carrier and J. R. Jeffrey, *External Migration: A Study of Available Statistics, 1815–1950* (General Register Office. Studies on Medical and Population Subjects, No. 6, H.M.S.O., 1953).
  The Census Explained (H.M.S.O., 1951).
  Guides to Official Sources: No. 2, Census Reports of Great Britain, 1801–1931 (H.M.S.O., 1951).
  B. R. Mitchell and P. Deane, *Abstract of British Historical Statistics* (Cambridge University Press, 1962).
  M. P. Newton and J. R. Jeffrey, *Internal Migration: Some Aspects of Population Movements within England and Wales* (General Register Office, Studies on Medical and Population Subjects, No. 5, H.M.S.O., 1951).
  Registrar General's Statistical Reviews.
  Report of the Royal Commission on Population (Cmd. 7695, H.M.S.O., 1949).

C. Manpower

1. Distribution of Employment by Industries, in Great Britain, 1871–1961

('000,000 persons)

	Agriculture, forestry, fishing	Mining and quarrying	Manufactures	Construction	Trade	Transport	Public and professional service	Domestic service	Total occupied population
1871	1·8	0·6	3·9	0·8	1·6	0·7	0·7	1·8	12·0
1881	1·7	0·6	4·2	0·9	1·9	0·9	0·8	2·0	13·1
1891	1·6	0·8	4·8	0·9	2·3	1·1	1·0	2·3	14·7
1901	1·5	0·9	5·5	1·3	2·3	1·3	1·3	2·3	16·7
1911	1·6	1·2	6·2	1·2	2·5	1·5	1·5	2·6	18·6
1921	1·4	1·5	6·9	0·8	2·6	1·4	2·1	1·3	19·3
1931	1·3	1·2	7·2	1·1	3·3	1·4	2·3	1·6	21·1
1951	1·1	0·9	8·8	1·4	3·2	1·7	3·3	0·5	22·6
1961	0·9	0·7	8·9	1·6	3·4	1·8	4·0	—	22·8

Sources: Deane and Cole.
Key Statistics.

II. Distribution of Employment by Sectors, Great Britain, 1801–1965
(Percentages)

	Primary Agriculture, forestry, and fishing	Secondary Mining, manu- facturing, and production	Tertiary Services, com- merce, transport, government
1801	36	30	34
1871	15	43	42
1901	9	46	45
1931	6	45	49
1951	5	49	46
1965	3·4	46·3	50·3

Source: Deane and Cole.

III. Unemployment (U.K.), 1870–1960
(Percentage unemployed of Unions making returns)

Year	%	Year	%	Year	%
1870	3·9	1901	3·3	1931	21·3
1	1·6	2	4·0	2	22·1
2	0·9	3	4·7	3	19·9
3	1·2	4	6·0	4	16·7
4	1·7	5	5·0	5	15·5
5	2·4	6	3·6	6	13·1
6	3·7	7	3·7	7	10·8
7	4·7	8	7·8	8	13·5
8	6·8	9	7·7	9	11·6
9	11·4	1910	4·7	1940	9·7
1880	5·5	1	3·0	1	6·6
1	3·5	2	3·2	2	2·4
2	2·3	3	2·1	3	0·8
3	2·6	4	3·3	4	0·7
4	8·1	5	1·1	5	1·2
5	9·3	6	0·4	6	2·5
6	10·2	7	0·6	7	3·1
7	7·6	8	0·8	8	1·8
8	4·9	9	2·1	9	1·6
9	2·1	1920	2·0	1950	1·5
1890	2·1	1	12·9	1	1·2
1	3·5	2	14·3	2	2·1
2	6·3	3	11·7	3	1·8
3	7·5	4	10·3	4	1·5
4	6·9	5	11·3	5	1·2
5	5·8	6	12·5	6	1·3
6	3·3	7	9·7	7	1·6
7	3·3	8	10·8	8	2·2
8	2·8	9	10·4	9	2·3
9	2·0	1930	16·0	1960	1·7
1900	2·5				

Sources: Mitchell and Deane.
Key Statistics.

IV. Unemployment by Regions, Inter-War Years
(Percentages)

	1912	1929	1932	1936	1955
Great Britain	3·9	10·5	22·2	10·8	2·6
South Britain	5·2	7·1	16·2	6·9	1·8
North and Wales	2·5	13·8	28·2	15·0	4·1
London	8·7	5·6	13·5	6·3 }	1·5
South-Eastern	4·7	5·6	14·3	6·7 }	
South-Western	4·6	8·1	17·1	7·8	2·1
Midlands	2·8	9·3	20·1	7·2	1·9
North-Eastern	2·5	13·7	28·5	11·0	5·0
North-Western	2·7	13·3	25·8	14·0	3·1
Northern Ireland	6·9	12·0	23·0	26·2	6·8
Scotland	1·8	12·1	27·7	15·9	4·8
Wales	3·1	19·3	36·5	22·3	3·7

Source: Beveridge.

V. Numbers and Membership of Trade Unions (U.K.), 1892–1965

	Total unions (Numbers)	Total members ('000)
1892	1,233	1,576
1895	1,340	1,504
1900	1,323	2,022
1905	1,244	1,997
1910	1,269	2,565
1915	1,299	4,359
1920	1,379	8,347
1925	1,170	5,506
1930	1,114	4,841
1935	1,049	4,867
1940	1,022	6,559
1945	777	7,813
1950	704	9,235
1955	704	9,741
1960	664	9,835
1965	580	10,180

Sources: Mitchell and Deane.
Annual Abstract of Statistics.

VI. Industrial Disputes (U.K.), 1888–1965

	A	B		A	B		A	B
1888	517		1914	972	9,878	1940	922	940
9	1,211		5	672	2,953	1	1,251	1,079
1890	1,040		6	532	2,446	2	1,303	1,527
1	906	6,809	7	730	5,647	3	1,785	1,808
2	700	17,382	8	1,165	5,875	4	2,194	3,714
3	615	30,468	9	1,352	34,969	5	2,293	2,835
4	929	9,529	1920	1,607	26,568	6	2,205	2,158
5	745	5,725	1	763	85,872	7	1,721	2,433
6	926	3,746	2	576	19,850	8	1,759	1,944
7	864	10,346	3	628	10,672	9	1,426	1,807
8	711	15,289	4	710	8,424	1950	1,339	1,389
9	719	2,516	5	603	7,952	1951		
1900	648	3,153	6	323	162,233	2		
1	642	4,142	7	308	1,174	3		
2	442	3,479	8	302	1,388	4		
3	387	2,339	9	431	8,287	5		
4	355	1,484	1930	422	4,399	6	2,648	2,083
5	358	2,470	1	420	6,983	7	2,859	8,412
6	486	3,029	2	389	6,488	8	2,629	3,462
7	601	2,162	3	357	1,072	9	2,093	5,270
8	399	10,834	4	471	959	1960	2,832	3,024
9	436	2,774	5	553	1,955	1	2,686	3,046
1910	531	9,895	6	818	1,829	2	2,449	5,798
1	903	10,320	7	1,129	3,413	3	2,068	1,755
2	857	40,915	8	875	1,334	4	2,524	2,277
3	1,497	11,631	9	940	1,356	1965	2,354	2,925

A = No. of industrial disputes with stoppages.
B = Working days lost each year as result of stoppages ('000).

Sources: Mitchell and Deane.
Annual Abstracts of Statistics.

VII. Explanation of Manpower Statistics

Manpower statistics are concerned with the employment of labour in economic activity: for example, the size of the labour force, its industrial distribution, the rate of unemployment, and the state of industrial relations. Information about manpower comes mainly from the Census of Population, the Census of Production, and from the publications of the Ministry of Labour. The Census of Population estimates the size of the employed population and its distribution between occupations; the Census of Production has detailed employment figures for those industries it covers (mining, manufacturing, construction, and gas, water, and electricity); the Ministry of Labour provides estimates of total working population, monthly figures of employment and unemployment, statistics of industrial disputes and working days lost through stoppages, indexes of weekly rates of wages, and statistics of cost of living, retail prices, and family budgets. Manpower statistics have been collected systematically since 1893, when the Labour Department of the Board of Trade was established; *The Labour Gazette* has been published monthly since 1893 and the *Annual Abstract of Labour Statistics* since 1894. Manpower statistics can be used for various purposes: to show how the structure of the economy changes over time (agriculture declining in importance, and industry and services increasing), how the structure of industry changes (with the decline of old and the growth of new industries), how unemployment fluctuates, both over time and between regions, and how industrial disputes also vary over time and in intensity.

Sources for Section C:
  Annual Abstract of Labour Statistics (General Statistical Office, H.M.S.O.), *Guides to Official Sources*: No. 1, *Labour Statistics* (H.M.S.O., 1958).
  Annual Abstract of Statistics (General Statistical Office, H.M.S.O.).
  W. H. Beveridge, *Full Employment in a Free Society* (London, Allen and Unwin, 1944).
  The British Economy, Key Statistics 1900–1966 (London, Times Newspapers Ltd., 1967).
  P. Deane and W. A. Cole, *British Economic Growth, 1688–1959* (Cambridge University Press, 1967).
  B. R. Mitchell and P. Deane, *Abstract of British Historical Statistics* (Cambridge University Press, 1962).

D. Production Statistics

I. Structure of the British National Product, 1801–1965

(Percentages)

	Primary Agriculture, forestry, and fishing	Secondary Mining, manu- facturing, and construction	Tertiary Services, com- merce, transport, government
1801	36	30	34
1881	13	43	44
1901	9	44	47
1931	6	45	49
1950	5·2	46·3	48·5
1965	3·4	46·3	50·3

Sources: Deane and Cole.
National Income Blue Book.

II. Censuses of Production: Output of Individual Industries, 1907–1963

(£'000,000)

	1907		1935		1963	
	Gross output	Net output	Gross output	Net output	Gross output	Net output
Food, drink, and tobacco	283	87	665	203	4,477	1,248
Chemicals and allied industries	90	27	206	89	2,692	1,016
Metal manufacture	147	45	245	88	2,479	826
Engineering and allied industries	—	—	710	357	9,013	4,388
Textiles, leather, and clothing	458	187	656	249	3,081	1,272
Other manufactures	—	—	413	237	3,771	1,955
Mining and quarrying	134	115	167	136	1,031	772
Construction	—	—	295	150	3,930	1,857
Gas, electricity, and water	51	32	181	128	1,593	982

Source: Annual Abstract of Statistics.

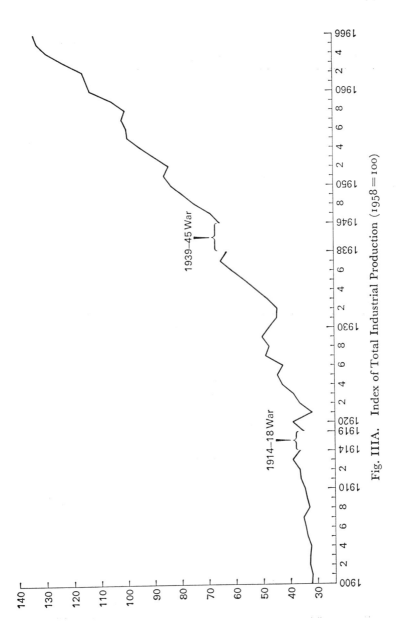

Fig. IIIA. Index of Total Industrial Production (1958 = 100)

III. Index of Total Industrial Production, 1900–1966
(Index Number, average 1958 = 100)

1900	31·9	1923	38·8	1949	78·6
1	31·8	4	43·2	1950	83·1
2	33·0	5	44·8	1	85·8
3	33·0	6	42·5	2	83·9
4	32·7	7	48·9	3	88·9
5	33·8	8	47·6	4	94·1
6	34·6	9	50·0	5	98·9
7	35·2	1930	47·8	6	99·3
8	33·3	1	44·8	7	101·1
9	34·1	2	44·5	8	100·0
1910	34·8	3	47·5	9	105·1
1	35·8	4	52·3	1960	112·5
2	36·3	5	56·2	1	113·9
3	39·1	6	61·3	2	115·1
1914	36·6	7	65·0	3	119·0
1919	35·1	8	63·2	4	128·2
1920	39·0	1946	64·9	5	131·9
1	31·7	7	68·4	1966	133·1
2	36·7	8	74·3		

Source: *Key Statistics.*

IV. Staple Industries
(i) Cotton Production

	Raw cotton consumption ('000,000 lbs.)	Exports piece goods ('000,000 yards)	Total cotton output ('000,000 yards)	
1870	1,078	3,267		
1900	1,737	5,032		
			1905	7,090
			1912	8,050
1913	2,178	7,075		
1920	1,726	4,643		
			1924	5,589
1930	1,272	2,491		3,179
			1937	3,640
1938	1,109	1,448		
1950	1,017	822		2,123
1965	506	385		1,015

Sources: *Annual Abstract of Statistics.*
Key Statistics.

(ii) Coal Production and Employment

	U.K. total output ('000,000 tons)	Numbers employed ('000)	Exports and bunkers ('000,000 tons)	Percentage coal cut by machine
1870	110·4	350·9	11·2	0
1900	222·5	780·1	44·1	1
1913	287·4	1,127·9	73·4	8
1920	229·5	1,248·2	24·9	13
1930	243·9	931·4	54·9	31
1938	227·0	790·9	35·9	59
1950	216·3	697·0	17·1	
1965	187·5	465·6	3·7	

Sources: Mitchell and Deane.
Key Statistics.
Annual Abstract of Statistics.

(iii) Iron and Steel Production

	Total output of pig iron ('000 tons)	Total output of steel ('000 tons)	Total iron and steel exports ('000 tons)
1871	6,627	329	3,169
1900	8,960	4,901	3,541
1913	10,260	7,664	4,969
1920	8,035	9,067	3,251
1930	6,192	7,326	3,160
1938	6,761	10,398	1,960
1950	9,633	16,293	3,250
1965	17,460	27,006	4,728

V. New Industries

(i) Cars and Commercial Vehicles
('000)

Year	Value	Year	Value	Year	Value
1910	14	1936	482	1951	734
1913	34	7	508	2	690
1922	73	8	445	3	834
3	95	9	402	4	1,018
4	147	1940	134	5	1,237
5	167	1	145	6	1,005
6	198	2	161	7	1,149
7	212	3	149	8	1,364
8	212	4	133	9	1,560
9	239	5	139	1960	1,811
1930	327	6	365	1	1,464
1	226	7	442	2	1,675
2	233	8	499	3	2,012
3	286	9	629	4	2,332
4	342	1950	784	1965	2,177
5	417				

Source: Key Statistics.

(ii) Electricity Generated
('000,000,000 Kwh)

Year	Value	Year	Value	Year	Value
1900	0·1	1934	15·6	1950	55·6
1913	1·3	5	17·7	1	60·7
1920	4·3	6	20·4	2	62·8
1	3·9	7	23·1	3	66·4
2	4·6	8	24·6	4	73·9
3	5·3	9	26·7	5	81·2
4	6·1	1940	29·0	6	88·3
5	6·7	1	32·7	7	92·2
6	7·1	2	36·0	8	99·8
7	8·5	3	37·4	9	106·6
8	9·4	4	38·8	1960	120·5
9	10·5	5	37·7	1	129·4
1930	11·0	6	41·7	2	143·6
1	11·5	7	43·1	3	156·0
2	12·3	8	47·0	4	164·6
3	13·7	9	49·7	1965	177·4

Source: Key Statistics.

VI. Transport

(i) Railways

	Length of road (Miles)	Passengers carried ('000)	Freight carried ('000 tons)
1871	15,376	375,221	169,365
1901	22,078	1,172,396	415,954
1911	23,417	1,326,317	523,577
1921	23,724	1,252,056	301,626
1931	20,395	1,156,353	268,380
1936	20,121	875,685	280,712
1950	19,500	981,700	281,300
1965	19,025	835,000	213,500

Sources: Annual Abstract of Statistics.
Mitchell and Deane.

(ii) Motor Vehicles in Use, Great Britain, 1904–1965 ('000)

	Private cars	Goods vehicles	Total (except trams)*
March 1904	8	4	—
1910	53	30	144
1913	106	64	306
1920	187	101	650
1930	1,056	348	2,274
1938	1,944	495	3,085
1950	2,258	900	4,414
1965	8,917	1,607	12,939

* Includes cycles, buses, coaches, taxis, vans, etc.

Sources: Mitchell and Deane.
Annual Abstract of Statistics.

VII. Agriculture

(i) Acreage of Crops, Great Britain, 1870–1960
('000 acres)

	1870	1900	1930	1960
Wheat	3,501	1,845	1,400	2,102
Barley	2,372	1,990	1,127	3,372
Oats	2,763	3,026	2,640	1,974
Other corn	912	474	493	224
Potatoes	588	561	548	829
Turnips and swedes	2,211	1,689	1,004	453
Mangolds	307	414	289	134
Sugar beet		(1920 = 3)	349	436
Cabbage, rape, etc.	144	196	180	49
Other green crops	361	320	303	366
Bare fallow	611	308	300	193
Rotation grasses {hay	2,069	2,202	2,006	3,171
{grazing	2,436	2,557	1,967	3,616
Permanent pasture {hay	3,067	4,373	5,222	3,372
{grazing	9,006	12,356	11,895	9,437
Orchards		232	248 }	286
Small fruit		74	74 }	
Hops	61	51	20	20
Total area of arable land	18,300	19,500	14,100	18,100

Sources: Mitchell and Deane.
Annual Abstract of Statistics.

(ii) Livestock Numbers

(a) Numbers of Sheep, 1871–1954
('000)

	England and Wales	Scotland
1871–5	21,629	7,161
1935–8	17,056	7,703
1954	14,513	7,428

(b) Numbers of Cows and Heifers in Milk and Total Cattle, 1871–1955
('000)

	Milk cows		Total cattle		Milk cows	Cattle
	England and Wales	Scotland	England and Wales	Scotland	Great Britain	Great Britain
1871–5	1,812	392	4,686	1,127	2,204	5,813
1835–8	3,067	497	6,604	1,309	3,564	7,913
1955	3,543	658	8,039	1,725	4,201	9,764

Source: H. T. Williams.

VIII. EXPLANATION OF PRODUCTION STATISTICS

Conventionally production is divided into three sections: *primary production* or *agricultural production*, which covers farming, fishing, and forestry; *secondary production* or *industrial production*, which covers manufacturing, mining, construction, and the supply of gas, electricity, and water; and *tertiary production* or *services*, which covers transport, commerce, distribution, professional services, public administration, education, entertainment, and domestic service. In an advanced economy like Great Britain, industrial production and services dominate total production, with agriculture contributing only a small and declining percentage (33 per cent in 1801 and 3·4 per cent in 1965).

Industrial Production. There are three sets of statistics for industrial production: the Census of Production, statistics of particular industries collected by trade organizations or government departments, and index numbers of total industrial production. The first Census of Production was taken in 1907; other censuses followed in 1912, 1924, 1930, 1935, and 1946; from 1948 a census has been made each year. The Census of Production covers manufacturing, mining, building and contracting, and public utilities; for example, in 1948 it covered about half the working population. The aim of the Census of Production is to have accurate statistics about the output of British industry: it includes statistics for *gross output* (the selling value of the goods made during the year by each industrial group in the census, and includes the values of all materials, fuels, and services bought from other industries) and *net output* (obtained by subtracting from the gross output of each industrial group the value of all materials and fuels used in producing that output). The Census of Production makes it possible to analyse the relationship between industries (by showing where the materials of an industry come from, and where the final products are sold), and also to analyse the distribution of output by size of individual producing units (and thus to show how unit size varies over time). Statistics for particular industries are too numerous and too varied to list here, but important examples are *Lloyds Register of Shipping* (a trade journal) and the annual *Statistical Digest of the Ministry* of *Fuel and Power* (a government publication). Most of the important information collected by government departments is summarized conveniently in the *Annual Abstract of Statistics*. Index numbers of production aim to give a continuous measure of changes in the volume of production, in particular industries or groups of industries, as well as for industry as a whole. An official monthly index of production has been calculated since 1952 by the Central Statistical Office, using 1,300 individual production series and measures changes in output compared with 1948. For the pre-war period, a number of indices have been devised, including the well-known indices of K. S. Lomax (from 1900) and W. G. Hoffman (from 1700), but little reliance can be placed on estimates for the period before 1900.

Agricultural Production. Agricultural statistics are collected by the government Agricultural Departments and are published both in the *Annual Abstract of Statistics* and in *Agricultural Statistics* (which gives details of acreage, farm incomes, livestock numbers, production, employment, machinery, and prices).

Services. The statistics about services are not collected together in the same way as those for agriculture or industry. There are, however, statistics for distribution and transport. Distribution statistics cover merchandise activities of all kinds, and have been collected systematically only since the first Census of Distribution in 1950. Statistics for retail trade go back to 1930, and these are important for the picture they give of consumption trends. Transport statistics, especially for railways and shipping, have a much longer history: the railways have long been required to publish regular statistics (*Railway Statistics*, monthly, and *Railway Returns*, yearly), but statistics for road transport are neither as detailed nor as comprehensive as railway statistics; shipping statistics, both for the size and type of ships, and for the work they do, go back to the nineteenth century (published annually in *Lloyds Register of Shipping* and in the *Annual Abstract of Statistics*).

Sources for Section D:
    Annual Abstract of Statistics (General Statistical Office, H.M.S.O.
    The British Economy. Key Statistics, 1900–1966.

P. Deane and W. A. Cole, *British Economic Growth, 1688–1959* (Cambridge University Press, 1967).

E. Devons, *An Introduction to British Economic Statistics*, op. cit. *Guides to Official Sources*: No. 4, *Agricultural and Food Statistics*; No. 6, *Census of Production Reports*.

B. R. Mitchell and P. Deane, *Abstract of British Historical Statistics* (Cambridge University Press, 1962).

National Income Blue Book.

H. T. Williams, *Principles for British Agricultural Policy* (Oxford University Press, 1960).

E. Foreign Trade

I. Directions of U.K. Trade, 1900–1965
(£'000,000)

	1900	1913	1920	1929	1937	1950	1965
Total Exports	291	525	1,335	729	521	2,160	4,724
To Western Europe	109	165	494	196	137	559	1,736
To rest of sterling area	86	171	413	287	222	1,009	1,645
To North America	28	55	122	83	61	240	700
To rest of world	67	134	306	163	101	352	642
Total Imports	523	769	1,933	1,221	1,028	2,598	5,763
From Western Europe	202	273	427	389	259	650	1,897
From rest of sterling area	87	160	460	305	309	972	1,812
From North America	163	175	666	246	208	392	1,132
From rest of world	77	161	380	281	252	584	922

Sources: Mitchell and Deane.
Key Statistics.

II. Composition of Domestic Exports, 1870–1966
(As percentages of all domestic exports)

	1870	1890	1910	1930	1950	1966
Staple Industries						
Textiles	56	43	37	26	19	7
Iron and steel	14	15	11	10	10	11
Engineering products	3	7	11	19	37	45
Total	73	65	59	55	66	63
Staple Products						
Coal	3	7	9	9	5	—
Woollens	13	10	9	7	7	—
Cottons	36	28	24	15	7	—
Vehicles	1	4	4	9	19	16
Value in £m.						
Domestic exports	200	264	430	571	2,171	5,042
Re-exports	45	65	104	87	85	194
Re-exports as percentage of domestic exports	23	25	24	15	4	4

Sources: Annual Abstract of Statistics.
Mitchell and Deane.
Deane and Cole.

III. Balance of Payments, U.K., 1870–1965

(£'000,000)

	Imports	Domestic exports	Re-exports	Overseas investment earning	Invisible trade	Overall balance on current account
1870	303·3	199·6	44·5	35·3	76·8	44·1
1880	411·2	223·1	63·4	57·7	96·4	35·6
1890	420·7	263·5	64·7	94·0	99·6	98·5
1900	523·1	291·2	63·2	103·6	109·1	37·9
1910	678·3	430·4	103·8	170·0	146·7	167·3
1920	1,932·6	1,334·5	222·8	200·0	395·0	252·0
1930	1,044·0	570·8	86·8	220·0	194·0	25·0
1937	1,027·8	521·4	75·1	210·0	176·0	−144·0
1950	2,608·2	2,171·3	84·8	237·0	357·0	221·0
1965	5,751·1	4,728·0	172·8	447·0	171·0	−110·0

Sources: Mitchell and Deane.
Annual Abstract of Statistics.
Key Statistics.

IV. Explanation of Foreign Trade Statistics

Foreign trade statistics are concerned with the economic relations between the United Kingdom and other countries. These relationships take the form of transactions of two main types: the importing of goods and services from other countries, and the exporting of goods and services to other countries; borrowings from, and loans to, other countries. The trade in goods is *merchandise trade*; the trade in services is *invisible trade*; the movements of capital are *capital exports* or *capital imports*. The net results of these transactions are important for an economy, and a summary of total international transactions gives the *balance of payments*. The balance of payments is summed up in a table showing three things: the *current account*, the difference between the total exports of goods and services and the total imports of goods and services (if positive, the *balance of trade* is *favourable*; if negative, *unfavourable*); the *capital account*, the net flow of loans and investment by both individuals and government (again positive or negative); and *gold movements* (which are treated separately because gold is the ultimate means of payment), with gold exports providing a positive element, and gold imports a negative element, to the balance of payments. One other commonly used term to help explain foreign trade movements is that which relates export prices to import prices—the *terms of trade*; if export prices rise relatively to import prices, the terms of trade are becoming favourable; if import prices rise relatively to export prices, the terms of trade are becoming unfavourable; the terms of trade are expressed as a ratio of export prices to import prices.

Historical series of trade statistics have existed for centuries, mainly because Customs were for long a main source of government revenue, and trade statistics were gathered as part of the administration of taxation. Today trade statistics appear monthly in the *Board of Trade Journal* and yearly in the *Annual Statement of Trade*. The *Annual Abstract of Statistics* also gives a balance of payments table, as well as details about size and direction of trade flows, of the commodity composition of trade, of the sources of imports and the destination of exports. However, whereas trade statistics have been accurately estimated over a long period, international capital movements have not, and for the period before 1939 we are dependent on individual estimates. In a trading nation like the U.K., the importance of the balance of payments is obvious: a negative balance of payments means indebtedness, while a positive balance (which the U.K. enjoyed for the hundred years or more before 1930) means an ability to consume more foreign goods and to be able to invest more abroad.

Sources for Section E:
 Annual Abstract of Statistics (General Statistical Office, H.M.S.O.).
 Board of Trade Journal.
 The British Economy. Key Statistics, 1900–1966.
 A. K. Cairncross, *Home and Foreign Investment, 1870–1913* (1953).
 E. Devons, *An Introduction to British Economic Statistics* (Cambridge University Press, 1961).
 B. R. Mitchell and P. Deane, *Abstract of British Historical Statistics* (Cambridge University Press, 1962).

INDEX

(Bold type means either a textual Sub-heading or a main source for a document.)

Aberfan disaster, 280–4
Advertising, 148, 235, 240, 320, 323
Aerial warfare, 201–2, 210–11
Aeroplane industry, 171, 332
Affluence, post-1945, 19, 231, 235, 240, 261, 309, 312, 322–4
African nationalism, 312–14
Age distribution, defined, 380
Agriculture, 7, 25, 27–9, 38, **39–41**, 134, 213, 233, 332, 337–9, 353; workers in, 28, 40–1, 59, 73–4, 156, 353
Agriculture, Report from the Select Committee on, 1968–9, **337–9**
Agriculture, Royal Commission on, 1893, 28
Andrews, P. W. S. and Brunner, E., *The Life of Lord Nuffield*, **192–3**
Annual Register, 1947, **243–5**
Appleton, Sir Edward, 266
Arch, Joseph, 27, 31
Army, the, 11, 124
Arnold, Dr., 72
Arnold, Matthew, 34
Artisans Dwelling Act, 1875, 29
Artisan, skilled, 31–2, 59, 205–6; *see also* Classes, working; Labour; workers; workmen, skilled
Ashley, W. J., *British Industries*, **116–18**
Asquith, H. H., 9, 100
Attlee, Clement, 10, 244–5; *As It Happened*, **309–10**
Auden, W. H., 136
Automation, 295–8, 335
Automation in Perspective, 1956, **295–8**

Badley, J. H., 36
Bagehot, Walter, *The English Constitution*, **89–90**
Balance of payments, 6, 19, 232, 234, 325–6, 327–9, 330, 359; defined, 397
Balance of trade, 222; defined, 397
Baldwin, Stanley, 161–3, **200–1**
Balfour Act, 1901, 138
Barnardo, Dr., 90
B.B.C., the, 139, 148–50, 244, 252; *see also* Broadcasting; Television
Beatles, the, 5
Beer, A Report on the Supply of, 1969, 340–2

Belfast, violence in, 168–71
Bell, Lady, *At the Works*, **77–80**, 246
Bennett, Arnold, 47
Bevan, Aneurin, *In Place of Fear*, **246–8**
Beveridge, William, 24, 31, 236, 261
Bevin, Ernest, 303
Birmingham, 80–2, 278
Birth control, 32, 263
Birth-rate, defined, 379
Blackett, Professor, 266
Booker, C., *The Neophiliacs*, **319–22**
Booth, Charles, 24, 30–1, 59, 163; *Life and Labour of the People of London*, 54, **84–6**
Booth, William, *In Darkest England and the Way Out*, 31, 90, **92–4**
Boothby, Robert, 195
Brassey, Lord, 'Introduction' to S. J. Chapman, *Work and Wages*, 119–20, **121–2**
Briggs, A., *The Birth of Broadcasting*, **148–9**
Bristol, 171–5
British Association, the, *Britain in Recovery*, **192–4**
British Gazette, the, 158, **160–1**
British Worker, the, 158, **159–60**
Broadcasting, 17, 19, 148–50, 208; *see also* B.B.C.; Television
Brontë, Charlotte, *The Professor*, 29
Brown, Joseph, Birmingham ironworker, 80–2
Buchanan, Professor, *Traffic in Towns*, 278
Burnham Committee, 1938, 138
Burt, Cyril, 140

Caird, James, 31, **39–41**
Capital, 37, 38, 61–2, 213, 223, 231–4, 309–10, 337, 356, 359; value of landowners', 1880, 39
Capitalism, 199–200, 240, 311
Car ownership, 235, 277–9
Carr-Saunders, A. M., 240
Carstairs, G. M., *This Island Now*, 1962 Reith Lectures, **317–19**
Castle, Mrs. Barbara, 348
Casual labour, 186–9, 303

INDEX

Chamberlain, Joseph, 13, 29
Chamberlain, Neville, 209
Chaplin, Charlie, 151
Chapman, S. J., *The Cotton Industry and Trade*, **112-14**, 119
Chemical industry, 143, 206-7, 214, 233, 234
Children, 64-5, 73, 136; at school, 138, 140, 204, 205, 246; post-1945, 239, 242, 255, 262-5, 294, 318
Church of England, 11, 33, 288-90; see also Religion
Churchill, Winston, 17, 309
Into Battle, 208, **210-11**
Cinema, the, 151-4, 167-8
'Circular flow', defined, 368
Civil Service, the, 11, 24, 239, 299-302, 334-5
Clapham, J. H., *The Woollen and Worsted Industry*, 112, **114-16**
Clarendon Report on the Public Schools, 1864, 35-6, 69, **71-2**
Classes, middle and upper, 2, 5, 8, 36, 46-7, 106, 135, 137-8, 142, 150, 176-7, 205-6, 239-40, 241, 315-17, 322-4; working, 4, 5, 8, 9-10, 15, 23, 27, 29, 30-2, 33, 35-6, 61-2, 80-2, 86-8, 90, 105, 112-16, 124, 127-9, 137, 142, 144, 148, 150, 154-7, 171-5, 179, 205-6, 235, 239-40, 241, 249-54, 261-5, 292-4, 316-17, 322; lower middle, 102-3, 150, 240, 316-17; social, 39-41, 176-9; see also Society
Clegg, H. A. and R. Adams, *The Employers' Challenge*, **343-5**
'climacteric, the', 1, 6
Clothing industry, 52, 109, 191-2
Club, working men's, 251, 252-4
Coal industry, 29, 37, 112, 142, 143, 154-6, 212, 215, 219-20, 240; unrest in, 135, 158, 160; see also Mineworkers
Coates, K., and R. Silburn, *Poverty: the Forgotten Englishmen*, **261-5**
Cole, G. D. H., 35
Collective bargaining, 127, **348-52**
Collectivism, 12, 94-5
Combination Movement, 116-18
Commercial Policy in the Interwar Period: International Proposals and National Policies, **225-7**
Common Market, 19, 233, 234, 312, 361
Commonwealth, the, 234, 310, 312-14, 330, 357-9, 361-2

Communism, 141, 142, 313-14
Competition, international, 119-23; see also Trade
Consumer durable, 17, 148, 235, 293
Consumption, mass, 4, 323-4, 327-8
Co-operative Independent Commission Report, 1958, **259-61**
Co-operative shops, 259-61
Cost-of-living, rise in, 154-7; see also Prices
Cotton industry, 25, 29, 112-14, 116-17, 121-3, 142, 216, 218-19, 290-2, 325, 353
County Councils Act, 1888, 62
Crime, 19, 140, 242
Crisis, financial, of 1931, 144, 222-5, 225-6
Crisis of 1947, 243-5
Crossman, R. H. S., *New Fabian Essays*, 309, **310-11**
Crowther Report, 1959, 266
Cutlery trade, 110-11

Daniel, W. W., *Racial Discrimination in England*, **271-7**
Darwin, Charles, *The Origin of Species*, 33
Death-rate, defined, 379
Defence, expenditure on, 124, 234, 327, 328
Depressed areas, 133, 214-15, 325-6, 353, 355
Devaluation, 232-3, 339
Developing countries, 356-60
Development areas, 355; see also Depressed areas
Devlin Report, '*Into Certain Matters Concerning the Port Transport Industry*', **303-8**
Dicey, A. V., *Law and Opinion in England*, 12, **94-5**
Dickens, Charles, 26
Dilke, C. W., 13; *Problems of Greater Britain*, 104, **107**
Disraeli, Benjamin, 8; *Selected Speeches of the late Rt. Hon. the Earl of Beaconsfield*, **95-7**
Distributive trades, 216; see also Retail trade; Service industries
Dock workers, 186-9, 303-8
Donovan Report, 348, 350, 352
Drink, 31, 52

Economics, 140-1; Keynesian, 195-7, 235; terms explained, 367-8

INDEX 401

Economy, defined, command, 367; free market, 367
Economy, British, pre-1914, 2, 36–8, 144; post-1945, 3, 231–5, 240, 325–6, 327–9, 329–33, 340; government-controlled, 5–7, 195–7, 300–2; American view of, 212–14
Education, 5, 7, 9, 12, 23, 34–6, 69–72, 94–5, 125, 138–9, 144, 203, 234, 236–8, 239, 240, 242, 266–70, 324; *see also* Schools, Universities
Education Acts, 1870, 25, 34, 94; 1902, 35; 1918, 36, 138; 1944, 236
Edward VIII, abdication of, 200–1
Edwardian age, 2, 14–15, 25–6, 75, 102–3, 104–7
Electrical goods, 216, 233, 234
Electricity, 143, 192–4, 234, 243–4, 335–7
Electricity (Supply) Act, 1926, 193
Eliot, T. S., 136
Emigration, 13, 38
Empire, the, 1, 2, 8, 13, 24, 107, 200, 213, 226, 312; ideas of, 95–9; *see* Commonwealth; Imperialism
Employment figures, 1901 Census, 109, 114
Engineering industry, Gantrymen, the; 154, 168, 214–16, 234, 343–5, 348
Escott, T. H. S., 27, 29, 30
European Free Trade Association (EFTA), 233, **361–4**; *see also* Common Market
Evacuation from cities, 1939, 140
Exchange controls, 226, 232
Export industries, 7, 112, 122–3, 142–3, 212–14, 218–20, 221, 232–4, 325–6, 331; *see also* Trade

Fabians, the, 12, 24, 100–1, 102, 195, 239
Factory and Workshop Act, 1901, **82–3**, 110
Factory system, 11–12, 50, 53, 108–9; *see also* Classes, working; Sweating; Worker, lower-paid; Workshops
Family allowances, 236, 258; *see also* Social services, Welfare State
Family income, 31, 54–8, 137, 140, 175, 249–50, 262–5
Family life, 254–6, 256–9, 293–4
Farm servants, 73–4
Farm workers, *see* Agriculture, workers in
Fascism, 141, 142, 201; British, 198–200

Food, consumption of, 60, 79–80, 171–3, 205–6; imports of, 25, 27–8, 107, 212–13; price of, 38, 55–7, 134, 172
Fords, industrial relations at, 345–7
Fortnightly, the, 1888, **75–6**
Foster, W. E., 34, 94
Franchise, 1, 8–10, 11, 23, 25, 139
Free speech, 334
Free trade, 1, 6, 29, 36, 119, 144, 225, 362–4; *see also* Laissez-faire; Tariffs; Trade
Fulton Report, 1968, *The Civil Service*, **299–302**
Fyfe, H., *Behind the Scenes of the General Strike*, 158, **161–3**
Fyfe, T. A., *Employers and Workmen under Munitions Acts*, **86–8**
Fyvel, T. R., *Intellectuals Today*, 322–4

Gaitskell, Hugh, 10, 135
Galbraith, J. K., 14
Gantrymen, the, 77–9
General Agreement on Tariffs and Trade (GATT), 233, 364
General Strike, 1926, 13, 135, 158–63
Germany, 1, 8, 36, 37, 201, 226; industry in, 116, 119–21, 122, 143
Gladstone, W. E., 12, 34
Gold standard, 6, 143, 144, 221–2
Golden Age, the, 104–7
Great Depression, 1873–96, 1, 37–8; 1929–34, 3, 12, 139, 141–3, 151, 198, 222
Green, T. H., 24
Gross national product (G.N.P.), 328, 338; defined, 374
Growth, economic, 6–7, 234, 327, 329–33, 368
Growth in the British Economy, P.E.P., **327–9**

Hadow report, 1926, 138, 236
Haggard, Rider, 31; *Rural England*, 73, 74
Hannington, Wal, *Unemployed Struggles, 1919–1936. My Life and Struggles amongst the Unemployed*, **168–71**
Health, 140, 144, 239; expenditure on, 125, 327; in factories, 82–3; services, 12, 136, 246–8; *see also* National Health Service
Henderson, H. D., *The Interwar Years and Other Papers*, **214–16**, **325–6**
Hire-purchase, 235
Hobhouse, L. T., 136

INDEX

Hobson, J. A., *Imperialism*, 13, 95, **97–9**
Hoggart, Richard, 239, 317; *The Uses of Literacy*, **251–2**
Holidays, 139, 144, 242
Hospitals, 246; *see also* National Health Service
Hours of work, 53, 77–8, 81, 86, 144, 157, 191, 241, 304, 345
Housing and Planning Act, 1919, 135
Housing conditions, 19, 135–6, 140, 143–4, 177, 238–9, 249–50; in Merseyside, 164–5
Hunger Marches, 168
Huxley, Aldous, 136

Ibsen, Henrik, 102
Immigrants, coloured, 238, 271–7
Imperialism, 1, 13–14, 95–9, 107, 202, 312–14
Import Duties Act, 1932, 144, 226
Imports, 25, 27, 37, 38, 142, 213, 232–3; *see also* Trade
Income distribution, 32, 137, 235, 239, 257
India, 96–9, 122, 218–19, 231, 312, 357
Industrial Population, Royal Commission on the Distribution of the, 214, **216–17**
Industrial powers, foreign, 24, 116; *see also* Germany, Japan, United States of America
Industrial relations, 345–7, 348; *see also* Strikes, Trades unions
Industrial Revolution, new, 214–17
Industrialization, 1, 29, 32, 59, 98–9, 212–13, 325
Industries, declining, 2, 6–7, 133, 142; light, 136, 285–7; new, 6–7, 17, 24, 130, 143, 144, 145, 192–4; small-scale, 80–2, 108–11, 353; staple, 38, 112–16, 142–3, 215–16, 218; research in, 206–7, 266–7
Inflation, 6, 154, 232, 242, 327–9
Insurance, national, 62, 180–2, 236, 246–8; *see also* National Health Service; Welfare State
Inter-war years, 2, 3, 10, 133–45, 195, 225, 235
Investment, 7, 37, 197, 213, 234, 278, 327, 328, 332, 357–9
Invisible trade, defined, 397
Ireland, 36, 168–71
Iron and steel industry, 77–9, 112, 117, 119–21, 142, 215, 292–4; *see also* Engineering industry, Gantrymen, the, Workshop of Joseph Brown

Jackson, Brian, 241; *Working Class Community*, 251, **252–4**
Japan, 1, 37, 123
Jenkins, Roy, 317
Jennings, Hilda, *Brynmawr*, **166–8**
Jewkes, J., *Ordeal by Planning*, **333–5**
Jews, 5, 52, 209
Jones, A. Creech, *Working Days*, **186–9**
Jones, D. Caradog, *The Social Survey of Merseyside*, **163–5**
Joyce, James, 136

Kebbel, T. E., *The Agricultural Labourer*, **73–4**
Keesings Contemporary Archives, **312–14**
Keynes, J. M., 2, 141, 146; *The Economic Consequences of the Peace*, 104, **106–7**; *The End of Laissez-Faire*, **195–7**
Kingsley, Charles, 101; *Alton Locke*, 26
Kipling, Rudyard, 95

Labour, skilled, 80–2, 178, 190, 290–2, 296–7; semi-skilled, 86–7, 274; unskilled, 55–8, 86–7, 112–14, 274; dilution of skilled, 87–8; need for movement of, 332; exploitation of, *see* Sweating
Labour governments, 136–7, 202, 222–5, 235, 237, 239–40, 243, 261, 309–11, 327, 329–33
Labour Party, 1, 9, 18, 23, 100, 142, 154, 168, 198, 201, 222, 231, 315
Laissez-faire, 1, 5, 141, 143–5, 195–7, 333
Landowners, 26, 39–40
Lansbury, G., 309, *My Quest for Peace*, **201–3**
Law, the, 84–6, 176, 333
Lewis, C. A., 149
Lewis, R. and Angus Maude, *Professional People*, **288–90**; *The English Middle Classes*, **315–17**
Liberal ministry, 1906–14, 38, 44; *see also* Lloyd George, David
Life, quality of, 14–16, 177–8
Life expectancy, defined, 380
Linen industry, 168
Literature, cheap, 46, 139
Liverpool, *see* Merseyside
Living standards, *see* Standards of living
Lloyd, G. I. H., *The Cutlery Trades*, 108–11
Lloyd George, David, 9, 11, 142, 156, 208, 246

INDEX 403

London, 30, 67–8, 84, 109, 139, 151, 210–11, 217, 238, 278, 307, 343; East End of, 50–3; world financial centre, 6, 37, 144, 213, 221, 232
London County Council, 67, 140
London, Jack, *On the Edge of the Abyss*, 31
London Life and Labour, New Survey of, 1928, 136, **151–2**

MacDonald, Ramsay, 161
Macmillan, Harold, 142, 195 231, 238, 243, 312, 361
Macrosty, H. W., 'The Trust Movement in Great Britain', chapter in W. J. Ashley, *British Industries*, **116–18**
Mallet, B., *British Budgets, 1887–88 to 1912–13*, **124–6**
Mann, Tom, 170
Manpower gap, estimated, 332
Marconi, 26
Marshall, A., 24
Marshall Aid, 243
Masterman, C. F. G., 322, *Condition of England*, 16, 31, **44–7**, **104–6**
Marx, Karl, 12, 89, 101
McKenna duties, 143
Means test, 142, 236
Mearn, Rev. Andrew, *Bitter Cry of Outcast London*, 31
Mellanby, Sir E., *Recent Advances in Medical Science*, **203–6**
Melly, George, 321
Merchandise trade, defined, 397
Merseyside, 1934, **163–5**, 238
Metropolitan Board of Works, Report of, 1888, 67–8
Middle classes, *see* Classes, middle
Milk-bar, the, 251–2
Mine-workers, 29, 189–91, 215; *see also* Coal Industry
Mobility, social, 32, 36, 317–18
Monarchy, the, 11, 16, 25–6, 89–90, 200–1
Monetary policy, 144, 222–5, 328
Mosley, Oswald, 142, 198–200
Motor industry, 143, 192–3, 214–16, 233, 345–7
'Municipal socialism', 11, 12, 67–8
Music halls, 151

National Government of the thirties, 17, 144
National Grid, 143, 192, 193; *see also* Electricity
National Health Service, 19, 236, 246–8; *see also* Social services; Welfare State
National Income, 179, 233, 234, 238; defined, 374
National Insurance Act, 1911, 25
National Plan, 1965, 329–33
National Wealth, increase in, 59–62, 104–6, 128–9
Nationalized Industries, Report from the Select Committee on, **335–7**
Net reproduction rate, defined, 380
New Towns Act, 1946, 239
Newsom Report, The, 1963, 266, **269–71**
Newspapers, 13, 15, 33, 139, 141, 235
Nuclear power, 335–7
Nuffield, Lord, 192

Occupation returns, 109, 384–5
Occupational changes, 215–16
Office work in the Twenties, 183–5
Oil, 218, 219, 233, 336
Old Age Pension Act, 1908, 25
Old age pensions, 32, 124, 126, 181–2, 261–2
Opportunity cost, defined, 367
Origin of Species, The, 33
Orr, John Boyd, 136
Orwell, George, 136, 137, 322; *Road to Wigan Pier*, 141; *The Lion and the Unicorn*, **176–9**
Output, defined, gross, 393; net, 393
Overseas Aid, **356–60**
Owen, Harold, *The Staffordshire Potter*, **47–50**

Pacifism, 201–3
P.E.P. (Political and Economic Planning), 136
'Permissive Society, the', 18, 317–19
Philip, J. C., 'The Chemist in the Service of the Community' in *What Science Stands For*, 203, **206–8**
Pickford, Mary, 151
Piecework, 81–2, 86–7, 111, 189, 191–2, 305–6, 343–4
Planning, attitude to, 333–5; *see also* National Plan
Police, clashes with, 134–5, 168–70
Pollock, M. A., *Working Days*, **189–91**
Pollution 19, 207–8, 238, 280
Poor Law Commission, 1909, 42, 62–6
Poor Law relief, 62, 66, 138, 166, 171, 179, 181, 236

Population, 197, 238, 379–80; changes in, 4, 26, 137, 139, 214–17, 256–9, 368; in towns, 31, 77, 278
Population, Royal Commission on, 1949, **256–9**
Potteries, The, 47–50, 215
Poverty, 30–2, 50–3, 54–8, 59, 104–6, 137, 140, 154–7, 171, 179–82, 236, 249, 261–5; *see also* Sweating; Worker, lower-paid; Workhouse
Powell, Enoch, 271, 333
Power for industry, 192, 219, 335–7; *see also* Coal industry; National Grid
Press, the, *see* Newspapers
Prices, changes in, 28, 37–8, 328; *see also* Cost-of-living, rise in; Wages, real
Priestley, J. B., *English Journey*, 136; *Angel Pavement*, 151, **152–4**, **183–5**
Problems Created by Modern Technology, **277–9**
Production, Committee on, 1917, 154–6
Production sectors, analysed, 393–4
Productivity, 19, 37, 295, 328–9, 331–2
Professions, the, 84–6, 144, 176, 333; and women, 139
Protection, 144, 225–7, 233; *see also* Tariffs
Public Health Act, 1875, 12, 25

Racial discrimination, 238, 271–7
Radio, *see* Broadcasting; B.B.C.
Railways, 121, 139, 279
Rationing, 231–2, 257–8
Reform Bill, 1884, 8–9
Regional changes in employment, 133, 214–16
Reith, Lord, 139, 149
Religion, 5, 23, 33–4, 70, 90–4, 242; *see also* Church of England
Rents, 55–7, 60, 238, 249–50
Report of the Committee on Finance and Industry, **221–2**
Report of Committee on Industry and Trade, Survey of Overseas Markets, 1925, **218–19**
Report of Proceedings at the Forty-fifth Annual Trades Union Congress, 1912, **127–9**
Report of Proceedings at the Forty-ninth Annual Trades Union Congress, 1917, 127, **129–30**
Research, chemical, 206–8; industrial, 203, 206–7, 266–7; medical, 204–6; scientific, 140–1, 203–8, 360
Retail trade, 235, 259–61, 353–4

Revolutions, European, 127, 129–30, 134
Roads, congestion on, 235, 277–9
Robbins Report, 1962, 237, 266
Robson, R., *The Cotton Industry in Britain*, **290–2**
Rousiers, Paul de, *The Labour Question in Britain*, **50–3**, **80–2**
Rowntree, Seebohm, 31, 59, 139, 163, 241, 261, 263, 264; *Poverty: a Study of Town Life*, **54–8**; *Poverty and Progress*, **179–82**; and G. R. Lavers, *Poverty and the Welfare State*, **249–50**
Russia, 3, 120, 141, 200, 209; Revolution in, 1917, 127, 129–30

Safety in factories, 82–3
Salvation Army, the, 90, 92–4
Schools, public, 69, 71–2, 239; village, 69–71; *see also* Education
Science, 15, 26, 133, 140–1, 176; and society, 203–8; *see also* Research, scientific; Technology
Scientists, need for, 266–9
Seeley, J. R., 13
Servants, domestic, 75–6, 139, 322, 323; farm, 73–4
Service industries, 7, 233, 239, 332
Shadwell, A., 13
Shaftesbury, Lord, 8
Shaw, G. B., 17, 26, 102, 141, 254
Shinwell, E., 243–5
Shipbuilding, 112, 117–18, 141–2, 155, 168, 215, 343–5, 353
Shirt-maker Machinist, 189, 191–2
Sidgwick, H., 24
Simon, John, *Three Speeches on the General Strike*, **158–9**
Sitwell, Osbert, *Laughter in the Next Room*, **146–8**
Smallholdings Act, 1892, 28
Snow, C. P., 241, 266
Snowden, P., *An Autobiography*, **222–5**
Social services, 32, 33, 124–6, 134, 137–8, 140, 179–82, 231, 236, 327, 334, 367–8; *see also* National Health Service, welfare state
Social Surveys, 8, 24; *see* Booth, C., Rowntree, Seebohm, Jones, Caradog
Socialism, 100–1, 102, 141, 196, 310
Society, 23–4, 36–7, 42–4; change in, 3, 17–19, 137–8, 176–9, 234–8, 240–1, 317–19, 322–4; *see also* Classes; Family life
South Wales, 166–8, 215

INDEX

Spencer, H., 24
Standard of living, 4, 17, 19, 25, 31, 59–62, 133–4, 136–7, 144, 179, 232, 234, 261, 285, 292–3; in Bristol, 1938, 171–5
State activity, 4, 5–6, 9–10, 12, 23–5, 32, 38, 100–1, 141–2, 143–5, 156–7, 195–7, 207, 232, 234, 279, 300–2, 333–5, 338–9
Statistical Appendix, 367–98
'Stop-go', 232, 234
Strachey, John, *The Menace of Fascism*, **198–200**
Strikes, 38, 127–9, 134–5, 158–63, 168–9, 242, 303, 343, 345, 348; wildcat, 346–7, 351
Subsidies, 338–9
Suburbans, the new, 44–7
Supermarkets, 235, 259–61
Survey of Listening to Sponsored Radio Programmes, 1938, 148, **149–50**
Sweating, 50–3, 110, 156–7, 189, 191–2

Tariffs, 7, 142–3, 144, 338, 361–3
Tawney, R. H., *The Acquisitive Society*, 35, 136, 195
Taxation, 6, 11, 23, 32, 33, 59, 124, 232, 258, 261
Technology, 2, 3, 5, 7, 145, 270, 277–9, 349; *see also* Research, Science
Television, 19, 235, 239, 240, 242, 244, 320; *see also* B.B.C.
Tenant farmers, 39–40
Terms of trade, analysed, 398
Theatres, 151–2
Thompson, F., *Lark Rise to Candleford*, **69–71**
Titmuss, R. M., 136, 241, 322; *Essays on the Welfare State*, **254–6**
Tout, Herbert, *The Standard of Living in Bristol, a Preliminary Report*, 136, **171–5**
Town and Country Planning, 238, 277
Towns, 26–7, 29–30, 67–8, 139, 140, 238, 277–9; *see also* Birmingham, Bristol, London, Merseyside, York
Trade, international, 3, 7, 16, 24–5, 37, 107, 119, 213–14, 218–20, 221, 225, 226–7, 232–3, 325–6, 397–8; cycles, 143; *see also* Imports, Export industries
Trade Disputes Act, 1927, 135
Trade unions, 9, 23–4, 28, 31, 38, 100, 127–30, 135, 154–7, 158–9, 161–3, 168, 303–8, 328, 343–4, 348–52

Transport, public, 139, 277–9
Turning-points, 3, 9, 280

Unemployment, 2, 4, 7, 10, 17, 38, 133–5, 137, 141–3, 154, 197, 214, 218, 223, 235, 244, 257, 265, 326; in South Wales, 1934, 166–8
United Kingdom, The. An Industrial, Commercial and Financial Handbook, **212–14**, 218, **219–20**
United States of America, 1, 3, 13, 36, 37, 144, 214, 222; industry in, 116, 119–21, 122–3, 143, 345
Universities, 34–5, 138, 203, 237, 240, 266–9

Valentino, Rudolph, 151
Victorian age, 15–16, 25, 102
Voluntary organizations, decline of, 333–4

Wages, 9, 81–2, 85–6, 129, 137, 234, 240, 249–50, 261–4, 293, 329; minimum, 156–7, 303, 353–5; real, 4, 26, 37–8, 61, 137, 144, 327; in shipbuilding and engineering, 343–5
Wages, Profits and Prices, A Report, 1921, **154–7**
War, First World, 2, 15, 16–17, 23–4, 32–3, 38, 86–8, 124, 142, 146–8, 201, 203, 312, 325; legacy of, 134, 222, 223
War, Second World, 2–3, 17, 143, 145, 203, 208–11, 234–5, 309, 312
Webb, Beatrice, 62, 141; *My Apprenticeship*, **42–3, 90–2, 100–1**
Welfare expenditure, growth of, 124–6
Welfare state, 1, 4, 124, 231, 235–6, 239–40, 309, 310–11, 367–8; *see also* National Health Service; Social services
Wells, D. A., *Recent Economic Change*, **59–62**
Wells, H. G., *Experiment in Autobiography*, **102–3**
Wheat growing, decline of, 6
Wheatley Act, 1924, 135
Wilkinson, Ellen, *The Town that was Murdered*, 141
Williams, E. W., *The Case for Protection*, 119, **120–1, 122–3**
Wilson, Harold, 10
Woman, Wife and Worker, Social Science Dept., L.S.E., **285–7**

Women, 23, 24, 33, 42–3, 64, 86, 102–3, 134, 139, 254–6, 263–5, 354, 355; in biscuit factory, 285–6; in domestic service, 75–6, 139, 322, 323; in sweated industries, 50–2; in textile industry, 112–14
Woolf, Leonard, *The Journey not the Arrival Matters*, **208–10**
Woolf, Virginia, 136
Woollen industry, 112, 114–16, 215–16, 353
Work, conditions of, 9, 82–3, 87–8, 186, 296–8, 303–8
Worker, lower-paid, 154–7, 353–5; *see also* Sweating
Workers, *see* Dockers; Labour; Mineworkers; Office work; Shirt-maker machinist; Sweating; Women, in biscuit factory
Workers' Educational Association, 35
Workhouse, 31, 62–5, 182, 246
Working classes, *see* Classes, working
Workshop, 109–11; of Joseph Brown, Birmingham, 80–2
'Workshop of the World', 119
Worsted industry, 115–16, 215
Wright brothers, 26

Young, G. M., *Portrait of an Age*, 14
Youth, post-1945, 18, 251–2, 318, 319–22

Zweig, F., *The Worker in an Affluent Society*, **292–4**